Power Systems

Electrical power has been the technological foundation of industrial societies for many years. Although the systems designed to provide and apply electrical energy have reached a high degree of maturity, unforeseen problems are constantly encountered, necessitating the design of more efficient and reliable systems based on novel technologies. The book series Power Systems is aimed at providing detailed, accurate and sound technical information about these new developments in electrical power engineering. It includes topics on power generation, storage and transmission as well as electrical machines. The monographs and advanced textbooks in this series address researchers, lecturers, industrial engineers and senior students in electrical engineering.

Power Systems is indexed in Scopus

Chih-Chun Yeh

AC Motor Control Loop Design

Practical Modeling, Delay Compensation, and Simulation Techniques Using MATLAB/SIMULINK

Springer

Chih-Chun Yeh
YEHSTALK.SCHOOL
Taoyuan, Taiwan

ISSN 1612-1287 ISSN 1860-4676 (electronic)
Power Systems
ISBN 978-981-95-3260-5 ISBN 978-981-95-3261-2 (eBook)
https://doi.org/10.1007/978-981-95-3261-2

Translation from the Chinese traditional language edition: “交流電機控制回路設計” by Chih-Chun Yeh,

This Springer imprint is published by the registered company Springer Nature Singapore Pte Ltd.
The registered company address is: 152 Beach Road, #21-01/04 Gateway East, Singapore 189721, Singapore

To my parents, with love and gratitude.
To my wife, Melissa Yu—your support and trust made this work possible.

Preface

The field of AC motor control lies at the heart of modern industrial automation, electric vehicles, and robotics. As industries advance toward higher efficiency, reliability, and intelligence, engineers are increasingly required to bridge the gap between classical control theory and the practical challenges of digital implementation. This book was written with that mission in mind.

During my years of teaching, research, and industrial collaboration, I repeatedly encountered the same issue: while textbooks on motor control thoroughly explain the theoretical foundations, they often fail to address the practical realities of digital systems. Modern controllers introduce inevitable delays through computation, sampling, and output stages. If these delays are ignored during control loop design, the resulting system will diverge significantly from its intended behavior, reducing stability margins and compromising performance. Recognizing this gap inspired me to develop the systematic approaches presented in this book.

The **Chinese edition** of this book was first published in Taiwan in May 2024. It was warmly received in the Chinese-speaking world and has been appreciated by many students and engineers. Most importantly, it helped readers overcome common bottlenecks in learning field-oriented control (FOC) technology. Encouraged by this positive response, I translated the book into English in August 2025 and entrusted its publication to Springer Nature, with the hope of helping a broader audience of engineers and students beyond the Chinese-speaking community.

This volume is designed for graduate students, practicing engineers, and researchers who wish to master AC motor control from both a theoretical and practical perspective. By integrating space vector modeling, field-oriented control (FOC), loop design with delay considerations, and modern modulation techniques, the book provides a complete framework for designing stable and high-performance control systems. The inclusion of MATLAB/SIMULINK-based simulations allows readers to build virtual prototypes that accurately mirror real-world dynamics, thereby reducing development costs and accelerating the learning curve.

The structure of the book reflects a gradual learning path:

- Chapter 1 introduces space vector modeling of AC machines, forming the mathematical foundation.
- Chapter 2 discusses current, speed, and position loop design, emphasizing delay-aware controllers.
- Chapter 3 explores inverter modulation techniques, including SVPWM, and integrates them into PMSM simulation.
- Chapter 4 demonstrates real-world validation using the ODrive platform.
- Chapter 5 addresses practical issues such as parameter identification, digital filters, and observer design.
- Appendix A provides a concise refresher on classical control theory, tailored for motor control engineers.

My hope is that readers will find this book not only a reference for technical knowledge but also a guide that fosters confidence in tackling complex control challenges. Whether you are developing your first motor control loop or refining advanced algorithms for industrial applications, this book aims to provide clear, systematic, and practical insights that can be immediately applied.

I would like to express my gratitude to the students, colleagues, and industry partners whose feedback and collaboration have shaped the material in these pages. Special thanks are due to the YEHSTALK learning community, whose enthusiasm and questions continue to inspire me to make motor control knowledge more accessible to the global engineering community.

I sincerely hope that this work will serve as both a valuable learning tool and a practical handbook for engineers striving to design the next generation of high-performance AC motor control systems. All example codes in this book can be downloaded from the following website: https://github.com/sn-code-inside/AC_Motor_Control_Loop_Design_En

I would also like to extend my heartfelt thanks to **Dr. Shuang Xu**, Assistant Professor at the University of New Brunswick, Canada, and **Dr. Pengxiang Huang**, Research Engineer at the National Renewable Energy Laboratory (NREL), USA, for serving as reviewers of this manuscript and providing thoughtful feedback that greatly improved the clarity and accuracy of the book. My sincere gratitude also goes to **Dr. Mengchu Huang**, Editor at Springer, and the entire Springer editorial team for their professional guidance and support throughout the publication process.

Chih-Chun Yeh
YEHSTALK.SCHOOL
Taoyuan, Taiwan

Competing Interests The author has no competing interests to declare that are relevant to the content of this manuscript.

Contents

About the Author

Dr. Chih-Chun Yeh is a leading motor control expert in Asia and an internationally recognized innovator in engineering education. He holds a Ph.D. in Electrical Engineering from National Taipei University of Technology and has authored several best-selling technical books in the Chinese-speaking world, including *AC Motor Control Loop Design*. With more than a decade of research and development experience, Dr. Yeh has published over ten peer-reviewed papers in leading journals and conferences and holds two granted patents in advanced motor control technology.

His expertise spans AC motor control, embedded systems, digital twin modeling, and multi-domain simulation. As the founder of YEHSTALK.SCHOOL—the largest Chinese-language learning platform for motor control—he has trained thousands of engineers and students by combining academic rigor with real-world industrial insights. Through his widely followed YouTube channel "YEHSTALK" and his BiliBili presence, Dr. Yeh has inspired over 30,000 dedicated followers, establishing himself as one of the most influential voices in electrical engineering education across the Chinese-speaking world. Currently, he is pioneering bilingual educational initiatives aimed at empowering engineers globally with practical and innovative motor control knowledge.

Chapter 1
Space Vector Modeling of AC Machines

In today's world, approximately **60% of all electrical energy** is consumed by electric motors, and among these, about **80% is used by induction motors**. In the early stages, due to the lack of applicable control theory [1], induction motors could not be effectively controlled. The foundation of **field-oriented control (FOC)**—also known as **vector control**—was first established by **K. Hasse** of the Technical University of Darmstadt in 1969 [2], who introduced the concept of **indirect field-oriented control**, and later by **F. Blaschke** of Siemens in 1972 [3], who proposed **direct field-oriented control**. These groundbreaking theories marked a turning point, allowing AC motor drives to become potential replacements for DC motor drives.

Despite the availability of sound theory, practical implementation was initially hindered by limitations in computational power and power electronics technology at the time. It wasn't until the **1980s**, with advancements in digital processors and semiconductor power devices, that **commercialization of AC motor drives** became possible. Prior to that, **separately excited DC motors** were the mainstream choice for high-performance motor drives. The key advantage of such motors is the ability to **decouple flux and torque**, enabling high-precision, high-performance control through well-established control theories.

However, DC motors inherently require **commutators and carbon brushes**, which are subject to wear and require regular maintenance. Additionally, due to the **sparking generated during mechanical commutation**, DC motors are unsuitable for **hazardous or explosive environments**. Despite these drawbacks, their excellent controllability—thanks to the independent control of flux and torque—set a high bar for what AC motor control systems would aspire to achieve.

The **objective of field-oriented control (FOC)** is to provide a means to **independently control the flux and torque** of a three-phase AC motor—specifically

---

Supplementary Information The online version contains supplementary material available at https://doi.org/10.1007/978-981-95-3261-2_1.

C-C. Yeh, *AC Motor Control Loop Design*, Power Systems,
https://doi.org/10.1007/978-981-95-3261-2_1

an induction motor—in a way that mimics the performance of a separately excited DC motor. This theoretical framework, combined with continuing progress in digital computation and power electronics, has paved the way for a transition from DC to AC motor systems. As this trend continues, **the use of AC motors is expected to grow even more widespread** in modern applications.

Before diving into the design of control loops for AC motors, this chapter provides a **conceptual foundation** for understanding field-oriented control. It reviews essential concepts such as the **principles of separately excited DC motors**, **space vector representation**, **coordinate transformations**, and **space vector-based modeling of AC machines**. The chapter focuses on the two major types of AC motors in practical use today: **induction motors** and **permanent magnet synchronous motors (PMSMs)**—offering detailed modeling and theoretical insights to support the advanced topics covered in later chapters.

1.1 Fundamentals of Separately Excited DC Motors

Before formally introducing AC motor control theory, it is essential to first understand the control principles of **separately excited DC motors**, as the mainstream method for AC motor control—**Field Oriented Control (FOC)**—was developed with the goal of replicating the control characteristics of separately excited DC motors [1, 2, 4].

A separately excited DC motor can be equivalently represented by **two independent circuits**: the **armature circuit** and the **field winding circuit**, as shown in Fig. 1.1 [1, 4]. The field winding circuit is stationary and located within the motor's stator. Its primary function is to provide a **stable and continuous magnetic field** for the armature. The **armature circuit**, on the other hand, is part of the rotating rotor. When an external voltage is applied to the armature, it produces a current which interacts electromagnetically with the stator's magnetic field to generate an **electromagnetic torque**, thus driving the motor to rotate.

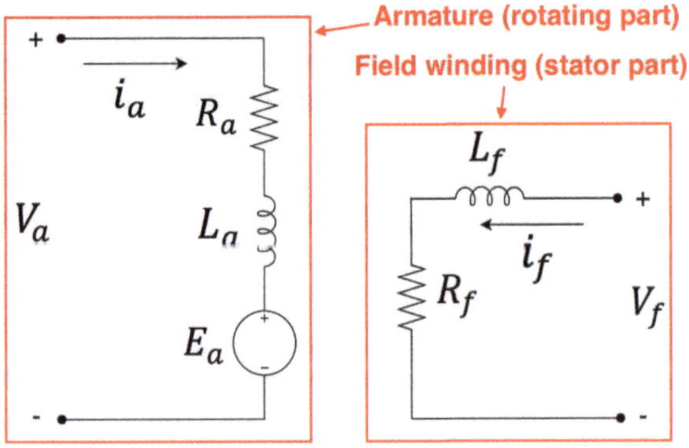

Fig. 1.1 Separately excited DC motor circuit diagram

The voltage equation of the armature circuit can be expressed as:

$$V_a = R_a i_a + L_a \frac{di_a}{dt} + E_a \tag{1.1.1}$$

where V_a is the armature voltage, i_a is the armature current, E_a is the armature back electromotive force (EMF), and R_a and L_a represent the armature resistance and inductance, respectively.

And the armature back electromotive force E_a can be expressed as:

$$E_a = K_e \lambda_f \omega_m \tag{1.1.2}$$

where K_e is the back EMF constant of the motor, λ_f is the flux linkage generated by the field winding, and ω_m is the mechanical angular velocity of the motor (in radians per second).

The voltage equation of the field winding circuit can be expressed as:

$$V_f = R_f i_f + L_f \frac{di_f}{dt} \tag{1.1.3}$$

where V_f is the field voltage, i_f is the field current, and R_f and L_f are the resistance and inductance of the field winding, respectively.

Under steady-state conditions, the steady-state field current I_f can be expressed as:

$$I_f = \frac{V_f}{R_f} \tag{1.1.4}$$

And the flux linkage λ_f generated by the field winding can be expressed as:

$$\lambda_f = L_f i_f \tag{1.1.5}$$

The electromagnetic torque of the separately excited DC motor can be expressed as:

$$T_e = K_t \lambda_f i_a \tag{1.1.6}$$

where K_t is the torque constant of the motor.

The mechanical equation of the motor is:

$$T_e = K_t \lambda_f i_a = J \frac{d\omega_m}{dt} + B\omega_m + T_L \tag{1.1.7}$$

where J is the total moment of inertia, with units of kg m^2; B is the friction coefficient, with units of N m/(rad/s); and T_L is the load torque, with units of N m.

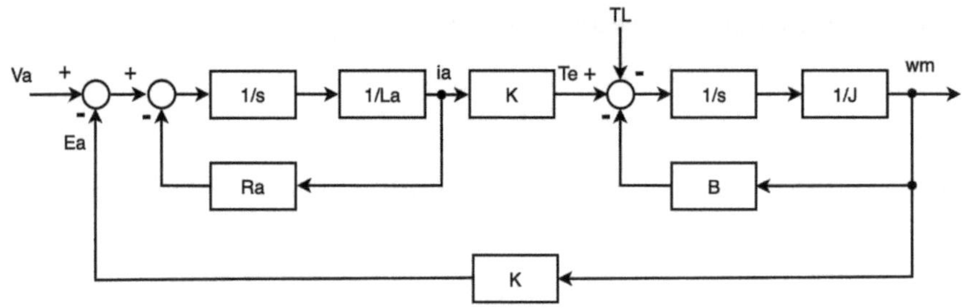

Fig. 1.2 Block diagram of the separately excited DC motor system [1]

The block diagram of the separately excited DC motor system is shown in Fig. 1.2, where the constant $K = K_e \lambda_f = K_t \lambda_f$.

Under the MKS (meter-kilogram-second) unit system, $K_e = K_t$, with K_e in units of V/(rad/s), and K_t in units of N m/A.

Explanation

For a separately excited DC motor, when the armature current flows into the back electromotive force (EMF), it indicates power input. Assuming that all the input power is converted into output mechanical power, the relationship can be expressed as $E_a i_a = T_e \omega_m$.

Expanding both sides: $K_e \omega_m i_a = K_t i_a \omega_m$.

Dividing both sides by $\omega_m i_a$, we obtain: $K_e = K_t$.

From the physical model of a separately excited DC motor, we can see that the **field winding is excited by a DC voltage** V_f, and under steady-state conditions, the **field current is given by** $I_f = \frac{V_f}{R_f}$. The **flux linkage generated by the field winding** is $\lambda_f = L_f i_f$, meaning that **controlling** i_f **is equivalent to controlling the magnetic flux** λ_f.

When the magnetic flux λ_f is held constant, the **electromagnetic torque** T_e can be **independently controlled by the armature current** i_a, according to

$$T_e = K_t \lambda_f i_a.$$

Thus, in a separately excited DC motor, the **magnetic field and torque can be independently controlled** via the **field current** i_f and **armature current** i_a, respectively. This **decoupled control structure** is precisely the target pursued by **Field Oriented Control (FOC)** theory. In essence, FOC is a control method that allows **induction motors (or AC motors in general)** to **achieve the same decoupling behavior**—independent control of flux and torque—as that of a separately excited DC motor.

To enable this kind of decoupled control for induction motors, the **traditional three-phase model** of the motor is not suitable for developing a flux–torque decoupling control strategy. Instead, the motor must be **re-modeled using the concept of space vectors**, resulting in a **space vector model of the induction motor**. Once this model is established, the **FOC strategy can be derived** based on the motor's space vector equations [1, 2, 5].

Tips

Referring to Fig. 1.2, from a control system perspective, R_a and B can be regarded as **damping elements**—where R_a represents **electrical damping**, and B represents **mechanical damping**. Although damping elements may reduce overall system efficiency, they play a crucial role in enhancing system stability. Without these damping components—i.e., when $R_a = 0$ and $B = 0$—the system would exhibit **undamped oscillations** [1].

1.2 Space Vector Representation and Coordinate Transformations

As mentioned earlier, it is difficult to develop decoupling control laws for flux and torque based on the traditional **three-phase model** of an induction motor. Therefore, in order to achieve flux–torque decoupling control, we must **reconstruct the induction motor model using the space vector approach**. By establishing the **space vector model** of the motor, we can then derive the **field-oriented control (FOC) strategy** based on this model [1, 2, 5].

A **space vector** refers to a synthesized physical quantity—such as voltage, current, or flux—formed by combining the three-phase windings' values into a **single rotating vector in space**.

Before using the space vector method, it is essential to ensure that the physical quantities of the three-phase windings satisfy the **balanced three-phase condition**, as described in Eq. (1.2.1).

$$f_a(t) + f_b(t) + f_c(t) = 0 \tag{1.2.1}$$

where $f_a(t)$, $f_b(t)$, and $f_c(t)$ represent the physical quantities (such as voltage, current, or flux) in the three-phase windings of the motor.

If the **three-phase balance condition** (as stated in Eq. (1.2.1)) is satisfied, then the **space vector** is defined as follows:

$$F_{abc} = \frac{2}{3}\left[f_a(t) + af_b(t) + a^2 f_c(t)\right] \tag{1.2.2}$$

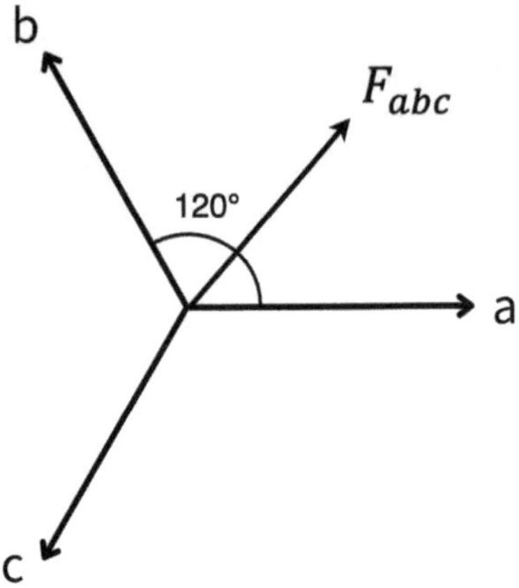

Fig. 1.3 Illustration of space vector in two-dimensional space

where $a = e^{j\frac{2\pi}{3}}, a^2 = e^{j\frac{4\pi}{3}}$.

Equation (1.2.2) represents the synthesis of the three-phase sinusoidal quantities into a space vector in a spatially symmetric manner. In other words,

$f_a(t)$ is placed along the spatial 0° axis (the *a*-axis in Fig. 1.3),
$f_b(t)$ is placed along the 120° axis (the *b*-axis), and
$f_c(t)$ is placed along the 240° axis (the *c*-axis).

Since $f_a(t)$, $f_b(t)$, and $f_c(t)$ are sinusoidal functions that vary with time, the resulting synthesized space vector F_{abc} **rotates in space**. The time it takes for the space vector to complete one full revolution corresponds exactly to **one period of the sinusoidal waveform**.

Tips

The **space vector** is a **physically realizable quantity** [5]. When balanced three-phase currents are applied to the windings of a three-phase induction motor, a **rotating magnetic field** is produced. The *angular frequency* ω_e of this rotating field—i.e., its rotational speed—is equal to the angular frequency of the input current. This angular frequency ω_e is also referred to as the **synchronous speed**.

Therefore, Equation (1.2.2) provides a mathematical vector-based representation of this physical phenomenon.

A **positive sequence** of three-phase sinusoidal waveforms is defined as **a–b–c**, meaning that phase *b* lags phase *a* by 120°, and phase *c* lags phase *b* by another 120°.

Here, we consider a set of **balanced three-phase voltages** in a **positive sequence (a–b–c)**, with a **peak amplitude of 311 V** and a **frequency of 50 Hz**. (*Note: 120° corresponds to* $\frac{2\pi}{3}$ *radians, and 240° corresponds to* $\frac{4\pi}{3}$ *radians.*)

$$v_a(t) = 311 \times \sin(2 \times \pi \times 50 \times t)$$
$$v_b(t) = 311 \times \sin(2 \times \pi \times 50 \times t - 120°)$$
$$v_c(t) = 311 \times \sin(2 \times \pi \times 50 \times t - 240°)$$

Tips

A **positive sequence** of three-phase sinusoidal waveforms is defined as **a–b–c**, meaning that phase *b* lags phase *a* by 120°, and phase *c* lags phase *b* by another 120°.

In contrast, a **negative sequence** is defined as **a–c–b**, where phase *c* lags phase *a* by 120°, and phase *b* lags phase *c* by another 120°.

We can use the following MATLAB code to plot the three-phase voltage waveforms in **positive sequence**, as shown in Fig. 1.4.

MATLAB m-file Example script: m1_2_1.m

```
t = linspace(0, 0.05, 100);
va = 220*1.414*sin(2*pi*50*t);
vb = 220*1.414*sin(2*pi*50*t-2*pi/3);
vc = 220*1.414*sin(2*pi*50*t-4*pi/3);
```

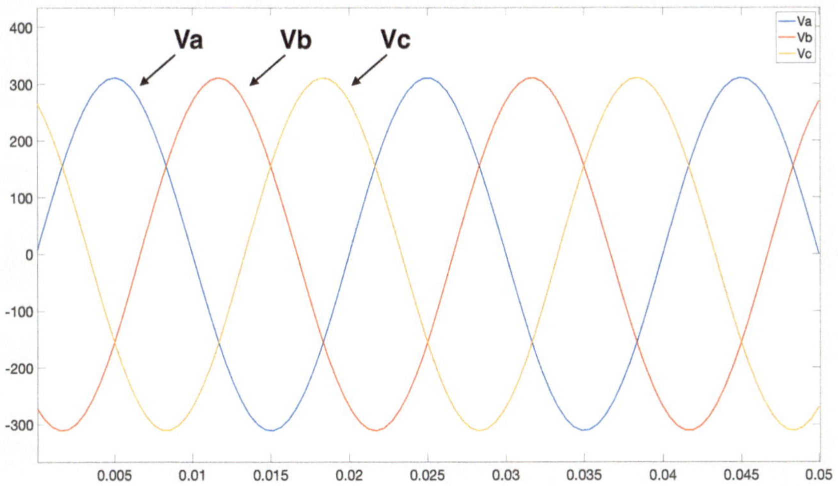

Fig. 1.4 Three-phase positive sequence voltage waveform

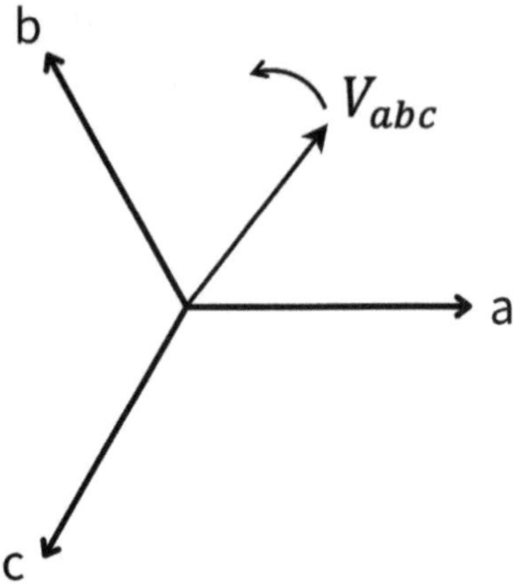

Fig. 1.5 Illustration of voltage space vector V_{abc} in two-dimensional space

```
plot(t, va, t, vb, t, vc);
legend('Va','Vb','Vc')
```

Next, by substituting $v_a(t)$, $v_b(t)$, and $v_c(t)$ into $f_a(t), f_b(t)$, and $f_c(t)$ in Eq. (1.2.2) respectively, we can obtain the **voltage space vector** V_{abc}.

$$V_{abc} = \frac{2}{3}\left[v_a(t) + e^{j\frac{2\pi}{3}} \times v_b(t) + e^{j\frac{4\pi}{3}} \times v_c(t)\right] \tag{1.2.3}$$

The space vector V_{abc} can be represented in two-dimensional space, as shown in Fig. 1.5. However, this space vector is **not stationary**—it rotates **counterclockwise** around the origin at the angular frequency of the sinusoidal waveform.

(Note: A ***positive sequence*** *of sinusoidal waveforms causes the space vector to rotate counterclockwise, while a* ***negative sequence*** *results in clockwise rotation.)*

The voltage space vector can be visualized using the following MATLAB code (**Example script: m1_2_2.m**), as shown in Fig. 1.6.

MATLAB m-file Example script: m1_2_2.m

```
t = linspace(0, 0.02, 100);
va = 220*1.414*sin(2*pi*50*t);
vb = 220*1.414*sin(2*pi*50*t-2*pi/3);
vc = 220*1.414*sin(2*pi*50*t-4*pi/3);
vabc = 2/3*(va + exp(1j*2*pi/3)*vb + exp(1j*4*pi/3)*vc);
polarplot(vabc);
hold on
```

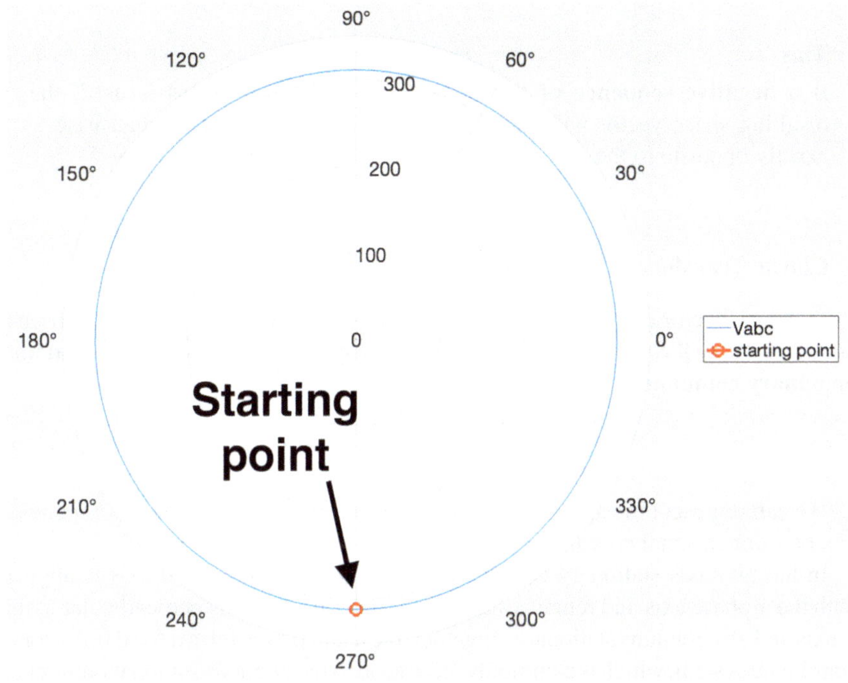

Fig. 1.6 Plot the voltage space vector in a polar coordinate diagram using MATLAB

```
t1 = 0;
va1 = 220*1.414*sin(2*pi*50*t1);
vb1 = 220*1.414*sin(2*pi*50*t1-2*pi/3);
vc1 = 220*1.414*sin(2*pi*50*t1-4*pi/3);
vabc1 = 2/3*(va1 + exp(1j*2*pi/3)*vb1 + exp(1j*4*pi/3)*vc1);
polarplot(vabc1,'-o');
legend('Vabc','starting point')
```

From Fig. 1.6, we can observe that the space vector V_{abc}, synthesized using Eq. (1.2.3), is a **rotating vector in space**. The magnitude of the vector is **311**, which matches the **peak value of the three-phase sinusoidal waveforms**.

At time t = 0, the space vector starts at the **− 90° position** and rotates **counter-clockwise**. The time required to complete one full revolution is exactly one period of the sinusoidal waveform, i.e., **0.02 s** (*since* 1/50 = 0.02). Therefore, the angular velocity of the space vector is

$$\omega = 2 \times \pi \times 50 = 314.16 \text{ (rad/s)}$$

which precisely matches the **angular frequency of the input sinusoid**.

Tips

If a **negative sequence** of three-phase sinusoidal waveforms is used, the resulting space vector will start at the **90° position** and rotate **clockwise**—exactly opposite to the space vector produced by a **positive sequence.**

- **Clarke Transformation (abc to αβ)**

Every space vector can be represented using a two-axis stationary reference frame, known as the **α–β coordinate system**, which corresponds to the sum of its **real and imaginary components**, as shown in Eq. (1.2.4).

$$F_{abc} = f_\alpha + jf_\beta \tag{1.2.4}$$

We can represent the space vector along with its real and imaginary components in a two-dimensional coordinate system, as shown in Fig. 1.7.

In the two-axis stationary coordinate system (α–β plane), the α-axis is aligned with the a-phase axis and remains stationary, while the β-axis is perpendicular to the α-axis and also remains stationary. Together, the α and β axes form a fixed orthogonal coordinate system, which is commonly referred to as the two-axis stationary reference frame.

Since F_{abc} is a rotating space vector, at any given moment it has corresponding projections on both the α-axis and β-axis, denoted as f_α and f_β, respectively. These projection components can be used to fully describe the magnitude and position of the space vector F_{abc}.

At this point, we know that the definition of a space vector is as follows:

$$F_{abc} = \frac{2}{3}\left[f_a(t) + e^{j\frac{2\pi}{3}} f_b(t) + e^{j\frac{4\pi}{3}} f_c(t)\right] \tag{1.2.5}$$

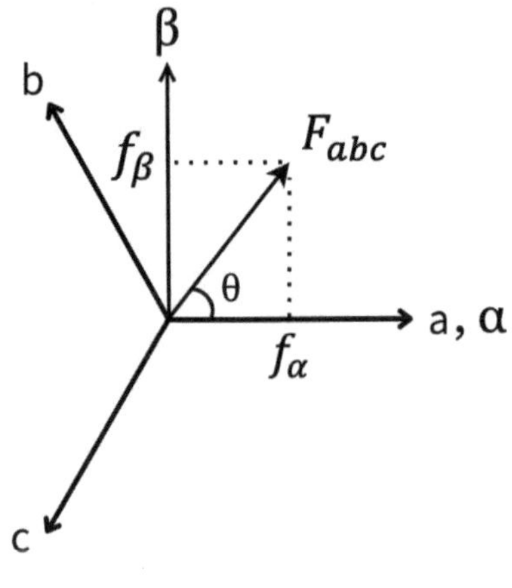

Fig. 1.7 Illustration of space vector in the two-axis stationary reference frame

Moreover, we also know that:

$$f_\alpha = \mathrm{Re}\left[\frac{2}{3}\left(f_a(t) + e^{j\frac{2\pi}{3}} f_b(t) + e^{j\frac{4\pi}{3}} f_c(t)\right)\right] \tag{1.2.6}$$

$$f_\beta = \mathrm{Im}\left[\frac{2}{3}\left(f_a(t) + e^{j\frac{2\pi}{3}} f_b(t) + e^{j\frac{4\pi}{3}} f_c(t)\right)\right] \tag{1.2.7}$$

Through derivation, we can obtain:

$$f_\alpha = \frac{2}{3}\left[f_a(t) - \frac{1}{2} f_b(t) - \frac{1}{2} f_c(t)\right] \tag{1.2.8}$$

$$f_\beta = \frac{2}{3}\left[\frac{\sqrt{3}}{2} f_b(t) - \frac{\sqrt{3}}{2} f_c(t)\right] \tag{1.2.9}$$

It can be expressed in matrix form as follows:

(Note: Here, $f_a(t), f_b(t)$, and $f_b(t)$ are denoted simply as f_a, f_b, and f_c for clarity.)

$$\begin{bmatrix} f_\alpha \\ f_\beta \end{bmatrix} = \frac{2}{3}\begin{bmatrix} 1 & \frac{-1}{2} & \frac{-1}{2} \\ 0 & \frac{\sqrt{3}}{2} & -\frac{\sqrt{3}}{2} \end{bmatrix}\begin{bmatrix} f_a \\ f_b \\ f_c \end{bmatrix} \tag{1.2.10}$$

Equation (1.2.10) is known as the Clarke Transformation, which converts from the a–b–c three-axis stationary frame to the α–β two-axis stationary frame.

Its inverse transformation is given by:

$$\begin{bmatrix} f_a \\ f_b \\ f_c \end{bmatrix} = \begin{bmatrix} 1 & 0 \\ -\frac{1}{2} & \frac{\sqrt{3}}{2} \\ -\frac{1}{2} & -\frac{\sqrt{3}}{2} \end{bmatrix}\begin{bmatrix} f_\alpha \\ f_\beta \end{bmatrix} \tag{1.2.11}$$

Here, we introduce the zero-sequence component f_0 to render the transformation matrix square. We can then modify Eq. (1.2.10) as follows:

$$\begin{bmatrix} f_\alpha \\ f_\beta \\ f_0 \end{bmatrix} = \frac{2}{3}\begin{bmatrix} 1 & -\frac{1}{2} & -\frac{1}{2} \\ 0 & \frac{\sqrt{3}}{2} & -\frac{\sqrt{3}}{2} \\ \frac{1}{2} & \frac{1}{2} & \frac{1}{2} \end{bmatrix}\begin{bmatrix} f_a \\ f_b \\ f_c \end{bmatrix} \tag{1.2.12}$$

The inverse transformation of Eq. (1.2.12) is:

$$\begin{bmatrix} f_a \\ f_b \\ f_c \end{bmatrix} = \begin{bmatrix} 1 & 0 & 1 \\ -\frac{1}{2} & \frac{\sqrt{3}}{2} & 1 \\ -\frac{1}{2} & -\frac{\sqrt{3}}{2} & 1 \end{bmatrix}\begin{bmatrix} f_\alpha \\ f_\beta \\ f_0 \end{bmatrix} \tag{1.2.13}$$

Tips

Under normal conditions, the zero-sequence component f_0 is zero. It becomes nonzero only in the case of three-phase imbalance.

Therefore, we have derived the Clarke transformation equations (Eq. 1.2.10 or 1.2.12) and their corresponding inverse Clarke transformations (Eq. 1.2.11 or 1.2.13).

Finally, we use the following MATLAB script (**Example script: m1_2_3.m**) to verify the functionality of the **Clarke transformation** and its **inverse**. The program first converts the three-phase sinusoidal quantities in the **a–b–c** frame into **α–β components** using the Clarke transformation. It then reconstructs the original **a–b–c components** from the α–β components using the inverse Clarke transformation. Lastly, the original a–b–c waveforms, the α–β components, and the reconstructed a–b–c waveforms are plotted for comparison, as shown in Fig. 1.8.

It can be observed that through the use of the **Clarke transformation** and its **inverse**, the three-phase a–b–c components can be **accurately and successfully reconstructed**. Additionally, we can see that the **α and β components** are sinusoidal in nature. As the space vector completes one full revolution, both the α and β components complete one full sinusoidal cycle.

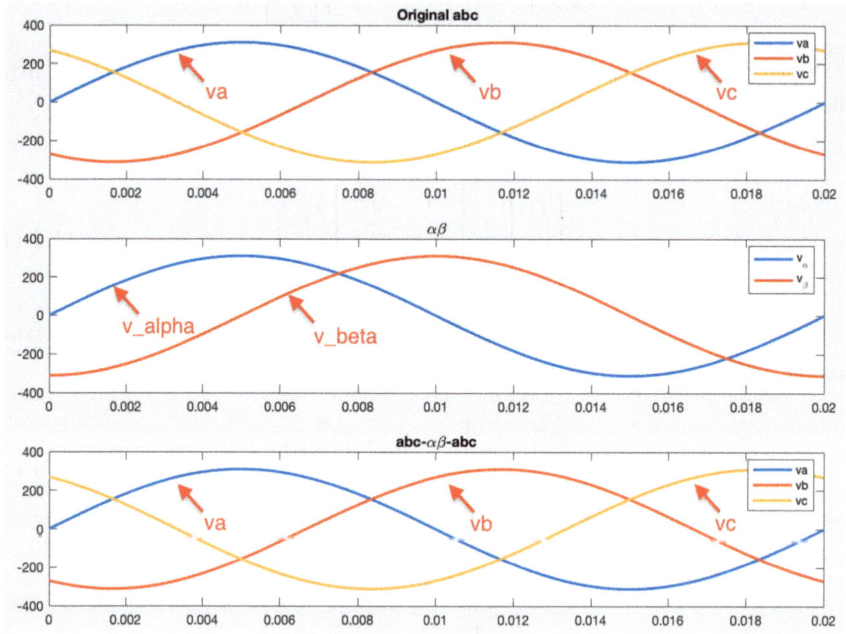

Fig. 1.8 Three-phase voltage waveforms, conversion into α–β stationary reference frame, and inverse transformation back to three-phase voltages

Since we are using a **positive-sequence three-phase sinusoidal input**, the space vector starts at **− 90°**. Therefore, at time zero, the **α component** begins increasing from zero, while the **β component** starts decreasing from its **negative peak**.

By understanding how a polar space vector projects onto the **α–β axes**, readers can gradually build a solid **geometric intuition** for how space vector projections behave in the two-axis stationary reference frame.

MATLAB m-file Example script: m1_2_3.m

```
t = linspace(0, 0.02, 100);
va = 220*1.414*sin(2*pi*50*t);
vb = 220*1.414*sin(2*pi*50*t-2*pi/3);
vc = 220*1.414*sin(2*pi*50*t-4*pi/3);
v_alpha = 2/3*(1*va - 0.5*vb - 0.5*vc);
v_beta = 2/3*(sqrt(3)/2*vb - sqrt(3)/2*vc);
v_zero = 2/3*(0.5*va + 0.5*vb + 0.5*vc);
va1 = v_alpha + v_zero;
vb1 = -0.5*v_alpha + sqrt(3)/2*v_beta + v_zero;
vc1 = -0.5*v_alpha - sqrt(3)/2*v_beta + v_zero;
subplot(3,1,1);
plot(t, va, t, vb, t, vc, 'LineWidth', 2);
legend('va', 'vb', 'vc');
title('Original abc');
subplot(3,1,2);
plot(t, v_alpha, t, v_beta, 'LineWidth', 2);
legend('v_\alpha', 'v_\beta');
title('\alpha\beta');
subplot(3,1,3);
plot(t, va1, t, vb1, t, vc1, 'LineWidth', 2);
legend('va', 'vb', 'vc');
title('abc-\alpha\beta-abc');
```

To further enhance your understanding of the **space vector projection method**, we will demonstrate a **geometric projection** using a voltage space vector. Assume a voltage vector $V_{abc} = 311\angle{-72^\circ}$. In this case, we can intuitively apply the **projection method** to determine its components in both the **a–b–c three-axis frame** and the **α–β two-axis frame**, as illustrated in Fig. 1.9.

If the angle θ is a simple and convenient value, the **projection components** on the a–b–c and α–β axes can be easily determined through visual inspection. However, in this example, the angle is **− 72°**, which is not easily handled analytically. Therefore, we can use the following **MATLAB script** (**Example script: m1_2_4.m**) to **accurately calculate the components** along each axis, as shown below:

Fig. 1.9 Representation of the voltage vector V_{abc} in two-dimensional space

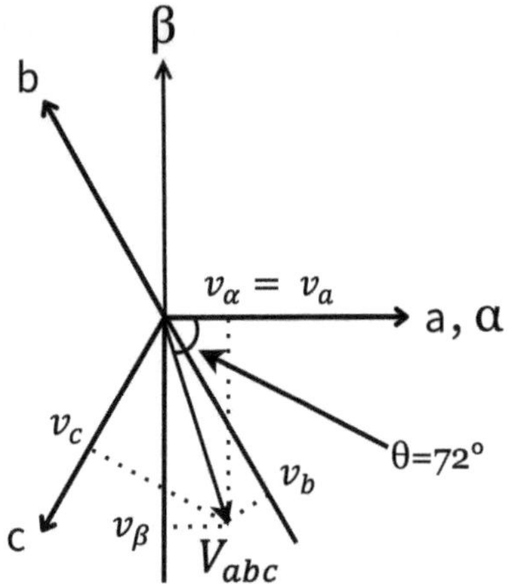

$$v_\alpha = 96.13$$
$$v_\beta = -295.8$$
$$v_a = 96.13$$
$$v_b = -304.3$$
$$v_c = 208.15$$

MATLAB m-file Example script: m1_2_4.m

```
t1 = 0.001;
va = 220*1.414*sin(2*pi*50*t1);
vb = 220*1.414*sin(2*pi*50*t1-2*pi/3);
vc = 220*1.414*sin(2*pi*50*t1-4*pi/3);
vabc1 = 2/3*(va + exp(1j*2*pi/3)*vb + exp(1j*4*pi/3)*vc);
rho = abs(vabc1);
theta = angle(vabc1)*180/pi;
v_alpha = 2/3*(1*va - 0.5*vb - 0.5*vc);
v_beta = 2/3*(sqrt(3)/2*vb - sqrt(3)/2*vc);
v_zero = 2/3*(0.5*va + 0.5*vb + 0.5*vc);
va1 = v_alpha + v_zero;
vb1 = -0.5*v_alpha + sqrt(3)/2*v_beta + v_zero;
vc1 = -0.5*v_alpha - sqrt(3)/2*v_beta + v_zero;
fprintf('The space vector Vabc: abs(aVbc) = %6.2f, angle(Vabc) = %6.2f\n',rho, theta)
fprintf('V_alpha of Vabc is %6.2f, V_beta of Vabc is %6.2f\n',v_alpha, v_beta)
```

```
fprintf('Va of Vabc is %6.2f, Vb of Vabc is %6.2f, Vc of Vabc is %6.2f\n',va1,
vb1, vc1)
```

- **Park Transformation (αβ to dq)**

The **Clarke transformation** converts the space vector's components from the **a–b–c three-axis stationary reference frame** to the **α–β two-axis stationary frame**. However, the resulting **α and β components remain sinusoidal**, because they are essentially the **instantaneous projections** of the **rotating space vector** onto the α and β axes.

As a result, the **angular frequency** ω of the α and β components is the **same as that of the space vector**, which is also the same as the **angular frequency of the original three-phase sinusoidal waveforms.**

Let us consider a scenario where, at time t = 0, a coordinate axis **d** is aligned with the **α-axis**, and the **q-axis** is perpendicular to the d-axis. For t > 0, the **d–q axes** begin rotating in the **same direction and at the same angular velocity** ω as the space vector. In this rotating reference frame, the projections observed on the **d and q axes**—denoted as f_d and f_q—become **DC quantities**, as illustrated in Fig. 1.10.

The **Park Transformation** is the coordinate transformation that converts the components f_α and f_β in the **stationary α–β frame** into the components f_d and f_q in the **rotating d–q frame**.

(Note: In this context, the angular velocity of the d–q frame is assumed to match that of the space vector.)

We can represent the space vector F_{abc} in the **d–q reference frame** as:

$$F_{abc} = f_d + jf_q \tag{1.2.14}$$

Through geometric relationships, Eq. (1.2.14) can be expressed as:

$$F_{abc} = \left[f_\alpha \cos(\gamma) + f_\beta \sin(\gamma)\right] + j\left[f_\beta \cos(\gamma) - f_\alpha \sin(\gamma)\right] \tag{1.2.15}$$

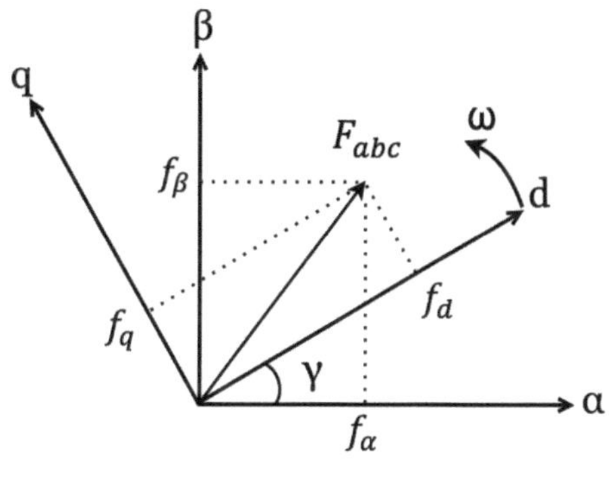

Fig. 1.10 Illustration of space vector in the two-axis synchronous rotating reference frame

Therefore, we can derive the **Park transformation** as:

$$F_{abc} = \begin{bmatrix} f_d \\ f_q \end{bmatrix} = \begin{bmatrix} \cos(\gamma) & \sin(\gamma) \\ -\sin(\gamma) & \cos(\gamma) \end{bmatrix} \begin{bmatrix} f_\alpha \\ f_\beta \end{bmatrix} \tag{1.2.16}$$

The **inverse Park transformation** is:

$$F_{abc} = \begin{bmatrix} f_\alpha \\ f_\beta \end{bmatrix} = \begin{bmatrix} \cos(\gamma) & -\sin(\gamma) \\ \sin(\gamma) & \cos(\gamma) \end{bmatrix} \begin{bmatrix} f_d \\ f_q \end{bmatrix} \tag{1.2.17}$$

Next, we can use the following MATLAB script (**Example script: m1_2_5.m**) to practice applying the **Park transformation** and its **inverse**, and to observe the results of the transformation through simulation.

MATLAB m-file Example script m1_2_5.m

```
t = linspace(0, 0.02, 100);
va = 220*1.414*sin(2*pi*50*t);
vb = 220*1.414*sin(2*pi*50*t-2*pi/3);
vc = 220*1.414*sin(2*pi*50*t-4*pi/3);
v_alpha = 2/3*(1*va - 0.5*vb - 0.5*vc);
v_beta = 2/3*(sqrt(3)/2*vb - sqrt(3)/2*vc);
v_zero = 2/3*(0.5*va + 0.5*vb + 0.5*vc);
gamma = 2*pi*50*t;
v_d = cos(gamma).*v_alpha + sin(gamma).*v_beta;
v_q = -sin(gamma).*v_alpha + cos(gamma).*v_beta;
subplot(3,1,1);
plot(t, va, t, vb, t, vc, 'LineWidth', 2);
legend('va', 'vb', 'vc');
title('Original abc');
subplot(3,1,2);
plot(t, v_alpha, t, v_beta, 'LineWidth', 2);
legend('v_\alpha', 'v_\beta');
title('\alpha-\beta');
subplot(3,1,3);
plot(t, v_d, t, v_q, 'LineWidth', 2);
ylim([-350 50])
legend('v_d', 'v_q');
title('d-q');
```

The result of running **Example script m1_2_5** is shown in Fig. 1.11. From the waveform, we can observe that the space vector V_{abc}, synthesized from three-phase sinusoidal voltages, produces **DC components** when projected onto the **d–q axis**: specifically, $v_d = 0$ and $v_q = -311$.

How do we explain this result? It's actually quite straightforward:

At $t \geq 0$, the space vector V_{abc} *starts at an angle of* ***−90°****, while the* ***d-axis*** *begins aligned with the* ***α-axis*** *(i.e., 0°). Because the* ***d–q frame rotates synchronously*** *with the space vector V_{abc}*, the **d-axis always leads the space vector by 90°**.

After t > 0, both the space vector and the d–q frame **rotate in the same direction and at the same angular speed**, maintaining this **constant 90° phase difference**. As a result:

- The **projection of V_{abc} onto the d-axis is always zero**, because the vector is always perpendicular to it.
- The **projection onto the q-axis is constant** and equals the **negative peak value**, i.e., $v_q = -311$.

This perfectly illustrates the core advantage of **Park transformation**: converting a sinusoidal space vector into steady DC quantities in a rotating reference frame, simplifying the control of AC machines.

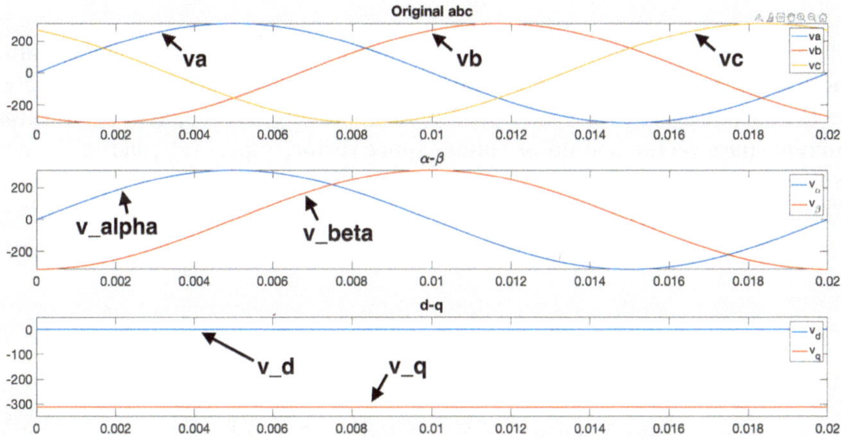

Fig. 1.11 Three-phase voltage waveforms, transformation into α–β stationary frame, and further transformation into d–q synchronous rotating frame

1.3 Space Vector Model of Three-Phase Squirrel-Cage Induction Motors

For **three-phase induction motors**, deriving a model directly using the state variables of the three-phase system results in a **time-varying model** with **strong coupling between variables**, which makes it difficult to use for control law development.

To obtain a **time-invariant** and **mathematically simplified model**, it is necessary to model the motor using the **space vector approach**. However, the derivation of the **space vector model** for a three-phase induction motor—especially for a squirrel-cage type—is relatively complex and lengthy [1, 4], and will not be detailed here. Interested readers are encouraged to refer to the references provided at the end of this chapter.

A fully derived **space vector model of a three-phase squirrel-cage induction motor** can be expressed as follows [1, 4]:

$$R_s I_{abcs} + L_s \frac{dI_{abcs}}{dt} + L_m \frac{dI_{abcr}}{dt} e^{j\theta_r} + j\omega_r L_m I_{abcr} e^{j\theta_r} = V_{abcs} \tag{1.3.1}$$

$$R_r I_{abcr} + L_r \frac{dI_{abcr}}{dt} + L_m \frac{dI_{abcs}}{dt} e^{-j\theta_r} - j\omega_r L_m I_{abcs} e^{-j\theta_r} = 0 \tag{1.3.2}$$

where R_s, R_r, L_s, L_r, and L_m represent the **stator resistance**, **rotor resistance**, **stator inductance**, **rotor inductance**, and **mutual inductance** of the motor, respectively.

The quantities I_{abcs}, I_{abcr}, and V_{abcs} denote the **stator current space vector**, **rotor current space vector**, and **stator voltage space vector**, respectively, and are defined as follows:

$$I_{abcs} = \frac{2}{3}\left[i_{as}(t) + e^{j\frac{2\pi}{3}} \times i_{bs}(t) + e^{j\frac{4\pi}{3}} \times i_{cs}(t)\right] \tag{1.3.3}$$

$$I_{abcr} = \frac{2}{3}\left[i_{ar}(t) + e^{j\frac{2\pi}{3}} \times i_{br}(t) + e^{j\frac{4\pi}{3}} \times i_{cr}(t)\right] \tag{1.3.4}$$

$$V_{abcs} = \frac{2}{3}\left[v_{as}(t) + e^{j\frac{2\pi}{3}} \times v_{bs}(t) + e^{j\frac{4\pi}{3}} \times v_{cs}(t)\right] \tag{1.3.5}$$

We can illustrate the space vectors I_{abcs}, I_{abcr}, and V_{abcs} in a two-dimensional vector space, as shown in Fig. 1.12.

In this diagram, **as–bs–cs** represents the **stator's stationary a–b–c reference frame** (*note: since the stator is fixed, it is referred to as the "stationary" three-phase frame*), which corresponds to the three-phase balanced windings of the stator. Relative to this stationary frame, **ar–br–cr** represents the **rotor's rotating a–b–c reference frame**, which rotates at an angular speed ω_r. The space vectors I_{abcs} and V_{abcs} are synthesized from the stator's as–bs–cs currents and voltages, while I_{abcr} is synthesized from the rotor's ar–br–cr three-phase currents.

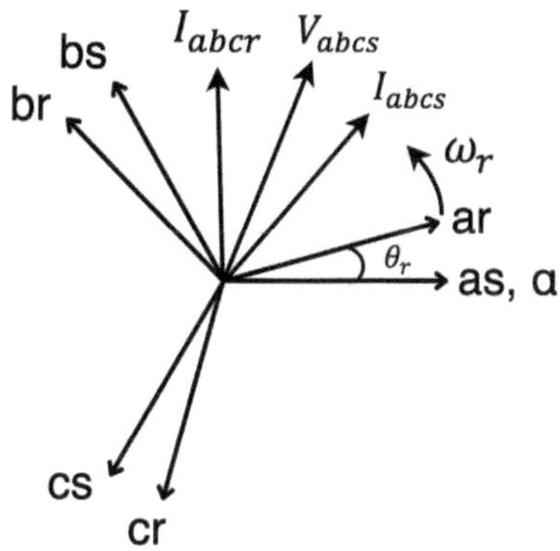

Fig. 1.12 Illustration of stator current vector, rotor current vector, and stator voltage vector in two-dimensional space

At this point, we introduce an **arbitrarily rotating two-axis coordinate system**, denoted as (d^a-q^a). Suppose this rotating frame rotates at an **angular velocity** ω, and the angle between the d^a**-axis** and the stationary **as-axis** is θ. In this **arbitrarily rotating reference frame**, the space vectors I_{abcs}, I_{abcr}, and V_{abcs} *are represented as* I^a_{abcs}, I^a_{abcr}, *and* V^a_{abcs}, respectively. (*Note: the superscript "a" denotes quantities expressed in the arbitrarily rotating reference frame.*) These vectors are related by the following expressions [1, 4]:

$$I_{abcs} = I^a_{abcs}e^{j\theta} \tag{1.3.6}$$

$$I_{abcr} = I^a_{abcr}e^{j(\theta-\theta_r)} \tag{1.3.7}$$

$$V_{abcs} = V^a_{abcs}e^{j\theta} \tag{1.3.8}$$

Explanation

For I_{abcr}, *which is the* **rotor-side current space vector**, *we must account for the fact that the rotor is also rotating and therefore introduces an angular displacement* θ_r. *The relationship between the rotor current space vector in the* ***stationary reference frame*** I_{abcr} and the **arbitrarily rotating frame** I^a_{abcr} is given by:

$$I^a_{abcr} = I^s_{abcr}e^{-j\theta} = I_{abcr}e^{j\theta_r}e^{-j\theta} = I_{abcr}e^{j(\theta_r-\theta)}.$$

Therefore, solving for I_{abcr}, we obtain:

$$I_{abcr} = I^a_{abcr}e^{j(\theta-\theta_r)}.$$

(Note: I_{abcr}^{s} denotes the value of I_{abcr} in the stationary reference frame, where the superscript "s" stands for "stationary".)

By substituting Eqs. (1.3.6) through (1.3.8) into Eqs. (1.3.1) and (1.3.2), and simplifying the expressions, we obtain the following equations:

$$R_s I_{abcs}^a + L_s \frac{dI_{abcs}^a}{dt} + j\omega L_s I_{abcs}^a + L_m \frac{dI_{abcr}^a}{dt} + j\omega L_m I_{abcr}^a = V_{abcs}^a \tag{1.3.9}$$

$$R_r I_{abcr}^a + L_r \frac{dI_{abcr}^a}{dt} + j(\omega - \omega_r) L_r I_{abcr}^a + L_m \frac{dI_{abcs}^a}{dt} + j(\omega - \omega_r) L_m I_{abcs}^a = 0 \tag{1.3.10}$$

Equations (1.3.9) and (1.3.10) represent the **fully derived space vector voltage equations** of a three-phase **squirrel-cage induction motor** expressed in an **arbitrarily rotating reference frame** (d^a-q^a).

Specifically:

- Equation (1.3.9) is the **stator space vector voltage equation**.
- Equation (1.3.10) is the **rotor space vector voltage equation**. (*Note: for squirrel-cage induction motors, the rotor windings are short-circuited, so the rotor voltage is zero.*)

Where ω is the **angular velocity of the arbitrarily rotating reference frame** (d^a-q^a), and ω_r is the **electrical angular velocity of the rotor**.

(Note: $\omega_r = \frac{P}{2}\omega_{rm}$, where P is the number of poles of the motor, and ω_{rm} is the ***mechanical angular velocity*** *of the rotor, in radians per second.)*

The space vectors expressed in the arbitrarily rotating reference frame—namely I_{abcs}^a, I_{abcr}^a, and V_{abcs}^a—can all be represented in the following **complex form** within that frame:

$$I_{abcs}^a = i_{ds}^a + j i_{qs}^a \tag{1.3.11}$$

$$I_{abcr}^a = i_{dr}^a + j i_{qr}^a \tag{1.3.12}$$

$$V_{abcs}^a = v_{ds}^a + j v_{qs}^a \tag{1.3.13}$$

By substituting Eqs. (1.3.11) through (1.3.13) into Eqs. (1.3.9) and (1.3.10), we obtain the following results:

$$R_s i_{ds}^a + L_s \frac{di_{ds}^a}{dt} - \omega L_s i_{qs}^a + L_m \frac{di_{dr}^a}{dt} - \omega L_m i_{qr}^a = v_{ds}^a \tag{1.3.14}$$

$$\omega L_s i_{ds}^a + R_s i_{qs}^a + L_s \frac{di_{qs}^a}{dt} + \omega L_m i_{dr}^a + L_m \frac{di_{qr}^a}{dt} = v_{qs}^a \tag{1.3.15}$$

$$L_m \frac{di_{ds}^a}{dt} - (\omega - \omega_r) L_m i_{qs}^a + R_r i_{dr}^a + L_r \frac{di_{dr}^a}{dt} - (\omega - \omega_r) L_r i_{qr}^a = 0 \tag{1.3.16}$$

$$(\omega - \omega_r) L_m i_{ds}^a + L_m \frac{di_{qs}^a}{dt} + (\omega - \omega_r) L_r i_{dr}^a + R_r i_{qr}^a + L_r \frac{di_{qr}^a}{dt} = 0 \tag{1.3.17}$$

Equations (1.3.14) to (1.3.17) represent the **stator and rotor voltage equations** of a three-phase squirrel-cage induction motor, expressed in terms of four state variables: i_{ds}^a, i_{qs}^a, i_{dr}^a, and i_{qr}^a.

(Note: v_{ds}^a and v_{qs}^a are not state variables; they are input signals.)

However, in practical applications, it is **very difficult to directly measure rotor currents**, making the model described by Eqs. (1.3.14)–(1.3.17) **difficult to implement**.

To address this issue, we introduce the following **flux linkage equations** to **eliminate the rotor current state variables**, expressing them instead in terms of **stator currents and magnetic fluxes**.

$$\Phi_{abcs}^a = L_s I_{abcs}^a + L_m I_{abcr}^a \tag{1.3.18}$$

$$\Phi_{abcr}^a = L_m I_{abcs}^a + L_r I_{abcr}^a \tag{1.3.19}$$

Equations (1.3.18) and (1.3.19) are the **stator and rotor flux linkage space vector equations**. These equations can be expanded and expressed in the form corresponding to an **arbitrarily rotating reference frame** (d^a–q^a) as follows:

$$\phi_{ds}^a = L_s i_{ds}^a + L_m i_{dr}^a \tag{1.3.20}$$

$$\phi_{qs}^a = L_s i_{qs}^a + L_m i_{qr}^a \tag{1.3.21}$$

$$\phi_{dr}^a = L_m i_{ds}^a + L_r i_{dr}^a \tag{1.3.22}$$

$$\phi_{qr}^a = L_m i_{qs}^a + L_r i_{qr}^a \tag{1.3.23}$$

We can use Eqs. (1.3.22) and (1.3.23) to substitute the rotor currents in terms of **stator currents** and **rotor flux linkages**, as follows:

$$i_{dr}^a = \frac{\phi_{dr}^a - L_m i_{ds}^a}{L_r} \tag{1.3.24}$$

$$i_{qr}^{a} = \frac{\phi_{qr}^{a} - L_m i_{qs}^{a}}{L_r} \quad (1.3.25)$$

By substituting Eqs. (1.3.24) and (1.3.25) for i_{dr}^{a} and i_{qr}^{a} in Eqs. (1.3.14) through (1.3.17), we can simplify and reorganize the equations to obtain the following results:

$$R_s i_{ds}^{a} + L_\sigma \frac{di_{ds}^{a}}{dt} - \omega L_\sigma i_{qs}^{a} + \frac{L_m}{L_r}\frac{d\phi_{dr}^{a}}{dt} - \frac{\omega L_m}{L_r}\phi_{qr}^{a} = v_{ds}^{a} \quad (1.3.26)$$

$$\omega L_\sigma i_{ds}^{a} + R_s i_{qs}^{a} + L_\sigma \frac{di_{qs}^{a}}{dt} + \omega \frac{L_m}{L_r}\phi_{dr}^{a} + \frac{L_m}{L_r}\frac{d\phi_{qr}^{a}}{dt} = v_{qs}^{a} \quad (1.3.27)$$

$$-R_r L_m i_{ds}^{a} + R_r \phi_{dr}^{a} + L_r \frac{d\phi_{dr}^{a}}{dt} - (\omega - \omega_r) L_r \phi_{qr}^{a} = 0 \quad (1.3.28)$$

$$-R_r L_m i_{qs}^{a} + (\omega - \omega_r) L_r \phi_{dr}^{a} + R_r \phi_{qr}^{a} + L_r \frac{d\phi_{qr}^{a}}{dt} = 0 \quad (1.3.29)$$

where $L_\sigma = L_s - \frac{L_m^2}{L_r}$ is the **leakage inductance** of the motor.

Equations (1.3.26) through (1.3.29) represent the **squirrel-cage induction motor model** using four state variables: i_{ds}^{a}, i_{qs}^{a}, ϕ_{dr}^{a} and ϕ_{qr}^{a}. This formulation is the **most commonly used motor model** in the development of **field-oriented control (FOC)** strategies for induction motors.

Based on Eqs. (1.3.26)–(1.3.29), we can draw the **d–q axis equivalent circuit** of a squirrel-cage induction motor in an **arbitrarily rotating reference frame** (d^a-q^a), as shown in Fig. 1.13 [1, 4].

To fully describe the dynamic behavior of a three-phase induction motor, we also need a **torque equation**. The **electromagnetic torque equation** for a three-phase induction motor can be expressed as follows:

$$T_e = \frac{3P}{4}\frac{L_m}{L_r}\left(i_{qs}^{a}\phi_{dr}^{a} - i_{ds}^{a}\phi_{qr}^{a}\right) \quad (1.3.30)$$

T_e is the **electromagnetic torque** produced at the motor shaft, with units of **N m**. The relationship between the electromagnetic torque T_e and the motor's mechanical angular velocity ω_{rm} can be described by the **mechanical equation** of the motor as follows:

$$T_e = J\frac{d\omega_{rm}}{dt} + B\omega_{rm} + T_L \quad (1.3.31)$$

where:

- J is the **total moment of inertia**, with units of **kg m^2**
- B is the **friction coefficient**, with units of **N m/(rad/s)**
- T_L is the **load torque**, with units of **N m**

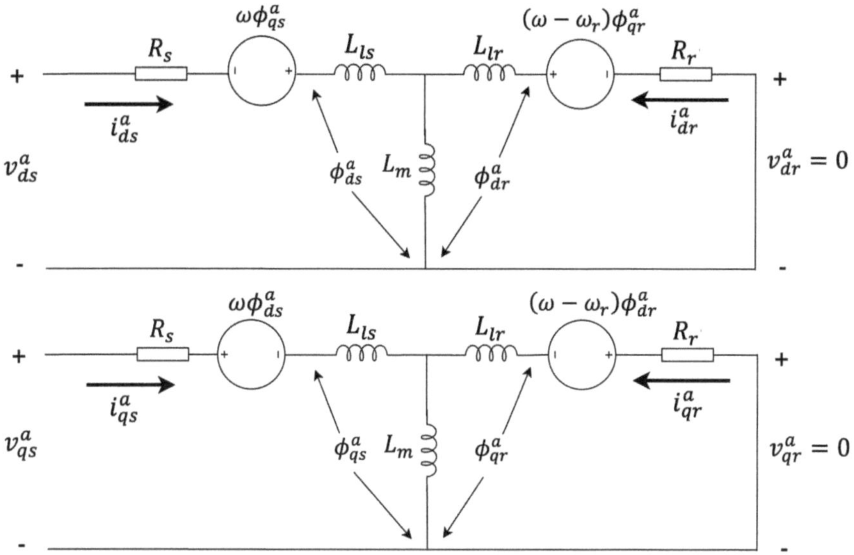

Fig. 1.13 d–q axis equivalent circuit diagram of a squirrel-cage induction motor

(Note: In the motor model, ω_r refers to the ***electrical angular velocity*** *of the motor. It is related to the* ***mechanical angular velocity*** *ω_{rm} by the equation:*
$\omega_r = \frac{P}{2}\omega_{rm}$. Both ω_r and ω_{rm} are expressed in ***rad/s****, where P is the number of poles of the motor.)*

Combining all of the above, Eqs. (1.3.26) through (1.3.30) form the **complete mathematical model** of a **three-phase squirrel-cage induction motor** in an **arbitrarily rotating reference frame** (d^a-q^a). When coupled with the **mechanical equation** (Eq. 1.3.31), this model fully describes the **dynamic behavior of the motor driving a load**.

- **Building an Induction Motor Subsystem Model Using MATLAB/SIMULINK**

Before implementing the model in **MATLAB/SIMULINK**, it is necessary to first simplify Eqs. (1.3.26) to (1.3.30). These equations describe the voltage model of an induction motor in an **arbitrarily rotating reference frame**.

By setting the angular velocity $\omega = 0$, we can obtain the **voltage model in the two-axis stationary reference frame (α–β)**. Under this condition, the variable substitutions are as follows:

$$i_{ds}^{a} = i_{s\alpha},\ i_{qs}^{a} = i_{s\beta},\ \phi_{dr}^{a} = \phi_{r\alpha},\ \phi_{qr}^{a} = \phi_{r\beta},\ v_{ds}^{a} = v_{s\alpha} \text{ and } v_{qs}^{a} = v_{s\beta}$$

We first reorganize Eqs. (1.3.28) and (1.3.29) under these assumptions as follows:

$$\frac{d\phi_{r\alpha}}{dt} = \frac{R_r L_m}{L_r} i_{s\alpha} - \frac{R_r}{L_r}\phi_{r\alpha} - \omega_r \phi_{r\beta} \tag{1.3.32}$$

$$\frac{d\phi_{r\beta}}{dt} = \frac{R_r L_m}{L_r} i_{s\beta} + \omega_r \phi_{r\alpha} - \frac{R_r}{L_r} \phi_{r\beta} \tag{1.3.33}$$

By substituting Eqs. (1.3.32) and (1.3.33) into Eqs. (1.3.26) and (1.3.27), we obtain:

$$\frac{di_{s\alpha}}{dt} = K_1 i_{s\alpha} + K_2 \phi_{r\alpha} + K_3 \omega_r \phi_{r\beta} + K_4 v_{s\alpha} \tag{1.3.34}$$

$$\frac{di_{s\beta}}{dt} = K_1 i_{s\beta} - K_3 \omega_r \phi_{r\alpha} + K_2 \phi_{r\beta} + K_4 v_{s\beta} \tag{1.3.35}$$

where $K_1 = \frac{-R_s L_r^2 - R_r L_m^2}{L_r w}$, $K_2 = \frac{R_r L_m}{L_r w}$, $K_3 = \frac{L_m}{w}$, $K_4 = \frac{L_r}{w}$, $w = L_r L_s - L_m^2$.

$\left(\text{Note: } L_\sigma = \frac{w}{L_r}\right)$

Next, we reorganize Eqs. (1.3.32) and (1.3.33) as follows:

$$\frac{d\phi_{r\alpha}}{dt} = K_5 i_{s\alpha} + K_6 \phi_{r\alpha} - \omega_r \phi_{r\beta} \tag{1.3.36}$$

$$\frac{d\phi_{r\beta}}{dt} = K_5 i_{s\beta} + \omega_r \phi_{r\alpha} + K_6 \phi_{r\beta} \tag{1.3.37}$$

where $K_5 = \frac{R_r L_m}{L_r}$ and $K_6 = -\frac{R_r}{L_r}$.

We now reorganize Eqs. (1.3.30) and (1.3.31) as follows:

$$\frac{d\omega_{rm}}{dt} = \frac{1}{J}(T_e - T_L - B\omega_{rm}) \tag{1.3.38}$$

$$T_e = \frac{3P}{4}\frac{L_m}{L_r}\left(i_{s\beta}\phi_{r\alpha} - i_{s\alpha}\phi_{r\beta}\right) \tag{1.3.39}$$

where $\omega_{rm} = \frac{2}{P}\omega_r$, and P is the number of poles of the motor.

Next, we will build the **induction motor model** in the **SIMULINK environment** using Eqs. (1.3.34) to (1.3.39), implemented within a dedicated **Subsystem block**.

Before constructing the **three-phase squirrel-cage induction motor model** in SIMULINK, we will first define the necessary **motor parameters** using a **MATLAB m-file script**, as shown in Table 1.1 [4]. These parameters will be used throughout the simulation model.

MATLAB m-file Example script: im_params.m

```
Rs = 0.8;
Rr = 0.6;
```

Table 1.1 Induction motor parameters

Parameter	Symbol	Value	Unit
Stator resistance	R_s	0.8	Ω
Rotor resistance	R_r	0.6	Ω
Stator inductance	L_s	0.085	H
Rotor inductance	L_r	0.085	H
Mutual inductance	L_m	0.082	H
Number of poles	P	4	–
Moment of inertia	J	0.033	Kg m^2
Friction coefficient	B	0.00825	N m s/rad

```
Ls = 0.085;
Lr = 0.085;
Lm = 0.082;
pole = 4;
J = 0.033;
B = 0.00825;
w = Ls*Lr - Lm^2;
Lsigma = w/Lr;
K1 = (-Rs*Lr^2-Rr*Lm^2)/(Lr*w);
K2 = (Rr*Lm)/(Lr*w);
K3 = Lm/w;
K4 = Lr/w;
K5 = Rr*Lm/Lr;
K6 = -Rr/Lr;
```

Once you have finished creating the **m-file**, you can **run the script** to load the motor parameters into the **MATLAB workspace**. These variables will then be available for use in your SIMULINK model.

Next, please use **SIMULINK** to construct the **Subsystem model** as shown in Fig. 1.14.

After completing the construction of the model, **select all blocks** (you can press **Ctrl + A**), then **right-click** on the selection and choose "**Create Subsystem from Selection.**" This will encapsulate all the selected blocks into a single **Subsystem block**, as shown in Fig. 1.15.

Rename this Subsystem to **im_model_is_phir** and **save the model file** for later use in your motor control simulations or system integration.

After building the induction motor model, we also need to construct SIMULINK Subsystem models for the **Clarke transformation, inverse Clarke transformation, Park transformation**, and **inverse Park transformation**, in order to develop a complete simulation program for the **field-oriented control (FOC) system** of the induction motor.

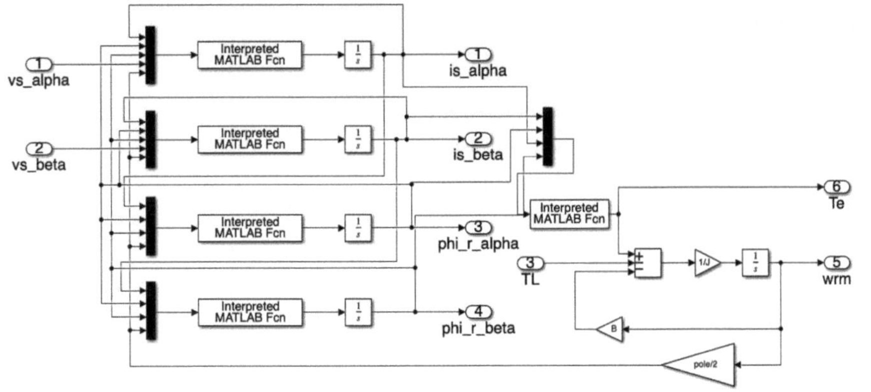

Fig. 1.14 SIMULINK model of a squirrel-cage induction motor, Example Model: im_model_is_phir.slx

Fig. 1.15 Subsystem diagram of a squirrel-cage induction motor, Example model: im_model_is_phir.slx

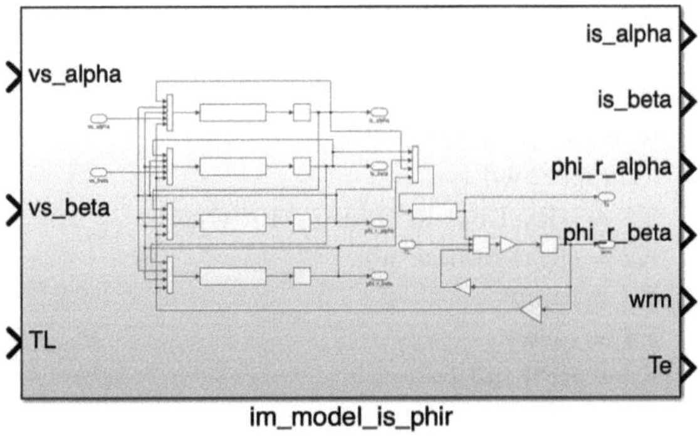

- **Building the SIMULINK Subsystem Models for Clarke Transformation and Inverse Clarke Transformation**

First, we will build the **Clarke transformation** based on Eq. (1.2.10).

Begin by creating a **new blank SIMULINK model**, and arrange the blocks as shown in Fig. 1.16.

Once the structure is complete, **select all blocks** (you can press **Ctrl + A**), then **right-click** on the selection and choose **"Create Subsystem from Selection"** to encapsulate them into a single **Subsystem block**, as shown in Fig. 1.17.

Name this Subsystem **Clarke**, and **save the file** for future use.

Next, we will build the **inverse Clarke transformation** based on Eq. (1.2.11). Start by creating a **new blank SIMULINK model**, and construct the blocks as shown in Fig. 1.18.

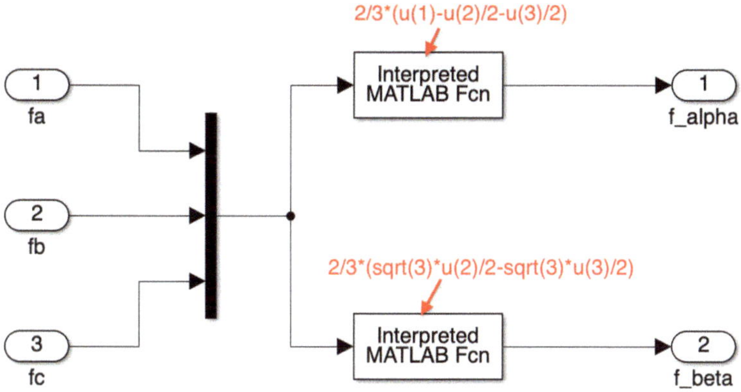

Fig. 1.16 SIMULINK model of Clarke transformation, Example model: clarke.slx

Fig. 1.17 Subsystem diagram of Clarke transformation, Example model: clarke.slx

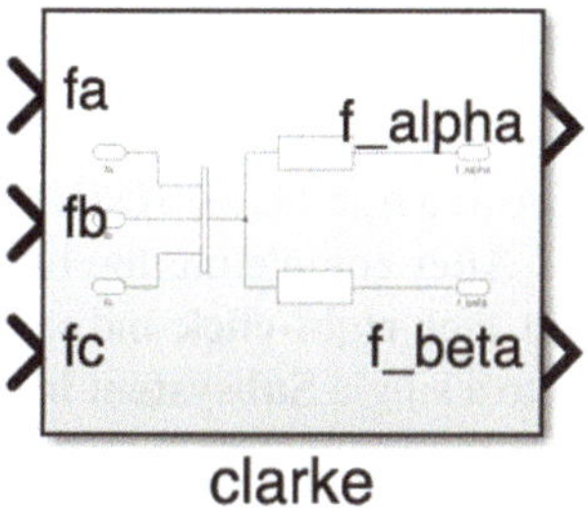

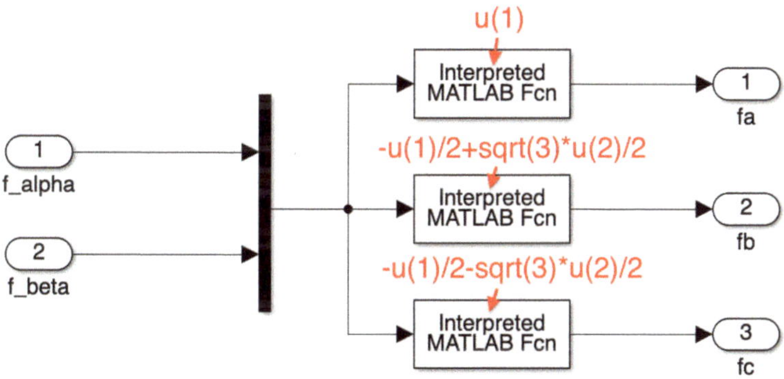

Fig. 1.18 SIMULINK model of inverse Clarke transformation, Example model: inv_clarke.slx

Once all blocks are in place, **select all of them** (by pressing **Ctrl + A**), then **right-click** on the selection and choose "**Create Subsystem from Selection**" to create a single **Subsystem block**, as shown in Fig. 1.19.

Rename this Subsystem to **inv_clarke**, and **save the file**.

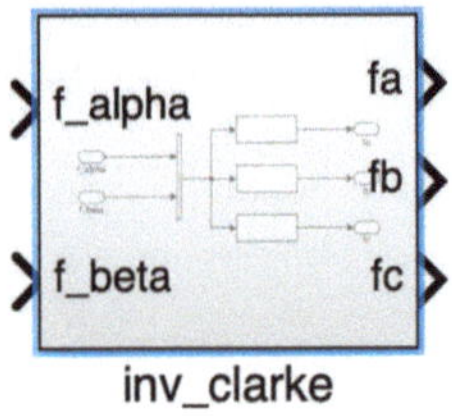

Fig. 1.19 Subsystem diagram of inverse Clarke transformation, Example model: inv_clarke.slx

- **Building the SIMULINK Subsystem Models for Park Transformation and Inverse Park Transformation**

Next, we will build the **Park transformation** based on Eq. (1.2.16).

Create a **new blank SIMULINK model**, and arrange the blocks according to Fig. 1.20. Once the block diagram is completed, **select all the blocks** (you can press **Ctrl + A**), then **right-click** and choose "**Create Subsystem from Selection**" to encapsulate them into a single **Subsystem block**, as shown in Fig. 1.21.

Name this Subsystem **park**, and **save the file** for later use.

Next, we will build the **inverse Park transformation** based on Eq. (1.2.17). Create a **new blank SIMULINK model**, and arrange the blocks as shown in Fig. 1.22.

After completing the block construction, **select all blocks** (by pressing **Ctrl + A**), then **right-click** and select **"Create Subsystem from Selection"** to group them into a single **Subsystem block**, as shown in Fig. 1.23.

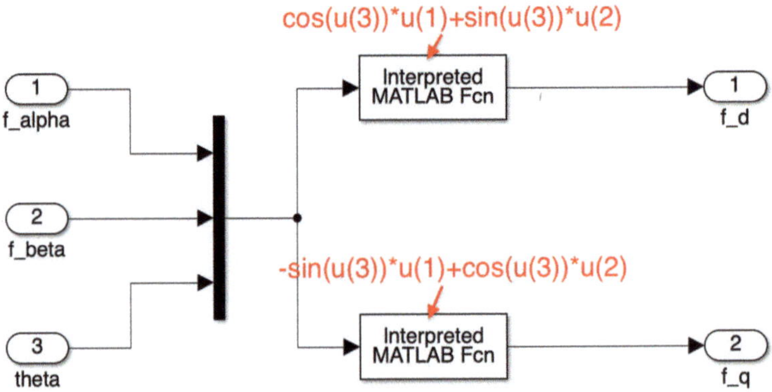

Fig. 1.20 SIMULINK model of Park transformation, Example model: park.slx

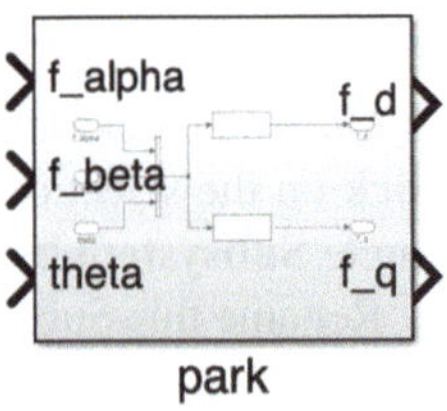

Fig. 1.21 Subsystem diagram of Park transformation, Example model: park.slx

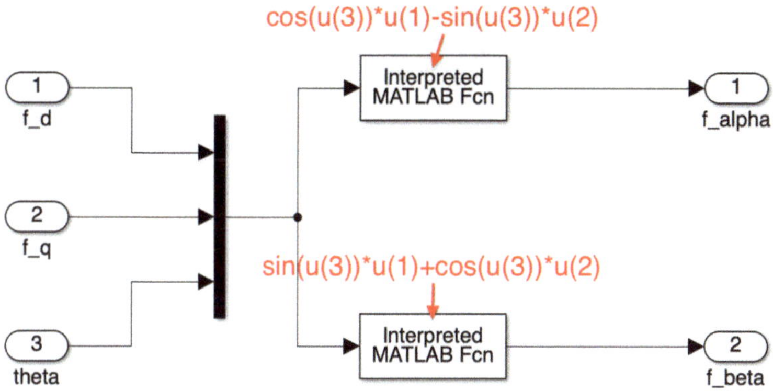

Fig. 1.22 SIMULINK model of inverse Park transformation, Example model: inv_park.slx

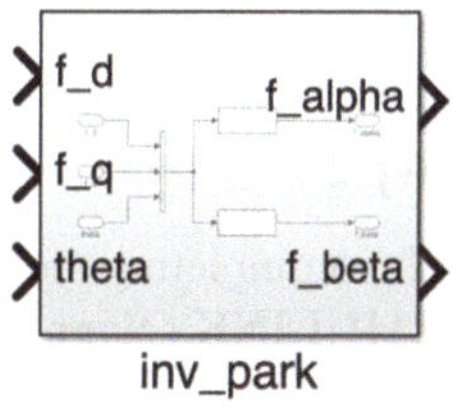

Fig. 1.23 Subsystem diagram of inverse Park transformation, Example model: inv_park.slx

Name this Subsystem **inv_park** (note: not inv_clarke), and **save the file** for later use.

- **Testing Coordinate Transformations (Clarke and Park) and the Induction Motor Model**

Step 1

Next, we will use the previously constructed **motor model** and **coordinate transformation subsystems** to simulate the dynamic behavior of an induction motor driven by a **three-phase voltage input**.

Set the input voltage to a **balanced three-phase system** in **positive sequence (a–b–c)**, with a **line voltage RMS of 220 V** and **frequency of 50 Hz**. The motor parameters are consistent with those defined in the im_params.m file, as listed in Table 1.1.

The load applied to the motor is a **constant torque load** of **8 N m**, which is applied **from the initial time**.

Please open a **blank SIMULINK model**, and connect the components as shown in Fig. 1.24.

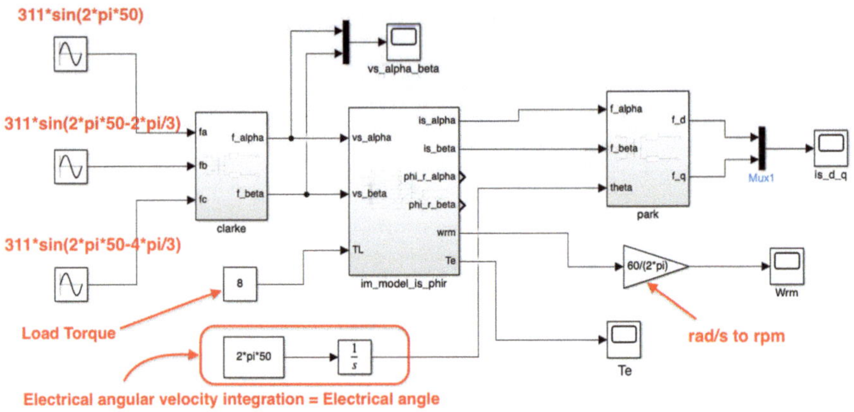

Fig. 1.24 SIMULINK model of a three-phase squirrel-cage induction motor directly driven by three-phase voltages, Example model: im_model_test1.slx

Step 2

After completing the construction of the SIMULINK block diagram, configure the **SIMULINK solver** settings as follows:

- Set the **solver type** to "**Variable-step**" with **Auto** solver selection
- Set the **maximum step size** to **0.0005**
- Set the **total simulation time** to **0.5 s** (Fig. 1.25).

Once the settings are configured, click **"Run"** to start the simulation.

Note: Before running the simulation, make sure to execute the im_params.m script in MATLAB. This script initializes the required **induction motor parameters**, and the simulation will not work properly without it.

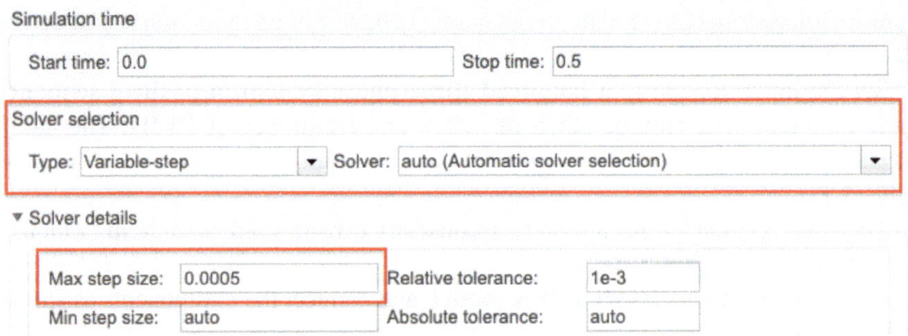

Fig. 1.25 Solver configuration block in SIMULINK

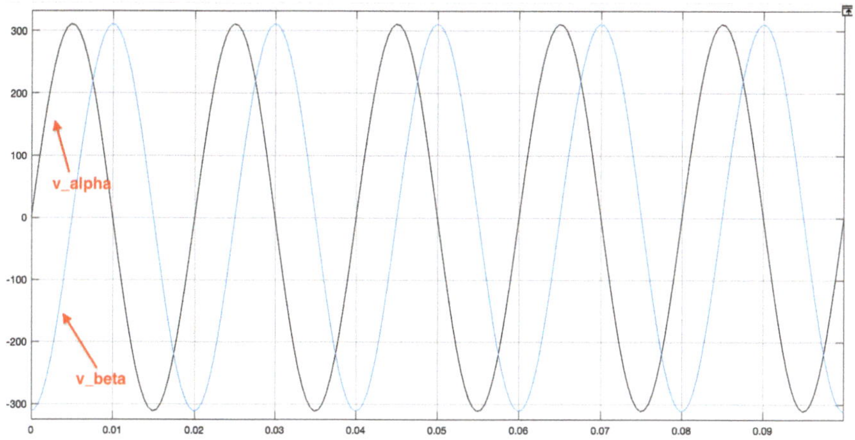

Fig. 1.26 Conversion of three-phase voltages into two-axis stationary reference frame voltage waveforms

Step 3

If the simulation completes successfully, first **double-click the vs_alpha_beta Scope block**, which displays the results of the **Clarke transformation** applied to the three-phase input voltages.

Observe the waveform from **0 to 0.1** s, as shown in Fig. 1.26. Since we applied a **positive-sequence three-phase sinusoidal voltage**, you should see that:

- v_β starts from its **negative peak** and decreases further
- v_α starts from **zero** and increases

This behavior matches the expected rotation of the **space vector under positive sequence excitation**, which is consistent with the result shown earlier in Fig. 1.8.

Therefore, the functionality of the **Clarke transformation Subsystem** has been successfully **verified**.

Step 4

Next, double-click the **Wrm scope block** to view the waveform shown in Fig. 1.27. You will observe that the motor speed stabilizes at approximately **1500 rpm** after **0.2** s. If you zoom in on the steady-state region, you'll find that the final steady-state speed is around **1490 rpm**, which is reasonable.

This is because, for a 4-pole induction motor supplied with **50 Hz** AC, the **electrical synchronous speed** is:

$$\text{Synchronous speed (RPM)} = \frac{120 \times f}{Poles} = 120 \times \frac{50}{4} = 1500 \text{ (RPM)}$$

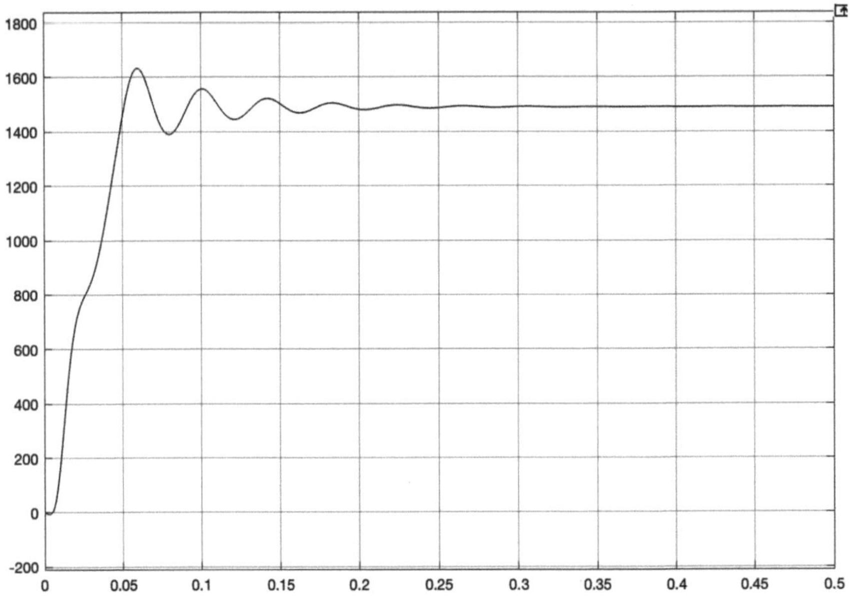

Fig. 1.27 Speed waveform of the induction motor

However, due to the **slip** in induction motors (i.e., the difference between synchronous speed and mechanical speed), the **actual mechanical speed** will always be **slightly lower** than the synchronous speed.

Next, double-click the **Te scope block**, and you'll see that the **steady-state electromagnetic torque** is approximately **9.2 Nm**, which is slightly higher than the **load torque of 8 Nm**. This is because a portion of the torque is also used to overcome **friction** (as modeled by the damping coefficient B).

This simulation result also **indirectly verifies** that the constructed **induction motor Subsystem model** is functioning properly.

Step 5

Next, double-click the **is_d_q scope block** to observe the stator current of the motor after **Park transformation**, as shown in Fig. 1.28. From the waveform, you can see that after approximately **0.25 s** of transient behavior, the stator current in the **synchronous rotating reference frame** stabilizes to **DC values** for both the **d- and q-axis components**.

*(Note: the electrical angular frequency input to the park transformation is **50 Hz**, which is synchronized with the input voltage.)*

This confirms that the **Park transformation block is functioning correctly**.

In this example:

- The **d-axis component** of the stator current stabilizes at approximately − **11.5**
- The **q-axis component** stabilizes at approximately − **3.4**.

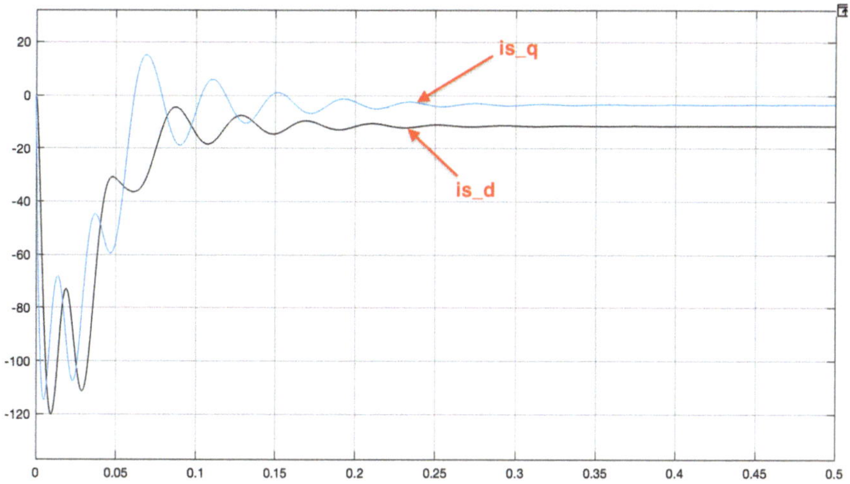

Fig. 1.28 Stator current waveforms in the two-axis synchronous rotating reference frame

We can calculate the **peak current magnitude** using the Euclidean norm:

$$\sqrt{(-11.5)^2 + (-3.4)^2} = 12$$

At this point, you can use the **inverse Park transformation** followed by the **inverse Clarke transformation** to reconstruct the three-phase stator current. This allows you to verify whether the actual peak current is indeed **12 A**, providing an additional check on the accuracy of the coordinate transformations and confirming the correctness of the implementation.

Step 6

Please add the previously created **inv_park** and **inv_clarke** subsystems to your **SIMULINK** model, as shown in Fig. 1.29.

After completing the integration, **run the simulation again** to verify the results.

After completing the simulation, double-click the is_abc scope block and zoom in on the steady-state sinusoidal currents, as shown in Fig. 1.30. The peak value of the three-phase current is approximately 12A, which matches the result calculated in Step 5. This indirectly verifies that the inv_park and inv_clarke blocks we created are functioning correctly.

- **Verifying the Power Factor Angle** φ

Step 1

First, we can plot the stator voltage space vector V_{abcs}, synthesized from the three-phase stator input voltages, and the stator current space vector I_{abcs} on a 2D coordinate plane (note: the subscript s denotes stator quantities), as shown in Fig. 1.31.

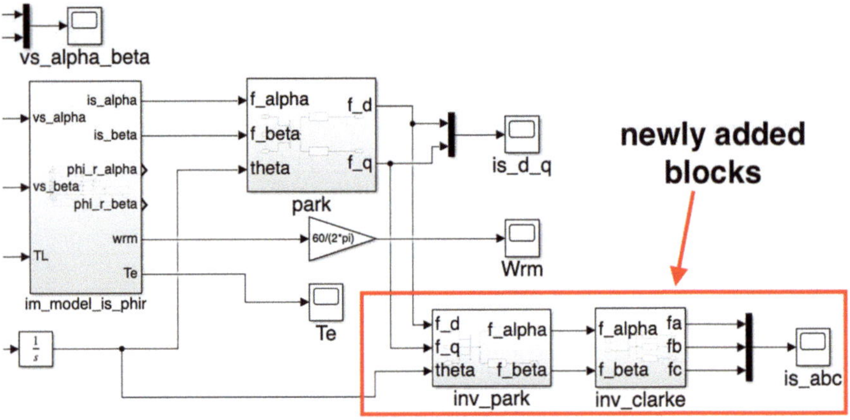

Fig. 1.29 Add the inverse park and inverse Clarke transformation blocks to the simulation system, Example model: im_model_test1.slx

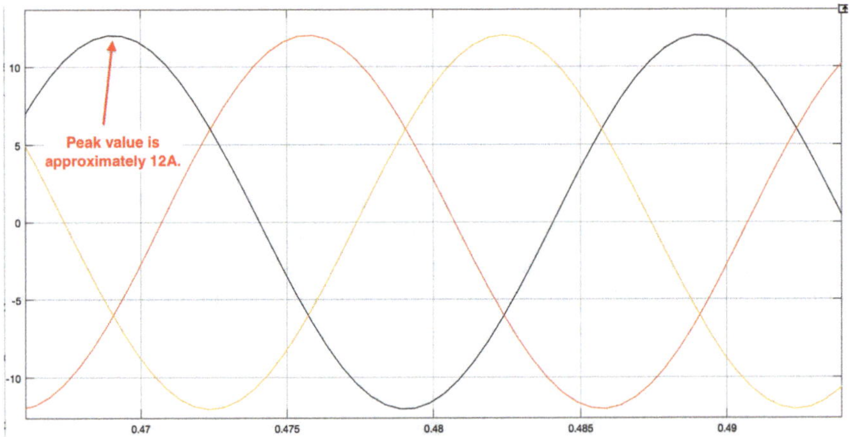

Fig. 1.30 Steady-state waveforms of the reconstructed three-phase stator currents

In this diagram, the d-q axes, the stator voltage space vector V_{abcs}, and the stator current space vector I_{abcs} all rotate synchronously at the electrical angular velocity ω_e, but there is a phase difference between them.

As previously verified in Sect. 1.2, for a positive-sequence voltage, its projection on the synchronously rotating d-q axes yields a **d-axis component of zero** and a **q-axis component equal to the negative peak value** (refer to the result in Fig. 1.4). This corresponds to the vector V_{abcs} shown in Fig. 1.31.

In this example, the d-axis component of the stator current is approximately − 11.5, and the q-axis component is about − 3.4. Thus, we can plot I_{abcs} on the coordinate plane as shown in Fig. 1.31.

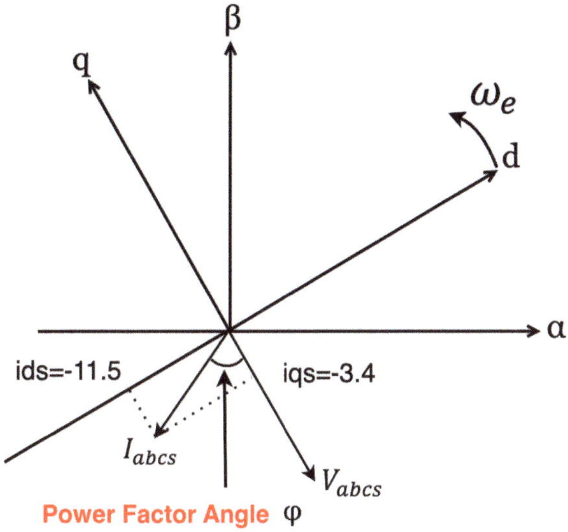

Fig. 1.31 Illustration of the power factor angle in two-dimensional space

The angle between V_{abcs} and I_{abcs} is the **power factor angle**, and it can be calculated using the following formula:

$$\varphi = \tan^{-1}\left(\frac{11.5}{3.4}\right) = 1.28 \text{ rad} = 73^\circ \tag{1.3.40}$$

By performing a simple trigonometric calculation, the power factor angle φ is found to be **73°**.

Step 2

To verify the correctness of the power factor angle, we can add another Scope block in the SIMULINK model to observe the waveforms of vsa (i.e., stator phase-a voltage) and isa (i.e., stator phase-a current), as shown in Fig. 1.32.

Step 2

After adding the blocks, run the simulation again. Once completed, double-click the **vsa_isa** Scope block to view the waveform, as shown in Fig. 1.33.

Next, zoom in on the waveform around **0.45 s** using the mouse. You will observe that vsa leads isa by approximately **0.004068 s** (note: this means the voltage leads the current by 0.004068 s).

We can then use the following formula to calculate the phase difference between voltage and current, which represents the **power factor angle**:

$$\varphi = 0.004068 \times 2 \times \pi \times 50 = 1.278 \text{ rad} \approx 73^\circ$$

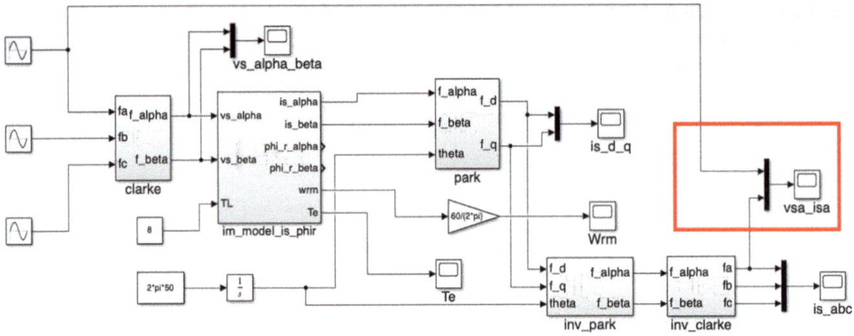

Fig. 1.32 Add an additional oscilloscope block to simultaneously display the phase-a voltage and phase-a current waveforms, Example model: im_model_test1.slx

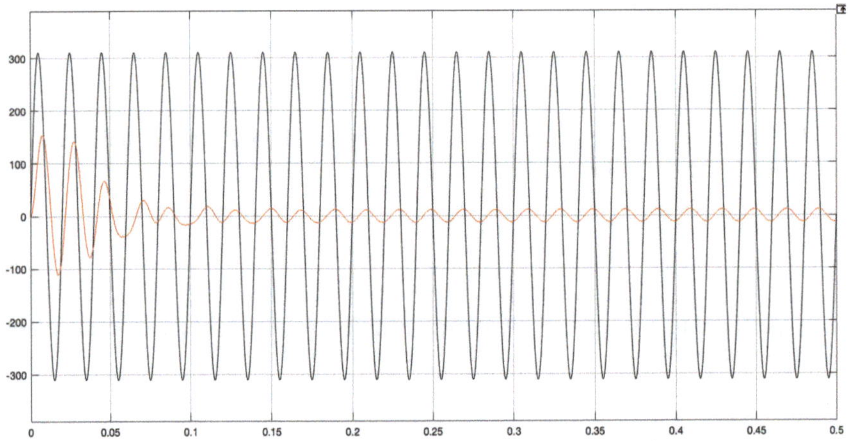

Fig. 1.33 Phase-a voltage and Phase-a current waveforms

So, the **power factor angle** φ is approximately **73°**, matching the previously calculated value.

1.4 Space Vector Model of Three-Phase Permanent Magnet Synchronous Motors (PMSMs)

Induction Motors (IM) and Permanent Magnet Synchronous Motors (PMSM) are two different types of AC motors, with their primary difference lying in the way the magnetic field is generated in the rotor.

In an **Induction Motor**, the rotor is made of conductive material. When AC current flows through the stator windings, it generates a rotating magnetic field. This rotating field induces a current in the rotor conductors, which in turn produces a magnetic field that opposes the stator field. The interaction between these magnetic fields generates torque, causing the rotor to rotate.

In contrast, a **Permanent Magnet Synchronous Motor** has permanent magnets mounted on or embedded in the rotor, which generate a constant magnetic field. The rotor's magnetic field stays precisely synchronized with the rotating magnetic field produced by the stator—hence the term "synchronous" motor. When AC current is applied to the stator, the interaction between the stator's and rotor's magnetic fields produces torque and drives the rotor.

Overall, PMSMs offer higher efficiency and power density due to the use of permanent magnets. They are typically used in high-performance and high-precision applications such as electric vehicles, robotics, and industrial drive systems. On the other hand, induction motors are generally used in applications where performance and precision are less critical, such as household appliances, pumps, and fans.

If the permanent magnets of a permanent magnet synchronous motor (PMSM) are mounted on the surface of the rotor, the motor is referred to as a **Surface-Mounted Permanent Magnet Synchronous Motor (SPMSM)**. If the magnets are embedded inside the rotor, the motor is known as an **Interior Permanent Magnet Synchronous Motor (IPMSM)**.

Since the rotor magnetic field speed ω_r (electrical speed) of a PMSM is synchronized with the stator magnetic field speed ω_e (also electrical speed), we can define an arbitrary rotating reference frame (called the d^a-q^a axis) attached to the rotor. By aligning the d^a-axis with the rotor magnetic pole direction and setting the rotation speed ω of the reference frame to the rotor magnetic field speed ω_r, we get:

$$\omega = \omega_r = \omega_e$$

This means the reference frame rotates synchronously with the rotor. Therefore, in PMSM systems, this synchronously rotating reference frame is also known as the **rotor reference frame**.

Due to the complexity of the derivation process for the space vector model of a three-phase Permanent Magnet Synchronous Motor (PMSM) [1, 4], it is omitted here. A fully derived space vector model of a **Surface-Mounted Permanent Magnet Synchronous Motor (SPMSM)** can be expressed as follows [1, 4]:

$$R_s I_{abcs} + L_s \frac{dI_{abcs}}{dt} + j\omega_e \lambda_f e^{j\theta_e} = V_{abcs} \tag{1.4.1}$$

where:

- I_{abcs} is the stator current space vector.
- V_{abcs} is the stator voltage space vector.
- λ_f is the flux linkage produced by the rotor (permanent magnets) in the stator windings.
- R_s is the stator resistance (per phase).

- L_s is the stator inductance (per phase).

The flux linkage λ_f generated by the rotor (permanent magnets) in the stator windings can be expressed as:

$$\lambda_f = L_m I_f \tag{1.4.2}$$

where L_m is the mutual inductance between the rotor and the stator windings. Since the permanent magnets in the rotor provide a constant magnetic field, they can be equivalently represented as a constant current source I_f, which interacts with the stator windings to produce the flux linkage λ_f.

Since we know that $I_{abcs} = i^r_{ds} + ji^r_{qs}$ and $V_{abcs} = v^r_{ds} + jv^r_{qs}$, Eq. (1.4.1) can therefore be expanded into the d-q axis form as follows:

$$v^r_{ds} = R_s i^r_{ds} + L_s \frac{di^r_{ds}}{dt} - \omega_r L_s i^r_{qs} \tag{1.4.3}$$

$$v^r_{qs} = R_s i^r_{qs} + L_s \frac{di^r_{qs}}{dt} + \omega_r \left(L_s i^r_{ds} + \lambda_f\right) \tag{1.4.4}$$

where $L_s = L_{ls} + L_m$, with L_{ls} being the stator leakage inductance and L_m the mutual inductance between the stator and rotor. The term $L_s i^r_{qs} = \lambda^r_{qs}$ represents the stator q-axis flux linkage, and $L_s i^r_{ds} + \lambda_f$ represents the stator d-axis flux linkage.

Additionally, the torque equation of a surface-mounted permanent magnet synchronous motor (SPMSM) can be expressed as:

$$T_e = \frac{3}{2}\frac{P}{2}\lambda_f i^r_{qs} \tag{1.4.5}$$

where P is the number of poles of the motor.

Furthermore, for an Interior Permanent Magnet Synchronous Motor (IPMSM), since the d- and q-axis inductances are not equal, the voltage Eqs. (1.4.3) and (1.4.4) need to be rewritten as Eqs. (1.4.6) and (1.4.7):

$$v^r_{ds} = R_s i^r_{ds} + L_d \frac{di^r_{ds}}{dt} - \omega_r L_q i^r_{qs} \tag{1.4.6}$$

$$v^r_{qs} = R_s i^r_{qs} + L_q \frac{di^r_{qs}}{dt} + \omega_r \left(L_d i^r_{ds} + \lambda_f\right) \tag{1.4.7}$$

Compared to SPMSMs, Interior Permanent Magnet Synchronous Motors (IPMSMs) have unequal d- and q-axis inductances, which results in the generation of additional reluctance torque. Therefore, their torque equation must be expressed as:

$$T_e = \frac{3}{2}\frac{P}{2}\left[\lambda_f i^r_{qs} + \left(L_d - L_q\right) i^r_{ds} i^r_{qs}\right] \tag{1.4.8}$$

For Interior Permanent Magnet Synchronous Motors (IPMSMs), the q-axis inductance is greater than the d-axis inductance, i.e., $L_q > L_d$. This characteristic is known as *saliency*. When the q-axis current is positive, applying a negative d-axis current can produce a positive reluctance torque. Typically, this additional reluctance torque can be harnessed when the IPMSM operates in Maximum Torque per Ampere (MTPA) mode or in the field-weakening region.

Explanation

For Surface-mounted Permanent Magnet Synchronous Motors (SPMSMs), the d-axis and q-axis inductances are equal, i.e., $L_d = L_q = L_s$.

Equations (1.4.6) and (1.4.7) can be represented as an equivalent circuit in the d-q axis reference frame, as shown in Fig. 1.34 [1, 6].

Tips

The mathematical model of an Interior Permanent Magnet Synchronous Motor (IPMSM) can be considered as the generalized form of a Permanent Magnet Synchronous Motor (PMSM) model. When setting $L_d = L_q = L_s$, this generalized model simplifies to that of a Surface-mounted Permanent Magnet Synchronous Motor (SPMSM), corresponding to Eqs. (1.4.3) and (1.4.4).

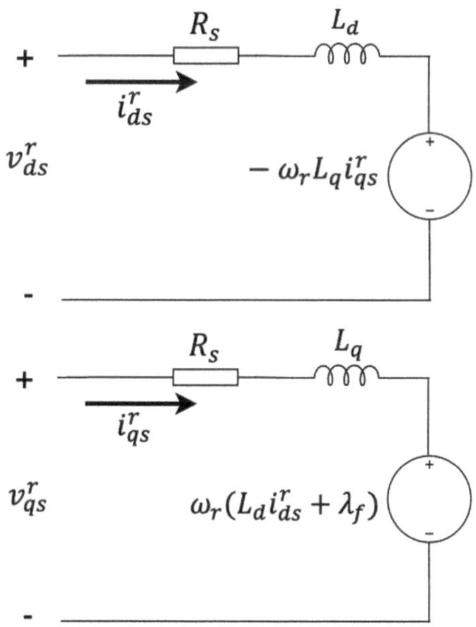

Fig. 1.34 d–q axis equivalent circuit diagram of a permanent magnet synchronous motor

- **Building an Permanent Magnet Synchronous Motor Subsystem Model Using MATLAB/SIMULINK**

Next, we will use MATLAB/SIMULINK to build a Permanent Magnet Synchronous Motor (PMSM) model. We will construct the mathematical model of an Interior Permanent Magnet Synchronous Motor (IPMSM) based on Eqs. (1.4.6) and (1.4.7). The reason for using the IPMSM model is its greater generality—when we set the parameters such that $L_d = L_q$, the IPMSM model simplifies into the form of a Surface-Mounted Permanent Magnet Synchronous Motor (SPMSM).

Before actually building the SIMULINK model, we will first use a MATLAB m-file to define all the parameters required for constructing the SIMULINK model of the Permanent Magnet Synchronous Motor (IPMSM), as listed in Table 1.2.

MATLAB m-file Example script: pm_params.m

```
Rs = 1.2;
Ld = 0.0057;
Lq = 0.0125;
Lamda_f = 0.03;
pole = 4;
J = 0.00016;
B = 0.0000028;
```

In the above, we set the inductance parameters L_d and L_q to be different in order to simulate the behavior of an Interior Permanent Magnet Synchronous Motor (IPMSM). After completing the creation of the m-file, please run the script to load the motor parameters into the MATLAB environment. Next, please use SIMULINK to build the Subsystem model as shown in Fig. 1.35.

After completing the model, select all the blocks (you can use CTRL + A), right-click, and choose "**Create Subsystem from Selection**" to create a single Subsystem block, as shown in Fig. 1.36. Name it "**pm_model_dq**" and save it. This completes the modeling of the three-phase Permanent Magnet Synchronous Motor (PMSM).

Table 1.2 Permanent magnet synchronous motor parameters [6]

Parameter	Symbol	Value	Unit
Stator resistance	R_s	1.2	Ω
Stator d-axis inductance	L_d	5.7	mH
Stator q-axis inductance	L_q	12.5	mH
Rotor flux linkage	λ_f	0.03	Wb
Number of poles	P	4	–
Moment of inertia	J	0.00016	Kg m^2
Friction coefficient	B	0.0000028	N m s/rad

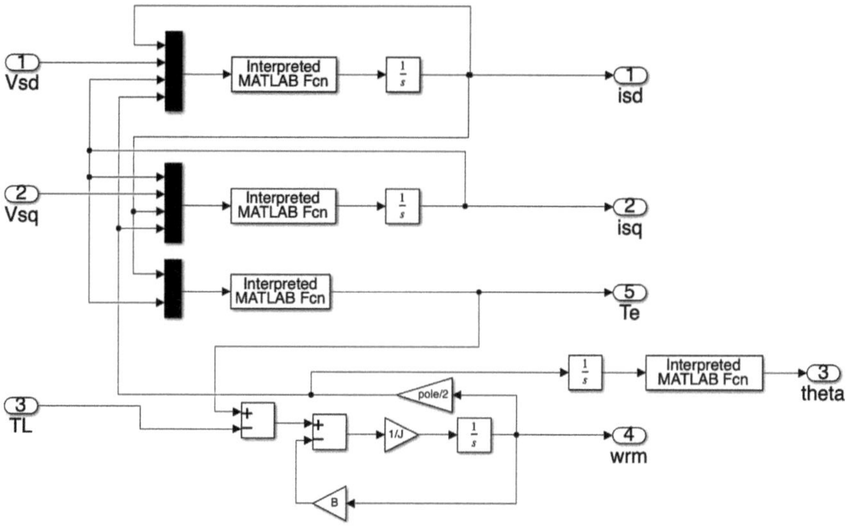

Fig. 1.35 SIMULINK model of a permanent magnet synchronous motor, Example model: pm_model_dq.slx

Fig. 1.36 Subsystem diagram of a permanent magnet synchronous motor, Example model: pm_model_dq.slx

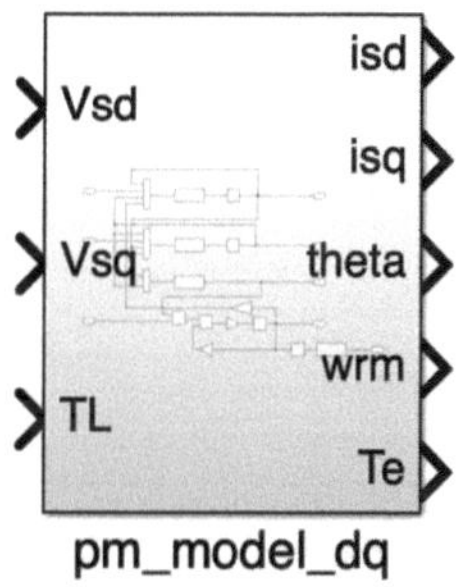

1.5 Summary

The key points of this chapter are summarized as follows:

- In the real world, induction motors can operate directly with three-phase AC input (as demonstrated in the simulation program of Sect. 1.3). However, permanent magnet synchronous motors (PMSMs) cannot be directly driven by three-phase AC, because they require rotor position feedback to decouple the d-q axis currents. Therefore, field-oriented control (FOC) must be applied to operate them correctly. In the next chapter, we will introduce the theory of field-oriented control for both induction motors and PMSMs, and use MATLAB/SIMULINK to simulate their field-oriented control systems.

- Figures 1.13 and 1.34 show the d-q axis equivalent circuits of induction motors and PMSMs, respectively. These are circuit representations of the motor's d-q axis models and have significant physical meaning. They are frequently used when developing motor control strategies or self-learning algorithms for motor parameters.
- The motor parameter values used in field-oriented control (e.g., those listed in Tables 1.1 and 1.2), such as stator resistance, rotor resistance, stator inductance, and rotor inductance, are all per-phase values of the motor's windings after being represented in an equivalent Y (wye) connection [1, 5, 6].

References

1. Sul, Seung-Ki, Control of Electric Machine Drive Systems, Wiley-IEEE Press, 2011.
2. K. Hasse, "On the Dynamics of Speed control of a Static AC Drive with a Squirrel-cage induction machine", PhD dissertation, Tech. Hochsch. Darmstadt, 1969.
3. F. Blaschke, "The principle of field orientation as applied to the NEW Transvector closed-loop control system for rotating-field machines," Siemens Rev., vol. 34, pp. 217–220, 1972.
4. N. Mohan, T. M. Undeland, and W. P. Robbins, Power Electronics: Converters, Applications and Design, Second ed. New York:Wiley, 1995.
5. Chang-Hwan Liu, AC Motor Control: Principles of Vector Control and Direct Torque Control, Taipei: Dong Hua Book Co., 2001.
6. Chapman, S. J. (2011). Electric machinery fundamentals (5th ed.). McGraw-Hill Education.

Chapter 2
Design of AC Motor Control Loops

In Chap. 1, we reviewed the fundamental concepts required for Field Oriented Control (FOC) of AC motors [1–3], including the principle of separately excited DC motors, the concept of space vectors, coordinate transformations, and the AC motor models derived using the space vector approach. We also discussed and modeled the two mainstream types of AC motors: induction motors and permanent magnet synchronous motors (PMSMs).

In this chapter, we will first explore the FOC strategies for both induction motors and PMSMs. Then, based on the FOC framework, we will proceed with the design of control loops for AC motor systems [4–6]. Since control loop design techniques are highly transferable, we will focus on the PMSM [3, 5, 7] as the primary target for control loop design. However, all the design concepts and techniques discussed in this chapter are equally applicable to the control of induction motors.

2.1 Field-Oriented Control (FOC) Strategies

2.1.1 FOC for Squirrel-Cage Induction Motors

The purpose of **Field-Oriented Control (FOC)** is to enable independent control of the **magnetic flux** and **torque** of a three-phase induction motor [1–3], thereby achieving a control effect similar to that of a **separately excited DC motor**.

Let us now revisit the **three-phase squirrel-cage induction motor model** established in Chap. 1, expressed in an **arbitrary rotating two-axis (d–q) reference frame** [1–3]:

Supplementary Information The online version contains supplementary material available at https://doi.org/10.1007/978-981-95-3261-2_2.

C-C. Yeh, *AC Motor Control Loop Design*, Power Systems,
https://doi.org/10.1007/978-981-95-3261-2_2

$$R_s i^a_{ds} + L_\sigma \frac{di^a_{ds}}{dt} - \omega L_\sigma i^a_{qs} + \frac{L_m}{L_r}\frac{d\phi^a_{dr}}{dt} - \frac{\omega L_m}{L_r}\phi^a_{qr} = v^a_{ds} \tag{2.1.1}$$

$$\omega L_\sigma i^a_{ds} + R_s i^a_{qs} + L_\sigma \frac{di^a_{qs}}{dt} + \omega \frac{L_m}{L_r}\phi^a_{dr} + \frac{L_m}{L_r}\frac{d\phi^a_{qr}}{dt} = v^a_{qs} \tag{2.1.2}$$

$$-R_r L_m i^a_{ds} + R_r \phi^a_{dr} + L_r \frac{d\phi^a_{dr}}{dt} - (\omega - \omega_r) L_r \phi^a_{qr} = 0 \tag{2.1.3}$$

$$-R_r L_m i^a_{qs} + (\omega - \omega_r) L_r \phi^a_{dr} + R_r \phi^a_{qr} + L_r \frac{d\phi^a_{qr}}{dt} = 0 \tag{2.1.4}$$

where $L_\sigma = L_s - \frac{L_m^2}{L_r}$.

When the rotational speed $\boldsymbol{\omega}$ of the arbitrary rotating two-axis coordinate frame (d^a-q^a) is equal to the **synchronous speed** (i.e., $\omega = \omega_e$), Eqs. (2.1.1) to (2.1.4) can be rewritten as:

$$R_s i^e_{ds} + L_\sigma \frac{di^e_{ds}}{dt} - \omega_e L_\sigma i^e_{qs} + \frac{L_m}{L_r}\frac{d\phi^e_{dr}}{dt} - \frac{\omega L_m}{L_r}\phi^e_{qr} = v^e_{ds} \tag{2.1.5}$$

$$\omega_e L_\sigma i^e_{ds} + R_s i^e_{qs} + L_\sigma \frac{di^e_{qs}}{dt} + \omega_e \frac{L_m}{L_r}\phi^e_{dr} + \frac{L_m}{L_r}\frac{d\phi^e_{qr}}{dt} = v^e_{qs} \tag{2.1.6}$$

$$-R_r L_m i^e_{ds} + R_r \phi^e_{dr} + L_r \frac{d\phi^e_{dr}}{dt} - (\omega_e - \omega_r) L_r \phi^e_{qr} = 0 \tag{2.1.7}$$

$$-R_r L_m i^e_{qs} + (\omega_e - \omega_r) L_r \phi^e_{dr} + R_r \phi^e_{qr} + L_r \frac{d\phi^e_{qr}}{dt} = 0 \tag{2.1.8}$$

In Eqs. (2.1.5)–(2.1.8), the superscript **e** indicates that the variables (such as stator current, rotor flux linkage, and stator voltage) are expressed in the **synchronously rotating two-axis reference frame** (d^e-q^e). In this reference frame, these quantities appear as **DC (steady-state) values** because we are observing them in a coordinate system that rotates at the **same angular speed** as their space vectors.

Explanation

Since the space vectors of stator current, stator voltage, and rotor flux linkage rotate in space at the synchronous speed ω_e, when observed in the **synchronously rotating reference frame**, they appear as **DC quantities**.

For the derivation of the FOC strategy, a crucial assumption must first be made (here we adopt the **direct rotor flux-oriented control method** [1, 3, 5, 6, 8, 9]):

It is assumed that at any given time, the **magnitude and position of the rotor flux linkage** are known. The **d-axis of the synchronous rotating reference frame** is aligned with the rotor flux linkage and rotates synchronously with it.

Under this condition, $\phi_{dr}^{e} = \left|\boldsymbol{\Phi}_{abcr}^{e}\right| = \boldsymbol{\Phi}_{r}$ and $\phi_{qr}^{e} = 0$ [1, 3, 5, 6, 8, 9]. Then, we can express Eqs. (2.1.5) to (2.1.8) as follows:

$$R_s i_{ds}^e + L_\sigma \frac{di_{ds}^e}{dt} - \omega_e L_\sigma i_{qs}^e + \frac{L_m}{L_r}\frac{d\phi_{dr}^e}{dt} - \frac{\omega L_m}{L_r} \times 0 = v_{ds}^e \tag{2.1.9}$$

$$\omega_e L_\sigma i_{ds}^e + R_s i_{qs}^e + L_\sigma \frac{di_{qs}^e}{dt} + \omega_e \frac{L_m}{L_r} \times \phi_{dr}^e + \frac{L_m}{L_r} \times 0 = v_{qs}^e \tag{2.1.10}$$

$$-R_r L_m i_{ds}^e + R_r \phi_{dr}^e + L_r \times \frac{d\phi_{dr}^e}{dt} - (\omega_e - \omega_r) L_r \times 0 = 0 \tag{2.1.11}$$

$$-R_r L_m i_{qs}^e + (\omega_e - \omega_r) L_r \phi_{dr}^e + R_r \times 0 + L_r \times 0 = 0 \tag{2.1.12}$$

It can be arranged as follows:

$$\frac{di_{ds}^e}{dt} = \left(-\frac{R_s}{L_\sigma} - \frac{1-\sigma}{\sigma \tau_r}\right) i_{ds}^e + \omega_e i_{qs}^e + \frac{1-\sigma}{\sigma \tau_r L_m}\Phi_r + \frac{v_{ds}^e}{L_\sigma} \tag{2.1.13}$$

$$\frac{di_{qs}^e}{dt} = -\frac{R_s}{L_\sigma} i_{qs}^e - \omega_e i_{ds}^e - \frac{(1-\sigma)}{\sigma L_m}\omega_e \Phi_r + \frac{v_{qs}^e}{L_\sigma} \tag{2.1.14}$$

$$\frac{d\Phi_r}{dt} = -\frac{R_r}{L_r}\Phi_r + R_r \frac{L_m}{L_r} i_{ds}^e \tag{2.1.15}$$

$$-R_r \frac{L_m}{L_r} i_{qs}^e + \omega_{sl}\Phi_r = 0 \tag{2.1.16}$$

Tips

To derive Eq. (2.1.13), we substitute Eq. (2.1.15) into Eq. (2.1.9), replacing the term $\frac{d\phi_{dr}^e}{dt}$ with $-\frac{R_r}{L_r}\Phi_r + R_r \frac{L_m}{L_r} i_{ds}^e$.

where $\sigma = 1 - \frac{L_m^2}{L_s L_r}$, $L_\sigma = \sigma L_s$, $\tau_r = \frac{L_r}{R_r}$,
and ω_{sl} (slip angular speed) $= \omega_e - \omega_r$.
Since $\phi_{qr}^e = 0$ and $\phi_{dr}^e = \Phi_r$, the torque Eq. (1.3.30) can be written as:

$$T_e = \frac{3P}{4}\frac{L_m}{L_r}\left(i_{qs}^e \phi_{dr}^e - i_{ds}^e \phi_{qr}^e\right) = \frac{3P}{4}\frac{L_m}{L_r}\left(i_{qs}^e \Phi_r\right) \tag{2.1.17}$$

From Eq. (2.1.15), it can be observed that the rotor flux linkage Φ_r is only related to the stator current i_{ds}^e in the d^e-axis. The motor torque Eq. (2.1.17) shows that the torque is proportional to the product of the rotor flux linkage Φ_r and the stator current i_{qs}^{e}. If the rotor flux linkage Φ_r is stably controlled, then the motor torque becomes proportional to i_{qs}^{e}. As a result, the flux and torque of the induction motor are successfully decoupled and can be independently controlled, achieving control performance similar to that of a separately excited DC motor.

Therefore, to effectively control the speed of an induction motor, the rotor flux linkage Φ_r must first be stably regulated. Since the rotor flux linkage Φ_r is controlled by the stator current i_{ds}^{e}, two control loops are required for field-oriented control:

- Rotor flux linkage Φ_r control loop
- Stator current i_{ds}^{e} control loop.

When the rotor flux linkage Φ_r is stably controlled, the motor torque can be regulated by the stator current i_{qs}^{e}. We can convert the torque command generated by the speed control loop into the q-axis stator current command i_{qs}^{e*}, which serves as the reference input for the stator current i_{qs}^{e} control loop. Therefore, to implement speed control, the following two additional control loops are required:

- Motor speed ω_{rm} control loop
- Stator current i_{qs}^{e} control loop.

Therefore, a complete FOC system for an induction motor requires a total of four control loops:

- **d-axis stator current i_{ds}^{e} control loop**
- **q-axis stator current i_{qs}^{e} control loop**
- **Rotor flux Φ_r control loop**
- **Motor speed ω_{rm} control loop.**

Next, we will design the controller parameters for each of these control loops i_{ds}^{e}, i_{qs}^{e}, Φ_r, and ω_{rm} in sequence. The design principles are as follows:

- **Inside-out design**: Design the inner loops first, followed by the outer loops.
- **Bandwidth separation**: The bandwidth of the inner loop must be at least **five times greater** than that of the outer loop.

Note: When the inner loop bandwidth is set to be five times or more than that of the outer loop, the inner loop can be approximated as a unity gain system during the outer loop controller design, simplifying the process [1, 4].

- **Induction Motor Field-Oriented Current Loop Design** [1, 4, 10]

In general, the design sequence of control loops follows an inside-out approach. To construct a complete field-oriented control (FOC) system for an induction motor, the first step is to establish the **d-axis** (d^e) **and q-axis** (q^e) **current control loops**, as these serve as the inner loops for both the flux and speed control loops.

By examining Eqs. (2.1.13) and (2.1.14), we observe that the differential equations of the stator d- and q-axis currents contain **nonlinear coupling terms**, as illustrated in Fig. 2.1.

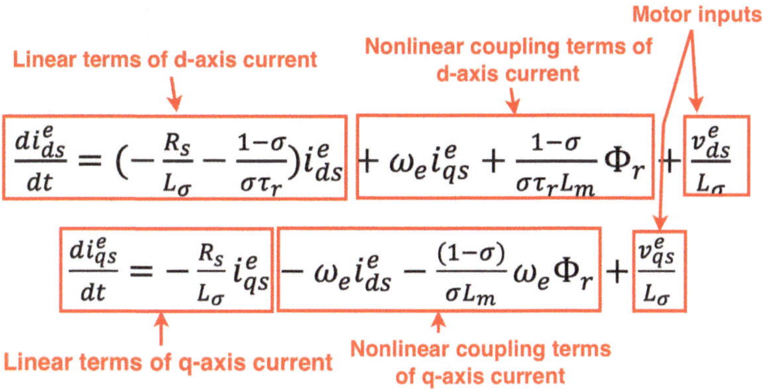

Fig. 2.1 Differential equations of the d- and q-axis currents of the induction motor

In Field-Oriented Control (FOC), the interaction between the stator current and rotor magnetic flux produces nonlinear coupling effects. These nonlinear coupling terms appear in the mathematical model of the induction motor as cross-coupling components, such as the product of the stator current and rotor speed. During actual control operation, these nonlinear coupling terms can degrade the performance of the current control loops.

To address this issue, compensation for the nonlinear coupling terms is required in FOC. Typically, a decoupling control strategy is employed to achieve this goal. Decoupling control is a method that introduces compensating terms to cancel out the effects of nonlinear coupling, thereby minimizing the mutual influence between flux-producing current and torque-producing current, and consequently improving overall control performance.

- **Design of the d-Axis Current Control Loop for Induction Motor**

To make it easier for readers to understand, we first assume that the nonlinear coupling terms have been perfectly compensated. Taking the d-axis as an example, if the nonlinear coupling terms are perfectly canceled, then the differential equation of the d-axis current can be written as:

$$\frac{di_{ds}^e}{dt} = \left(-\frac{R_s}{L_\sigma} - \frac{1-\sigma}{\sigma\tau_r}\right)i_{ds}^e + \frac{v_{ds}^e}{L_\sigma} \tag{2.1.18}$$

At this point, we take the Laplace transform of Eq. (2.1.18) to obtain:

$$I_{ds}^e(s) = V_{ds}^e(s) \cdot \frac{\frac{1}{L_\sigma}}{s + \left(\frac{R_s}{L_\sigma} + \frac{1-\sigma}{\sigma\tau_r}\right)} \tag{2.1.19}$$

Assuming a PI controller is designed for the d-axis current, with its output being $V_{ds}^e(s)$, the d-axis current control loop can be represented as shown in Fig. 2.2.

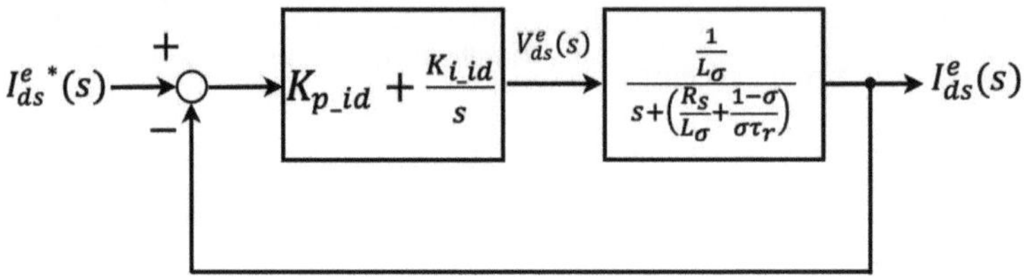

Fig. 2.2 d-axis current control block diagram of an induction motor with perfect nonlinearity compensation

Where $I_{ds}^{e*}(s)$ is the d-axis current reference, the proportional and integral gains of the d-axis current PI controller are K_{p_id} *and* K_{i_id} *respectively. The controlled plant for the d-axis current of the induction motor is given by* $\frac{\frac{1}{L_\sigma}}{s+\left(\frac{R_s}{L_\sigma}+\frac{1-\sigma}{\sigma\tau_r}\right)}$.

When the nonlinear coupling term in the d-axis has been perfectly compensated, the d-axis current loop of the induction motor can be represented by the control structure shown in Fig. 2.2. Under this structure, we can design the proportional gain K_{p_id} *and the integral gain* K_{i_id} of the d-axis current PI controller to meet the desired bandwidth specifications.

Let $\frac{1}{L_\sigma} = N$ and $\left(\frac{R_s}{L_\sigma} + \frac{1-\sigma}{\sigma\tau_r}\right) = D$, then the controlled plant of the d-axis current can be written as:

$$\frac{\frac{1}{L_\sigma}}{s+\left(\frac{R_s}{L_\sigma}+\frac{1-\sigma}{\sigma\tau_r}\right)} = \frac{N}{s+D} \tag{2.1.20}$$

We can express the d-axis current PI controller in the following form:

$$K_{p\_id} + \frac{K_{i\_id}}{s} = K_{p\_id}\left(\frac{s+\frac{K_{i\_id}}{K_{p\_id}}}{s}\right) \tag{2.1.21}$$

Assuming that ω_d is the desired bandwidth specification for the d-axis current control loop, we can use the pole-zero cancellation method to design the parameters of the d-axis PI current controller as follows:

$$K_{p\_id} = \frac{\omega_d}{N},\ K_{i\_id} = \frac{D\cdot\omega_d}{N} \tag{2.1.22}$$

By substituting the PI controller parameters, the open-loop transfer function of the d-axis current control loop, G_{d_open}, becomes:

$$G_{d\_open} = \frac{\omega_d}{N}\left(\frac{s+D}{s}\right)\times\frac{N}{s+D} = \frac{\omega_d}{s} \tag{2.1.23}$$

Therefore, the closed-loop transfer function of the d-axis current control loop, G_{d_close}, can be designed as a first-order low-pass filter with a cutoff frequency of ω_d, as shown in Eq. (2.1.23). This completes the design of the d-axis current control loop.

$$G_{d\_close} = \frac{G_{d\_open}}{1 + G_{d\_open}} = \frac{\omega_d}{s + \omega_d} \tag{2.1.24}$$

- **Decoupling Compensation for the d-Axis Current Control Loop of the Induction Motor**

In the previously discussed design of the d-axis current control loop for the induction motor, we assumed that the nonlinear coupling terms on the d-axis had been perfectly compensated. Now, we go back to address the compensation of these nonlinear coupling terms [1, 3]. We can rewrite the differential equation of the d-axis current for the induction motor as follows:

$$\frac{di_{ds}^e}{dt} = \left(-\frac{R_s}{L_\sigma} - \frac{1-\sigma}{\sigma\tau_r}\right)i_{ds}^e + \omega_e i_{qs}^e + \frac{1-\sigma}{\sigma\tau_r L_m}\Phi_r + \frac{v_{ds}^e}{L_\sigma} \tag{2.1.25}$$

Since the output of the PI controller is v_{ds}^e, and before v_{ds}^e is applied to the motor, a feedforward compensation term f_d is added as follows:

$$f_d = -L_\sigma\left(\omega_e i_{qs}^e + \frac{1-\sigma}{\sigma\tau_r L_m}\Phi_r\right) \tag{2.1.26}$$

Therefore, the actual d-axis voltage applied to the motor is given by $v_{ds}^{e*} = v_{ds}^e + f_d$. Substituting v_{ds}^{e*} into Eq. (2.1.25), we obtain:

$$\frac{di_{ds}^e}{dt} = \left(-\frac{R_s}{L_\sigma} - \frac{1-\sigma}{\sigma\tau_r}\right)i_{ds}^e + \frac{v_{ds}^e}{L_\sigma} \tag{2.1.27}$$

From Eq. (2.1.27), we can observe that the added feedforward control term f_d effectively cancels out the internal nonlinear coupling term of the motor. This transforms the originally nonlinear d-axis current differential equation into the linear form shown in (2.1.27). Therefore, we can adopt the PI controller structure shown in Fig. 2.2 to design the d-axis current loop as a first-order low-pass filter, as described by Eq. (2.1.24).

With the feedforward compensation included, the complete d-axis current loop can be represented by the structure shown in Fig. 2.3.

- **Design of the q-Axis Current Control Loop for Induction Motor**

Next, we proceed with the design of the q-axis current control loop. Here, we also assume that the nonlinear coupling terms in the q-axis have been perfectly compensated. Under this assumption, the q-axis current differential equation can be written

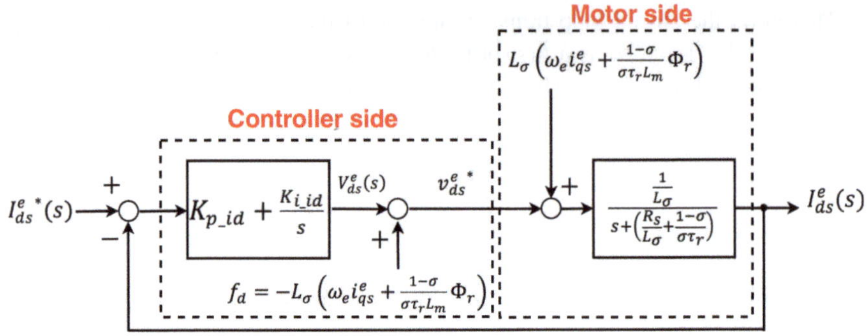

Fig. 2.3 d-axis current control block diagram of an induction motor including nonlinearity compensation

as:

$$\frac{di_{qs}^e}{dt} = -\frac{R_s}{L_\sigma} i_{qs}^e + \frac{v_{qs}^e}{L_\sigma} \tag{2.1.28}$$

At this point, we take the Laplace transform of Eq. (2.1.28) and obtain:

$$I_{qs}^e(s) = V_{qs}^e(s) \times \frac{\frac{1}{L_\sigma}}{s + \frac{R_s}{L_\sigma}} \tag{2.1.29}$$

Assuming a PI controller is designed for the q-axis current loop, with its output being $V_{qs}^e(s)$, the q-axis current control loop can be represented as shown in Fig. 2.4.

Where $I_{qs}^{e*}(s)$ is the q-axis current command, the proportional gain of the q-axis current PI controller is K_{p_iq}, and the integral gain is K_{i_iq}. The plant (controlled system) for the q-axis current loop of the induction motor is $\frac{\frac{1}{L_\sigma}}{s+\frac{R_s}{L_\sigma}}$.

When the nonlinear coupling terms in the q-axis have been perfectly compensated, the q-axis current loop of the induction motor can be represented using the control structure shown in Fig. 2.4. Under this structure, we can design the proportional gain

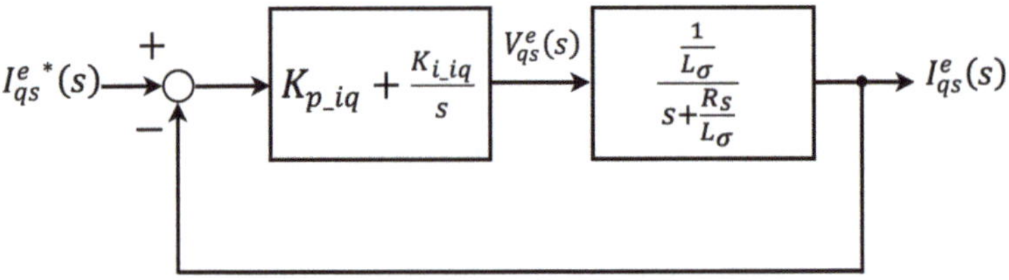

Fig. 2.4 q-axis current control block diagram of an induction motor with perfect nonlinearity compensation

K_{p_iq} and integral gain K_{i_iq} of the q-axis current PI controller to satisfy the desired bandwidth specification for the q-axis current loop.

Let $\frac{1}{L_\sigma} = N$ and $\frac{R_s}{L_\sigma} = D$. Then, the plant (controlled system) for the q-axis current loop can be expressed as:

$$\frac{\frac{1}{L_\sigma}}{s + \frac{R_s}{L_\sigma}} = \frac{N}{s + D} \tag{2.1.30}$$

We can express the q-axis current PI controller in the following form:

$$K_{p\_iq} + \frac{K_{i\_iq}}{s} = K_{p\_iq}\left(\frac{s + \frac{K_{i\_iq}}{K_{p\_iq}}}{s}\right) \tag{2.1.31}$$

Assuming that ω_q is the desired bandwidth specification for the q-axis current loop, we can use the **pole-zero cancellation method** to design the PI controller parameters for the q-axis current loop as follows:

$$K_{p\_iq} = \frac{\omega_q}{N},\ K_{i\_iq} = \frac{D \times \omega_q}{N} \tag{2.1.32}$$

By substituting the PI controller parameters, the open-loop transfer function of the q-axis current loop, G_{q_open}, becomes:

$$G_{q\_open} = \frac{\omega_q}{N}\left(\frac{s + D}{s}\right) \times \frac{N}{s + D} = \frac{\omega_q}{s} \tag{2.1.33}$$

Therefore, the closed-loop transfer function of the q-axis current loop, G_{q_close}, can be designed as a first-order low-pass filter with a cutoff frequency of ω_q, as shown in Eq. (2.1.34). This completes the design of the q-axis current control loop.

$$G_{q\_close} = \frac{G_{q\_open}}{1 + G_{q\_open}} = \frac{\omega_q}{s + \omega_q} \tag{2.1.34}$$

- **Decoupling Compensation for the q-Axis Current Control Loop of the Induction Motor**

In the q-axis current loop design for the induction motor described above, we assumed that the nonlinear coupling terms on the q-axis had been perfectly compensated. Now, we return to address the compensation of these nonlinear coupling terms [1, 3]. We can rewrite the differential equation of the induction motor's q-axis current as follows:

$$\frac{di_{qs}^e}{dt} = -\frac{R_s}{L_\sigma} i_{qs}^e - \omega_e i_{ds}^e - \frac{(1 - \sigma)}{\sigma L_m} \omega_e \Phi_r + \frac{v_{qs}^e}{L_\sigma} \tag{2.1.35}$$

Since the output of the PI controller is v_{qs}^e, a feedforward control term f_q is added before v_{qs}^e enters the motor, as follows:

$$f_q = L_\sigma\left(\omega_e i_{ds}^e + \frac{(1-\sigma)}{\sigma L_m}\omega_e \Phi_r\right) \tag{2.1.36}$$

Therefore, the actual q-axis voltage applied to the motor is $v_{qs}^{e*} = v_{qs}^e + f_q$. Substituting v_{qs}^{e*} into Eq. (2.1.35) in place of v_{qs}^e, we obtain:

$$\frac{di_{qs}^e}{dt} = -\frac{R_s}{L_\sigma}i_{qs}^e + \frac{v_{qs}^e}{L_\sigma} \tag{2.1.37}$$

From Eq. (2.1.37), we can observe that the added feedforward compensation term f_q effectively cancels out the nonlinear coupling terms inside the motor. As a result, the originally nonlinear q-axis current differential equation becomes a linear differential equation as shown in (2.1.37). This linearization enables the use of the PI controller architecture depicted in Fig. 2.4 to design the q-axis current control loop as a first-order low-pass filter, as described in Eq. (2.1.34). After incorporating the feedforward compensation, the q-axis current loop can be represented by the structure shown in Fig. 2.5.

- **Field-Oriented Flux Control Loop for Induction Motors**

After completing the current loop design, we now proceed to design the **flux control loop** for the induction motor. By rewriting Eq. (2.1.15), the relationship between the rotor flux linkage Φ_r and the stator d-axis current i_{ds}^e is given as:

$$\frac{d\Phi_r}{dt} = -\frac{R_r}{L_r}\Phi_r + R_r\frac{L_m}{L_r}i_{ds}^e \tag{2.1.38}$$

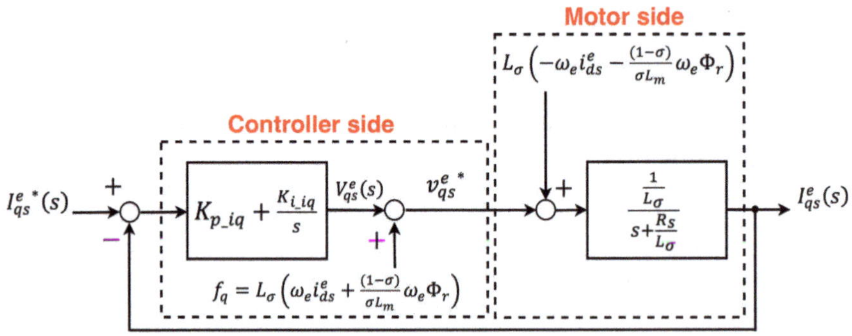

Fig. 2.5 q-axis current control block diagram of an induction motor including nonlinearity compensation

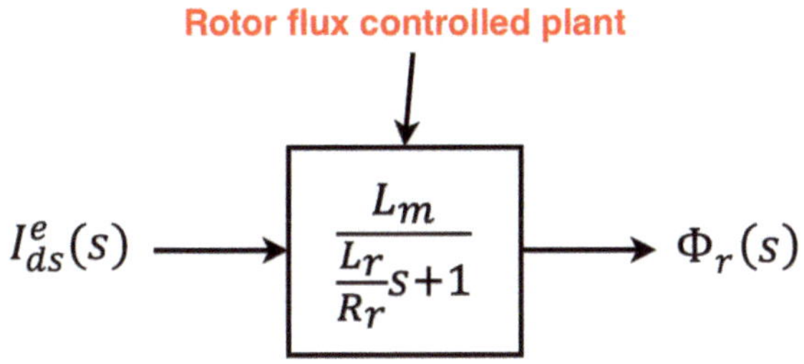

Fig. 2.6 Relationship between the d-axis current and rotor flux of an induction motor

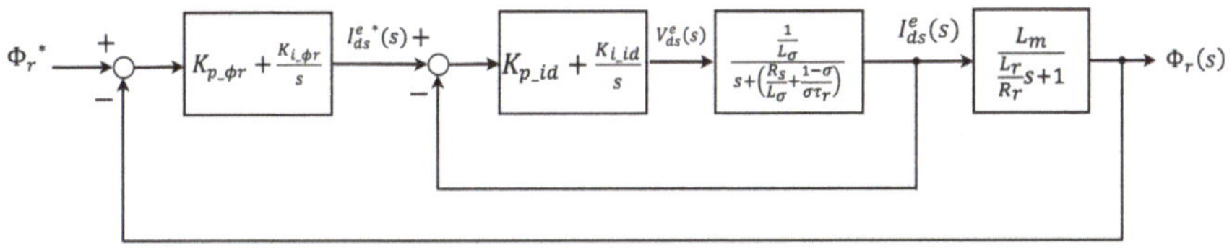

Fig. 2.7 Rotor flux control block diagram of an induction motor

The Laplace transform of Eq. (2.1.38) yields:

$$\Phi_r(s) = i_{ds}^e(s) \times \frac{L_m}{\frac{L_r}{R_r}s + 1} \tag{2.1.39}$$

From Eq. (2.1.39), it can be seen that the rotor flux linkage Φ_r depends solely on the stator d-axis current and can be generated by the d-axis current, as shown in Fig. 2.6.

To stably control the rotor flux linkage Φ_r, we need a rotor flux control loop, which serves as the outer loop of the d-axis stator current control loop, as shown in Fig. 2.7.

Here, Φ_r^* is the rotor flux reference, and the PI controller for rotor flux has a proportional gain of $K_{p\_\phi r}$ and an integral gain of $K_{i\_\phi r}$. The controlled plant for the rotor flux linkage of the induction motor is given by $\frac{L_m}{\frac{L_r}{R_r}s+1}$.

We assume that the bandwidth of the d-axis stator current loop is at least five times higher than that of the rotor flux loop. Therefore, when designing the PI controller for rotor flux, we can simplify the closed-loop transfer function of the d-axis stator current loop as 1. Under this control architecture, we can design the proportional gain $K_{p\_\phi r}$ and integral gain $K_{i\_\phi r}$ of the rotor flux PI controller so that the rotor flux loop meets the required bandwidth specification.

In this case, the controlled plant for the rotor flux linkage can be represented as:

$$\frac{L_m}{\frac{L_r}{R_r}s + 1} = \frac{L_m\frac{R_r}{L_r}}{s + \frac{R_r}{L_r}} \tag{2.1.40}$$

Let $L_m \frac{R_r}{L_r} = N$ and $\frac{R_r}{L_r} = D$, then the rotor-flux controlled plant of the induction motor can be written as:

$$\frac{L_m \frac{R_r}{L_r}}{s + \frac{R_r}{L_r}} = \frac{N}{s + D} \tag{2.1.41}$$

We can express the rotor flux PI controller in the following form:

$$K_{p\_\phi r} + \frac{K_{i\_\phi r}}{s} = K_{p\_\phi r} \left(\frac{s + \frac{K_{i\_\phi r}}{K_{p\_\phi r}}}{s} \right) \tag{2.1.42}$$

Assuming that ω_ϕ is the desired bandwidth specification for the rotor flux control loop, we can design the PI controller parameters for the rotor flux using the pole-zero cancellation method as follows:

$$K_{p\_\phi r} = \frac{\omega_\phi}{N}, \; K_{i\_iq} = \frac{D \cdot \omega_\phi}{N} \tag{2.1.43}$$

Substituting the PI controller parameters, the open-loop transfer function of the rotor flux control loop, $G_{\phi r\_open}$, becomes:

$$G_{\phi r\_open} = \frac{\omega_\phi}{N} \left(\frac{s + D}{s} \right) \cdot \frac{N}{s + D} = \frac{\omega_\phi}{s} \tag{2.1.44}$$

Therefore, the closed-loop transfer function of the rotor flux control loop, $G_{\phi r\_close}$, can be designed as a first-order low-pass filter with a cutoff frequency of ω_ϕ, as shown in Eq. (2.1.45). This completes the design of the rotor flux control loop.

$$G_{\phi r\_close} = \frac{G_{\phi r\_open}}{1 + G_{\phi r\_open}} = \frac{\omega_\phi}{s + \omega_\phi} \tag{2.1.45}$$

- **Field-Oriented Speed Control Loop for Induction Motors**

After completing the design of the d-axis and q-axis current loops as well as the rotor flux loop, the final step is to design the speed control loop for the induction motor. According to Eq. (2.1.17), we know that the electromagnetic torque T_e of the induction motor has a linear relationship with the product of the stator q-axis current i_{qs}^e and the rotor flux linkage Φ_r, as shown in Eq. (2.1.46).

$$T_e = \frac{3P}{4} \frac{L_m}{L_r} \left(i_{qs}^e \Phi_r \right) \tag{2.1.46}$$

When the rotor flux linkage Φ_r is stably controlled, the electromagnetic torque T_e will have a linear relationship with the stator q-axis current.

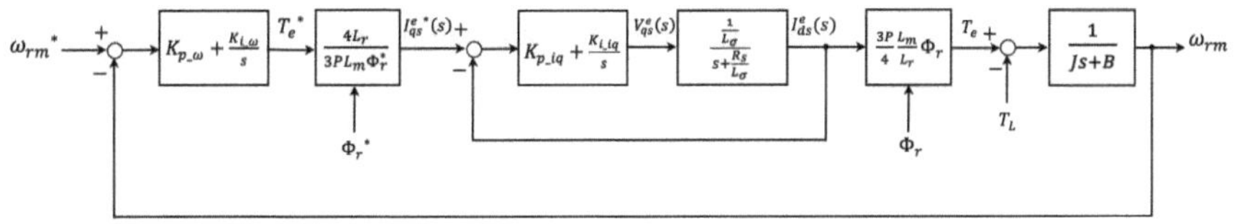

Fig. 2.8 Induction motor speed control block diagram

In addition, referring to the motor's mechanical equation (as given in Eq. (1.3.31)), it can be rearranged as follows:

$$T_e - T_L = J\frac{d\omega_{rm}}{dt} + B\omega_{rm} \tag{2.1.47}$$

Taking the Laplace transform of Eq. (2.1.48), we obtain:

$$\omega_{rm} = (T_e - T_L) \times \frac{1}{Js + B} \tag{2.1.48}$$

From Eq. (2.1.48), we can see that the electromagnetic torque T_e, after overcoming the load torque T_L, is applied to the motor's mechanical plant $\frac{1}{Js+B}$ to produce the motor's mechanical speed ω_{rm}.

Therefore, based on the above information, the speed control loop of the induction motor can be constructed as shown in Fig. 2.8.

Here, ω_{rm}^* is the speed command, and T_e^* is the torque command generated by the speed loop PI controller. The gain $\frac{4L_r}{3PL_m\Phi_r^*}$ *is responsible for converting the torque command into the q-axis current command, functioning similarly to the inverse of the torque constant K_T in permanent magnet motors. The output q-axis current from the q-axis current loop must then be multiplied by the gain $\frac{3P}{4}\frac{L_m}{L_r}\Phi_r$ to convert it into the motor's electromagnetic torque T_e. The gain $\frac{3P}{4}\frac{L_m}{L_r}\Phi_r$ functions similarly to the torque constant K_T in permanent magnet motors. When the rotor flux Φ_r of the induction motor is stably controlled—i.e., $\Phi_r^* = \Phi_r$*—the product of the gain $\frac{4L_r}{3PL_m\Phi_r^*}$ and the gain $\frac{3P}{4}\frac{L_m}{L_r}\Phi_r$ equals 1.

Here, it is assumed that the bandwidth of the q-axis stator current loop is at least five times greater than that of the speed loop. Therefore, when designing the PI controller for the speed loop, the closed-loop transfer function of the q-axis current loop can be simplified to 1. Additionally, assuming that the rotor flux linkage is stably controlled, the product of the gain $\frac{4L_r}{3PL_m\Phi_r^*}$ and the gain $\frac{3P}{4}\frac{L_m}{L_r}\Phi_r$ equals 1. Under this structure, we can design the proportional gain K_{p_ω} and integral gain K_{i_ω} of the speed loop PI controller to meet the required bandwidth specification. The mechanical plant of the motor can be represented as follows:

$$\frac{1}{Js + B} = \frac{\frac{1}{J}}{s + \frac{B}{J}} \tag{2.1.49}$$

Let $\frac{1}{J} = N$ and $\frac{B}{J} = D$, then the rotor mechanical plant of the induction motor can be expressed as:

$$\frac{\frac{1}{J}}{s + \frac{B}{J}} = \frac{N}{s + D} \tag{2.1.50}$$

We can express the speed loop PI controller in the following form:

$$K_{p\_\omega} + \frac{K_{i\_\omega}}{s} = K_{p\_\omega}\left(\frac{s + \frac{K_{i\_\omega}}{K_{p\_\omega}}}{s}\right) \tag{2.1.51}$$

Assuming that ω_s is the bandwidth specification required for the speed control loop, and that the motor's mechanical parameters J (moment of inertia) and B (viscous friction coefficient) can be accurately obtained, we can design the PI controller parameters for the speed loop using the pole-zero cancellation method as follows:

$$K_{p\_\omega} = \frac{\omega_s}{N}, K_{i\_\omega} = \frac{D \cdot \omega_s}{N} \tag{2.1.52}$$

Substituting the PI controller parameters, the forward open-loop transfer function of the speed control loop, G_{ω_open}, becomes:

$$G_{\omega\_open} = \frac{\omega_s}{N}\left(\frac{s + D}{s}\right) \cdot \frac{N}{s + D} = \frac{\omega_s}{s} \tag{2.1.53}$$

Therefore, the closed-loop transfer function of the speed control loop, G_{ω_close}, can be designed as a first-order low-pass filter with a cutoff frequency of ω_s, as shown in Eq. (2.1.54). This completes the design of the speed control loop.

$$G_{\omega\_close} = \frac{G_{\omega\_open}}{1 + G_{\omega\_open}} = \frac{\omega_s}{s + \omega_s} \tag{2.1.54}$$

- **MATLAB/SIMULINK Simulation Verification**

Next, we will use the induction motor parameters and bandwidth design specifications in Table 2.1 to perform simulation verification.

The PI controller parameters for the d-axis current loop can be calculated using the example program below. You may also adjust the values of N, D, and ω_d in the program as needed to compute the PI controller parameters for each control loop. Table 2.2 lists the PI controller parameter values for each control loop.

Table 2.1 Induction motor parameters and bandwidth design specifications

Parameter	Symbol	Value	Unit
Stator resistance	R_s	0.8	Ω
Rotor resistance	R_r	0.6	Ω
Stator inductance	L_s	0.085	H
Rotor inductance	L_r	0.085	H
Mutual inductance	L_m	0.082	H
Number of poles	P	4	–
Moment of inertia	J	0.033	kg m^2
Friction coefficient	B	0.00825	N m s/rad
Bandwidth of d- and q-axis current control loops	ω_d, ω_q	500	rad/s
Flux loop bandwidth	ω_ϕ	50	rad/s
Speed loop bandwidth	ω_s	50	rad/s

Table 2.2 PI controller parameter values for induction motor vector control loops

PI controller parameters	Value
$Kp_{\_id}$	2.95
$Ki_{\_id}$	679.2
$Kp_{\_iq}$	2.95
$Ki_{\_iq}$	400
$Kp_{\_\phi r}$	86.38
$Ki_{\_\phi r}$	609.76
$Kp_{\_\omega}$	1.65
$Ki_{\_\omega}$	0.41

MATLAB Example Script: m2_1_1.m

```
Rs=0.8; Rr=0.6; Ls=0.085; Lr=0.085; Lm=0.082;
pole=4; J=0.033; B=0.00825;
w = Ls*Lr - Lm^2; Lsigma = w/Lr;
sigma = 1 - Lm^2/(Ls*Lr); Tr = Lr/Rr;
N = 1/Lsigma; D = Rs/Lsigma + (1-sigma)/(sigma*Tr);
wd = 500;
Kp_id = wd/N
Ki_id = D*wd/N
```

Next, use SIMULINK to build the transfer function blocks for the induction motor, and input the PI controllers for each control loop. The completed SIMULINK transfer function control system is shown in Fig. 2.9.

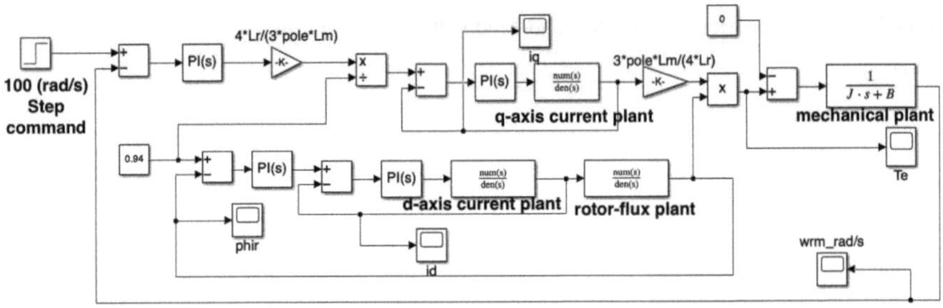

Fig. 2.9 Induction motor speed control transfer function block diagram, Example model: im_foc_loops.slx

Run the SIMULINK simulation to obtain the motor speed response, as shown in Fig. 2.10.

We can also use the vector control model of the induction motor developed in Chap. 1 to perform system simulation. Please first run the example script im_param.m from this section to load the parameters required for the induction motor vector control model. The completed induction motor vector control simulation system is shown in Fig. 2.11.

By running the SIMULINK simulation, the motor speed simulation result can be obtained as shown in Fig. 2.12.

It can be observed that the simulation result using the transfer function matches the result from the induction motor vector control model. This consistency is due to the implementation of current nonlinear decoupling compensation in the d- and q-axis current loops within the simulation system shown in Fig. 2.11. As a result, the vector control system simulation aligns with the transfer function simulation.

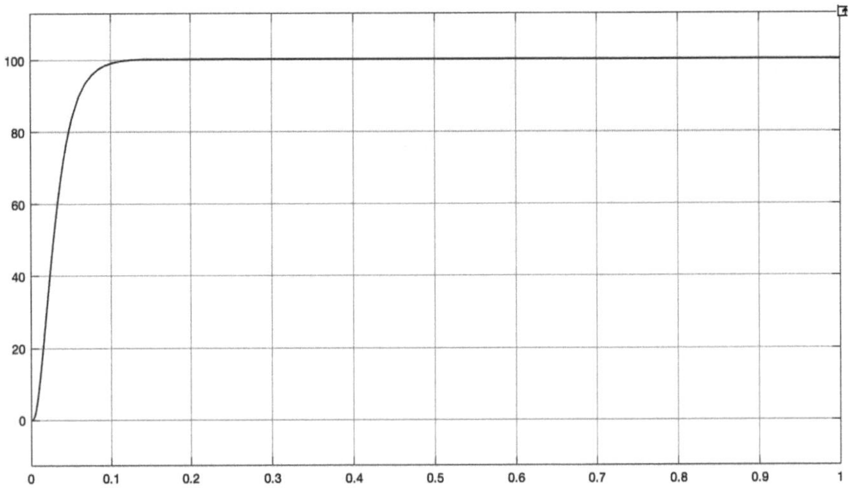

Fig. 2.10 Speed response waveform (oscilloscope block: wrm_rad/s)

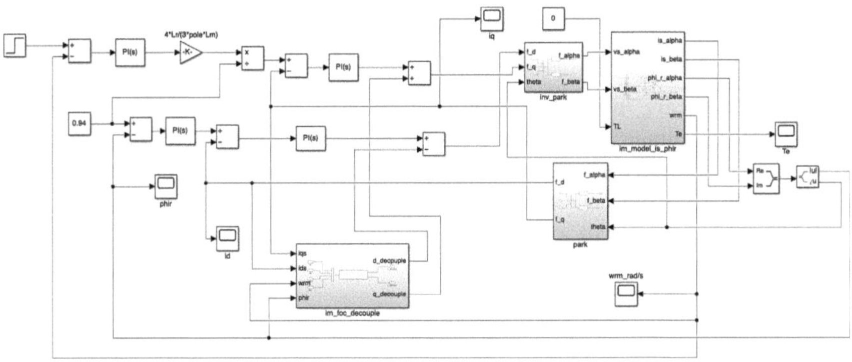

Fig. 2.11 Field-oriented control (FOC) block diagram of the induction motor, Example model: im_foc_models_decouple.slx

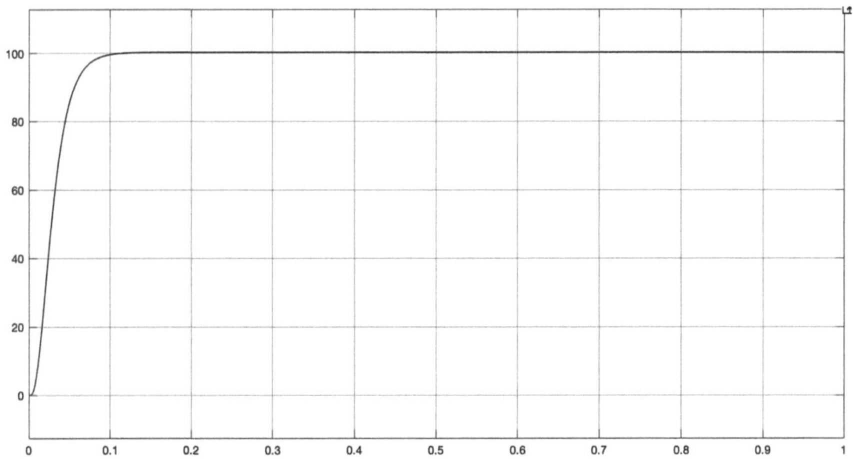

Fig. 2.12 Speed response waveform (oscilloscope block: wrm_rad/s)

2.1.2 FOC for Permanent Magnet Synchronous Motors (PMSMs)

The FOC method can effectively achieve near-perfect decoupling between torque and magnetic flux in a three-phase induction motor, resulting in control performance comparable to that of a separately excited DC motor. This has been fully demonstrated in the previous section through the derivation and simulation of the FOC strategy for induction motors.

For three-phase PMSMs, which are also AC machines, the FOC method is equally applicable [1, 3, 5, 7]. In fact, to some extent, the stator structures of three-phase

PMSMs and squirrel-cage induction motors are almost identical. The key difference lies in the rotor structure:

- In a squirrel-cage induction motor, the rotor consists of cage-shaped conductors that induce voltage and current from the stator's magnetic field. Since the rotor current is induced, the rotor speed must be lower than the synchronous speed—this necessitates the presence of slip.
- In contrast, the rotor of a PMSM uses permanent magnets and does not rely on induced voltage to generate current. As a result, no slip is needed, and the rotor speed equals the synchronous speed.

Let us revisit the mathematical model of the three-phase Interior Permanent Magnet Synchronous Motor (IPMSM) established in this chapter under the two-axis rotor reference frame:

$$v_{ds}^r = R_s i_{ds}^r + L_d \frac{di_{ds}^r}{dt} - \omega_r L_q i_{qs}^r \tag{2.1.55}$$

$$v_{qs}^r = R_s i_{qs}^r + L_q \frac{di_{qs}^r}{dt} + \omega_r \left(L_d i_{ds}^r + \lambda_f \right) \tag{2.1.56}$$

The q-axis flux linkage of the stator is given by $\lambda_{qs}^r = L_q i_{qs}^r$, and the d-axis flux linkage of the stator is given by $\lambda_{ds}^r = L_d i_{ds}^r + \lambda_f$, where λ_f is the permanent magnet flux linkage.

The torque equation of an Interior Permanent Magnet Synchronous Motor (IPMSM) can be expressed as:

$$T_e = \frac{3}{2}\frac{P}{2}\left[\lambda_f i_{qs}^r + \left(L_d - L_q \right) i_{ds}^r i_{qs}^r \right] \tag{2.1.57}$$

For a permanent magnet synchronous motor (PMSM), the significance of using a two-axis rotating reference frame aligned with the rotor is that the rotor position must be known at all times, and the d-axis of the synchronous rotating reference frame must be aligned with the rotor and rotate synchronously with it.

This alignment ensures that:

- The **d-axis** is aligned with the rotor's magnetic field (produced by the permanent magnets).
- The **q-axis** is orthogonal to the d-axis and represents the torque-producing component.

From Eq. (2.1.57), it can be seen that if the stator d-axis current i_{ds}^r is controlled to zero, the reluctance torque term $\left(L_d - L_q \right) i_{ds}^r i_{qs}^r$ will also become zero. The stator d-axis flux linkage will then consist only of λ_f, which is the flux linkage produced by the rotor's magnetic field in the stator windings (as shown in Eq. 1.4.2), and is a constant. Consequently, the torque will be proportional to the stator q-axis current i_{qs}^r, thus achieving the goal of decoupling the magnetic field and torque.

We can rearrange Eqs. (2.1.55) and (2.1.56) into the following form:

$$\frac{di_{ds}^r}{dt} = -\frac{R_s}{L_d}i_{ds}^r + \omega_r\frac{L_q}{L_d}i_{qs}^r + \frac{v_{ds}^r}{L_d} \tag{2.1.58}$$

$$\frac{di_{qs}^r}{dt} = -\frac{R_s}{L_q}i_{qs}^r - \omega_r\frac{\left(L_d i_{ds}^r + \lambda_f\right)}{L_q} + \frac{v_{qs}^r}{L_q} \tag{2.1.59}$$

Equations (2.1.58), (2.1.59), (2.1.57), and the motor mechanical Eq. (1.3.31) indicate that in order to effectively control the speed of a three-phase Permanent Magnet Synchronous Motor (PMSM), three control loops are required: the d-axis stator current i_{ds}^r control loop, the q-axis stator current i_{qs}^r control loop, and the rotor speed ω_{rm} control loop.

Compared to the Field-Oriented Control (FOC) of an induction motor, the PMSM control system omits the rotor flux control loop. This is because the permanent magnets on the PMSM rotor provide a stable magnetic field. Through mutual inductance with the stator, this creates an almost constant flux linkage λ_f. As long as the d-axis stator current i_{ds}^r is controlled to zero, the motor torque becomes proportional to the q-axis stator current i_{qs}^r, thereby achieving the decoupling of magnetic flux and torque.

Next, we will sequentially design the controller parameters for each control loop in the field-oriented control (FOC) system of the Permanent Magnet Synchronous Motor (PMSM), namely the i_{ds}^r, i_{qs}^r, and $\omega_{rm}\}$ control loops. The design principles are as follows:

- **From inner to outer loops**: The inner control loops should be designed first, followed by the outer loops.
- **Bandwidth separation**: The bandwidth of each inner loop should be at least five times greater than that of its corresponding outer loop.

(Note: The purpose of designing the inner loop with a bandwidth at least five times greater than that of the outer loop is to allow the inner loop transfer function to be approximated as 1 when designing the outer loop controller parameters, thus simplifying the design process) [1, 4].

- **PMSM Field-Oriented Current Loop Design** [1, 4, 10]

In general, the control loop design follows a "**from inner to outer**" approach. To build a complete field-oriented control (FOC) system for a Permanent Magnet Synchronous Motor (PMSM), the first step is to establish the d^r**- and** q^r**-axis stator current control loops**.

By examining Eqs. (2.1.57) and (2.1.58), we observe that the differential equations governing the stator d^r- and q^r-axis currents contain **nonlinear coupling terms**, as illustrated in Fig. 2.13.

In field-oriented control, the interaction between the stator current and the magnetic field generates **nonlinear coupling effects**. These nonlinear coupling terms appear in the mathematical model of a permanent magnet synchronous motor (PMSM) as **cross terms**, such as the product of the stator current and rotor speed.

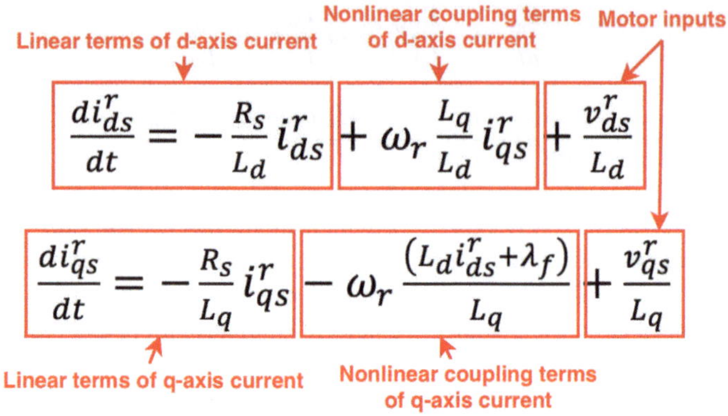

Fig. 2.13 d–q axis current differential equations of the PMSM

In practical control applications, these nonlinear coupling terms can **degrade the performance** of the current control loops.

To address this issue, it is necessary to **compensate for the nonlinear coupling terms** in field-oriented control. This is typically achieved using a **decoupling control strategy**. Decoupling control is a method that introduces compensation terms to cancel out the nonlinear coupling effects, thereby **minimizing the interaction between the flux-producing current and the torque-producing current**, which in turn **enhances the control performance**.

- **Design of the d-Axis Current Control Loop for PMSM**

To help readers better understand, we will first assume that the nonlinear coupling terms have been perfectly compensated. Taking the **d-axis** as an example, if the nonlinear coupling terms have been perfectly canceled, then the **d-axis current differential equation** can be written as:

$$\frac{di_{ds}^r}{dt} = -\frac{R_s}{L_d} i_{ds}^r + \frac{v_{ds}^r}{L_d} \tag{2.1.60}$$

At this point, applying the Laplace transform to Eq. (2.1.60), we obtain:

$$I_{ds}^r(s) = V_{ds}^r(s) \times \frac{\frac{1}{L_d}}{s + \frac{R_s}{L_d}} \tag{2.1.61}$$

Assuming that a PI controller is designed for the d-axis current, with its output denoted as $V_{ds}^r(s)$, the d-axis current control loop can then be represented as shown in Fig. 2.14.

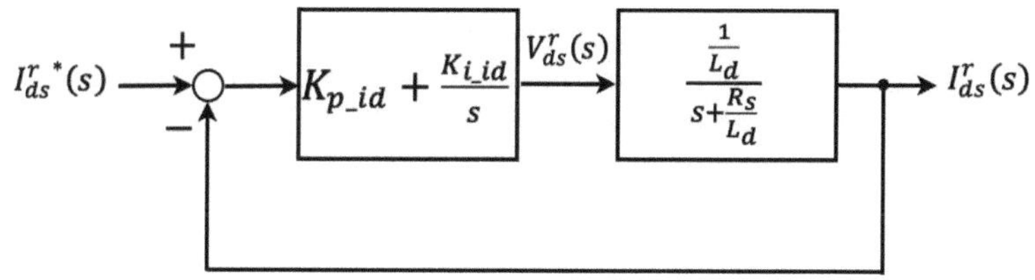

Fig. 2.14 PMSM d-axis current control block diagram with perfect nonlinear term compensation

Here, $I_{ds}^{r*}(s)$ is the d-axis current reference, and the PI controller for the d-axis current has a proportional gain K_{p_id} *and an integral gain* K_{i_id}. The controlled plant of the d-axis current in the synchronous motor is given by $\frac{\frac{1}{L_d}}{s+\frac{R_s}{L_d}}$.

When the nonlinear coupling terms on the d-axis have been perfectly compensated, the d-axis current loop of the permanent magnet synchronous motor (PMSM) can be represented by the control structure shown in Fig. 2.14. Under this structure, we can design the proportional gain K_{p_id} *and integral gain* K_{i_id} of the d-axis current PI controller to meet the desired bandwidth specifications.

Let $\frac{1}{L_d} = N$ and $\frac{R_s}{L_d} = D$. Then the controlled plant of the PMSM d-axis current loop can be written as:

$$\frac{\frac{1}{L_d}}{s+\frac{R_s}{L_d}} = \frac{N}{s+D} \tag{2.1.62}$$

We can express the d-axis current PI controller in the following form:

$$K_{p\_id} + \frac{K_{i\_id}}{s} = K_{p\_id}\left(\frac{s+\frac{K_{i\_id}}{K_{p\_id}}}{s}\right) \tag{2.1.63}$$

Assuming that ω_d is the desired bandwidth specification for the d-axis current control loop, we can apply the **pole-zero cancellation method** to design the PI controller parameters as follows:

$$K_{p\_id} = \frac{\omega_d}{N}, K_{i\_id} = \frac{D\cdot\omega_d}{N} \tag{2.1.64}$$

Substituting the PI controller parameters into the d-axis current loop, the open-loop transfer function G_{d_open} becomes:

$$G_{d\_open} = \frac{\omega_d}{N}\left(\frac{s+D}{s}\right)\cdot\frac{N}{s+D} = \frac{\omega_d}{s} \tag{2.1.65}$$

Therefore, the closed-loop transfer function of the d-axis current loop, G_{d_close}, can be designed as a first-order low-pass filter with a cutoff frequency of ω_d, as shown in Eq. (2.1.66). With this, the design of the d-axis current loop is complete.

$$G_{d\_close} = \frac{G_{d\_open}}{1 + G_{d\_open}} = \frac{\omega_d}{s + \omega_d} \tag{2.1.66}$$

- **Decoupling Compensation for the d-axis Current Control Loop of the PMSM**

In the above design of the d-axis current loop for the PMSMs, we assumed that the nonlinear coupling terms in the d-axis had been perfectly compensated. Here, we return to address the compensation of these nonlinear coupling terms [1, 3]. The differential equation of the PMSM d-axis current can be rewritten as follows:

$$\frac{di_{ds}^r}{dt} = -\frac{R_s}{L_d} i_{ds}^r + \omega_r \frac{L_q}{L_d} i_{qs}^r + \frac{v_{ds}^r}{L_d} \tag{2.1.67}$$

Since the output of the PI controller is v_{ds}^r, a feedforward compensation term f_d is added before it enters the motor, as follows:

$$f_d = -L_d \left(\omega_r \frac{L_q}{L_d} i_{qs}^r \right) \tag{2.1.68}$$

Therefore, the actual d-axis voltage applied to the motor is $v_{ds}^{r*} = v_{ds}^r + f_d$. Substituting v_{ds}^{r*} into Eq. (2.1.67), we obtain:

$$\frac{di_{ds}^r}{dt} = -\frac{R_s}{L_d} i_{ds}^r + \frac{v_{ds}^r}{L_d} \tag{2.1.69}$$

From Eq. (2.1.69), we can observe that the introduced feedforward control term f_d effectively cancels out the nonlinear coupling term inside the motor. This transforms the originally nonlinear d-axis current differential equation into the linear form shown in Eq. (2.1.69). As a result, we can apply the PI controller structure illustrated in Fig. 2.14 to design the d-axis current loop as a first-order low-pass filter, as shown in Eq. (2.1.66). With the addition of the feedforward compensation, the d-axis current loop can be represented as shown in Fig. 2.15.

- **Design of the q-Axis Current Control Loop for PMSM**

Next, we proceed to design the q-axis current control loop. Here, we also assume that the nonlinear coupling terms in the q-axis have been perfectly compensated. Under this assumption, the q-axis current differential equation can be expressed as follows:

$$\frac{di_{qs}^r}{dt} = -\frac{R_s}{L_q} i_{qs}^r + \frac{v_{qs}^r}{L_q} \tag{2.1.70}$$

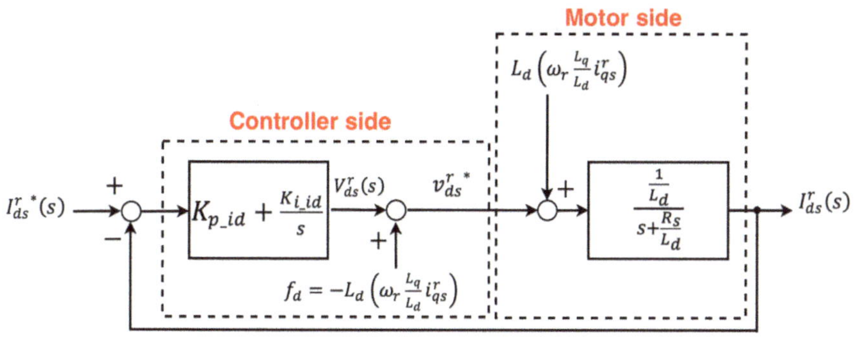

Fig. 2.15 PMSM d-axis current control block diagram with nonlinear term compensation

Taking the Laplace transform of Eq. (2.1.28), we obtain:

$$I_{qs}^r(s) = V_{qs}^r(s) \times \frac{\frac{1}{L_q}}{s + \frac{R_s}{L_q}} \tag{2.1.71}$$

Assuming a PI controller is designed for the q-axis current loop, with its output denoted as $V_{qs}^r(s)$, the q-axis current control loop can then be represented as shown in Fig. 2.16.

Here, $I_{qs}^{r*}(s)$ is the q-axis current command. The proportional gain and integral gain of the q-axis current PI controller are denoted by K_{p_iq} and K_{i_iq}, respectively. The controlled plant of the q-axis current loop for the permanent magnet synchronous motor (PMSM) is given by $\frac{\frac{1}{L_q}}{s+\frac{R_s}{L_q}}$.

When the nonlinear coupling terms of the q-axis have been perfectly compensated, the q-axis current loop of the permanent magnet synchronous motor (PMSM) can be represented using the control structure shown in Fig. 2.16. Under this structure, we can design the proportional gain K_{p_iq} and the integral gain K_{i_iq} of the q-axis current PI controller so that the current loop meets the desired bandwidth specification.

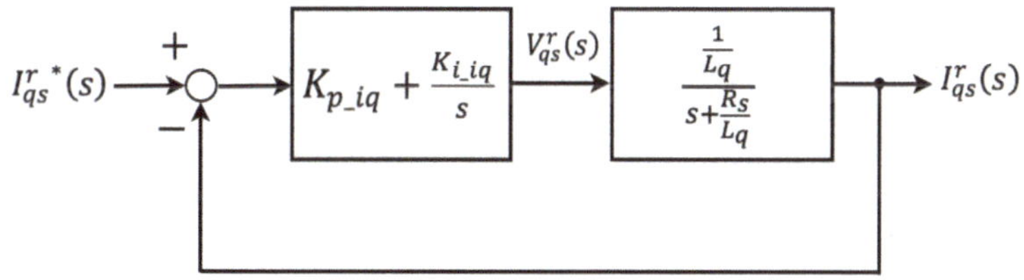

Fig. 2.16 PMSM q-axis current control block diagram with perfect nonlinear term compensation

Let $\frac{1}{L_q} = N$ and $\frac{R_s}{L_q} = D$. Then, the controlled plant of the PMSM q-axis current loop can be expressed as:

$$\frac{\frac{1}{L_q}}{s + \frac{R_s}{L_q}} = \frac{N}{s + D} \tag{2.1.72}$$

We can express the q-axis current PI controller in the following form:

$$K_{p\_iq} + \frac{K_{i\_iq}}{s} = K_{p\_iq}\left(\frac{s + \frac{K_{i\_iq}}{K_{p\_iq}}}{s}\right) \tag{2.1.73}$$

Assuming that ω_q is the desired bandwidth specification for the q-axis current control loop, and applying the pole-zero cancellation method, the PI controller parameters for the q-axis current loop can be designed as follows:

$$K_{p\_iq} = \frac{\omega_q}{N},\ K_{i\_iq} = \frac{D \times \omega_q}{N} \tag{2.1.74}$$

By substituting the PI controller parameters, the open-loop transfer function of the q-axis current loop G_{q_open} becomes:

$$G_{q\_open} = \frac{\omega_q}{N}\left(\frac{s + D}{s}\right) \times \frac{N}{s + D} = \frac{\omega_q}{s} \tag{2.1.75}$$

Therefore, the closed-loop transfer function of the q-axis current loop, G_{q_close}, can be designed as a first-order low-pass filter with a cutoff frequency of ω_q, as shown in Eq. (2.1.76). This completes the design of the q-axis current loop.

$$G_{q\_close} = \frac{G_{q\_open}}{1 + G_{q\_open}} = \frac{\omega_q}{s + \omega_q} \tag{2.1.76}$$

- **Decoupling Compensation for the q-Axis Current Control Loop of the PMSM**

In the above design of the q-axis current loop for the permanent magnet synchronous motor (PMSM), we assumed that the nonlinear coupling terms on the q-axis had been perfectly compensated. Now, we return to address the compensation of these nonlinear coupling terms [1, 3]. We can rewrite the q-axis current differential equation of the PMSM as follows:

$$\frac{di_{qs}^r}{dt} = -\frac{R_s}{L_q} i_{qs}^r - \omega_r \frac{\left(L_d i_{ds}^r + \lambda_f\right)}{L_q} + \frac{v_{qs}^r}{L_q} \tag{2.1.77}$$

Since the output of the PI controller is v_{qs}^r, a feedforward control term f_q is added before v_{qs}^r enters the motor, as shown below:

$$f_q = L_q\left(\omega_r \frac{(L_d i_{ds}^r + \lambda_f)}{L_q}\right) \tag{2.1.78}$$

Therefore, the actual q-axis voltage entering the motor is $v_{qs}^{r*} = v_{qs}^r + f_q a$. Substituting v_{qs}^{r*} into v_{qs}^r in Eq. (2.1.77), we obtain:

$$\frac{di_{qs}^r}{dt} = -\frac{R_s}{L_q} i_{qs}^r + \frac{v_{qs}^r}{L_q} \tag{2.1.79}$$

From Eq. (2.1.79), it can be seen that the feedforward compensation term f_q effectively cancels out the nonlinear coupling terms inside the motor. This transforms the original nonlinear q-axis current differential equation into the linear form shown in Eq. (2.1.79). As a result, the PI controller structure illustrated in Fig. 2.16 can be applied to design the q-axis current loop as a first-order low-pass filter, as described by Eq. (2.1.76). With the inclusion of feedforward compensation, the q-axis current loop can be represented by the control architecture shown in Fig. 2.17.

- **Field-Oriented Speed Control Loop for PMSMs**

After completing the design of the **d-axis and q-axis current control loops**, the final step is to design the **speed control loop** for the permanent magnet synchronous motor (PMSM). According to Eq. (2.1.57), the **electromagnetic torque** T_e of the PMSM has a **linear relationship** with the **stator q-axis current** i_{qs}^r, as shown in Eq. (2.1.80).

$$T_e = \frac{3}{2}\frac{P}{2}\left[\lambda_f i_{qs}^r\right] \tag{2.1.80}$$

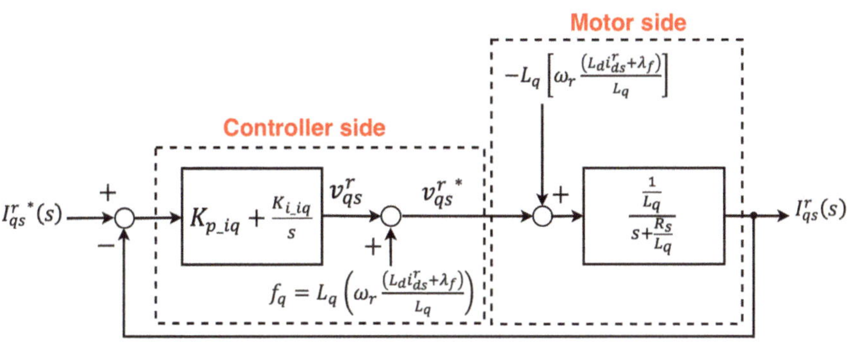

Fig. 2.17 PMSM q-axis current control block diagram with nonlinear term compensation

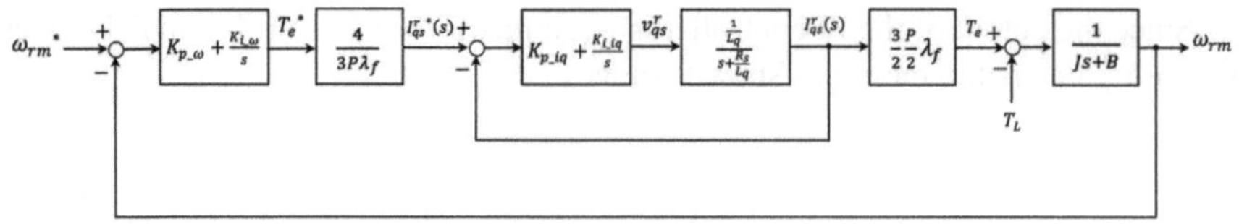

Fig. 2.18 PMSM speed control block diagram

In addition, referring to the motor mechanical equation (Eq. 1.3.31), it can be rearranged as follows:

$$T_e - T_L = J\frac{d\omega_{rm}}{dt} + B\omega_{rm} \tag{2.1.81}$$

Taking the Laplace transform of Eq. (2.1.81), we obtain:

$$\omega_{rm} = (T_e - T_L) \times \frac{1}{Js + B} \tag{2.1.82}$$

From Eq. (2.1.82), we can see that the electromagnetic torque T_e, after overcoming the load torque T_L, is applied to the motor mechanical plant $\frac{1}{Js+B}$, resulting in the motor's mechanical speed ω_{rm}.

Therefore, based on the above information, the speed control loop of the permanent magnet synchronous motor (PMSM) can be constructed as shown in Fig. 2.18.

Here, ω_{rm}^* is the speed command, and T_e^* is the torque command generated by the PI controller in the speed loop. The gain $\frac{4}{3P\lambda_f}$ is the inverse of the torque constant K_T of the permanent magnet synchronous motor (PMSM), responsible for converting the torque command into the q-axis current command.

The q-axis current output from the q-axis current loop is then multiplied by the gain $\frac{3}{2}\frac{P}{2}\lambda_f$ to convert it into the motor's electromagnetic torque T_e. This gain, $\frac{3}{2}\frac{P}{2}\lambda_f$, is the torque constant K_T of the PMSM. In this case, the product of $\frac{4}{3P\lambda_f}$ and $\frac{3}{2}\frac{P}{2}\lambda_f$ equals 1.

Assuming that the bandwidth of the q-axis stator current loop is at least five times higher than that of the speed loop, the closed-loop transfer function of the q-axis current loop can be simplified to 1 when designing the PI controller for the speed loop.

Under this architecture, we can design the proportional gain K_{p_ω} and integral gain K_{i_ω} of the speed loop PI controller to meet the desired bandwidth specification. In this case, the mechanical plant of the motor can be expressed as follows:

$$\frac{1}{Js + B} = \frac{\frac{1}{J}}{s + \frac{B}{J}} \tag{2.1.83}$$

Let $\frac{1}{J} = N$ and $\frac{B}{J} = D$, then the mechanical plant of the permanent magnet synchronous motor (PMSM) can be written as:

$$\frac{\frac{1}{J}}{s+\frac{B}{J}} = \frac{N}{s+D} \tag{2.1.84}$$

We can express the PI controller for the speed control loop in the following form:

$$K_{p\_\omega} + \frac{K_{i\_\omega}}{s} = K_{p\_\omega}\left(\frac{s+\frac{K_{i\_\omega}}{K_{p\_\omega}}}{s}\right) \tag{2.1.85}$$

Assuming that ω_s is the desired bandwidth of the speed control loop, and that the motor mechanical parameters J (moment of inertia) and B (viscous friction coefficient) are accurately known, we can design the PI controller parameters for the speed loop using the pole-zero cancellation method as follows:

$$K_{p-\omega} = \frac{\omega_s}{N} \cdot K_{i-\omega} = \frac{D \times \omega_s}{N} \tag{2.1.86}$$

Substituting the PI controller parameters, the forward open-loop transfer function of the speed control loop, G_{ω_open}, becomes:

$$G_{\omega\_open} = \frac{\omega_s}{N}\left(\frac{s+D}{s}\right) \times \frac{N}{s+D} = \frac{\omega_s}{s} \tag{2.1.87}$$

Therefore, the closed-loop transfer function of the speed control loop, G_{ω_close}, can be designed as a first-order low-pass filter with a cutoff frequency of ω_s, as shown in Eq. (2.1.88). This completes the design of the speed control loop.

$$G_{\omega\_close} = \frac{G_{\omega\_open}}{1+G_{\omega\_open}} = \frac{\omega_s}{s+\omega_s} \tag{2.1.88}$$

- **MATLAB/SIMULINK Simulation Verification**

Next, we will use the permanent magnet synchronous motor (PMSM) parameters and bandwidth design specifications listed in Table 2.3 to perform simulation verification.

Using the following example program, the PI controller parameters for the d-axis current loop can be calculated. You may also modify the values of N, D, and ω_d in the program to determine the PI controller parameters for each control loop. Table 2.4 lists the PI controller parameter values for each control loop.

Table 2.3 PMSM parameters and bandwidth design specifications

Parameter	Symbol	Value	Unit
Stator resistance	R_s	1.2	Ω
Stator d-axis inductance	L_d	5.7	mH
Stator q-axis inductance	L_q	12.5	mH
Rotor flux linkage	λ_f	0.03	Wb
Number of poles	P	4	–
Moment of inertia	J	0.00016	kg m^2
Friction coefficient	B	0.0000028	N m s/rad
Bandwidth of d- and q-axis current control loops	ω_d, ω_q	1000	rad/s
Speed loop bandwidth	ω_s	100	rad/s

Table 2.4 PI controller parameter values for PMSM vector control loops

PI controller parameters	Value
K_{p_id}	5.7
K_{i_id}	1200
K_{p_iq}	12.5
K_{i_iq}	1200
K_{p_ω}	0.016
K_{i_ω}	2.8e−4

MATLAB Example Script: m2_1_2.m

```
Rs=1.2; Ld=0.0057; Lq=0.0125;
Lamda_f=0.03;
pole=4; J=0.00016; B=0.0000028;
N = 1/Ld; D = Rs/Ld;
wd = 1000;
Kp_id = wd/N
Ki_id = D*wd/N
```

Next, use SIMULINK to build the transfer function control blocks for the permanent magnet synchronous motor (PMSM), and input the PI controllers for each control loop. The completed SIMULINK transfer function control system is shown in Fig. 2.19.

By running the SIMULINK simulation, the motor speed simulation result can be obtained as shown in Fig. 2.20.

We can also use the vector control model of the permanent magnet synchronous motor (PMSM) developed in Chap. 1 to perform system simulation. The completed PMSM vector control simulation system is shown in Fig. 2.21.

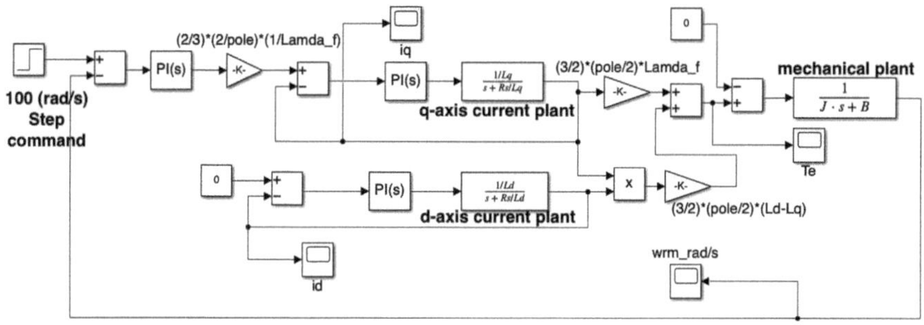

Fig. 2.19 PMSM speed control transfer function block diagram, Example model: pm_foc_loops.slx

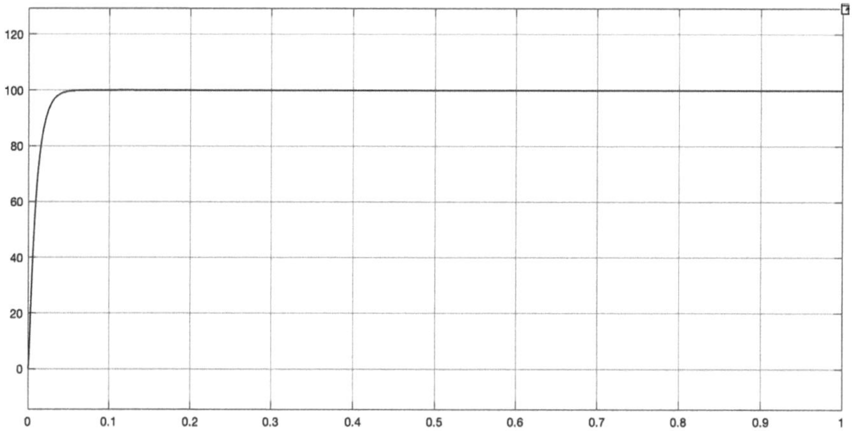

Fig. 2.20 Speed response waveform (oscilloscope block: wrm_rad/s)

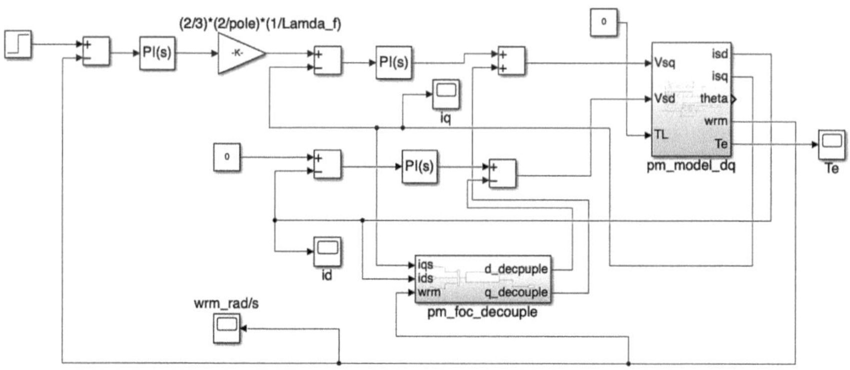

Fig. 2.21 PMSM FOC control block diagram, Example model: pm_foc_models_decouple.slx

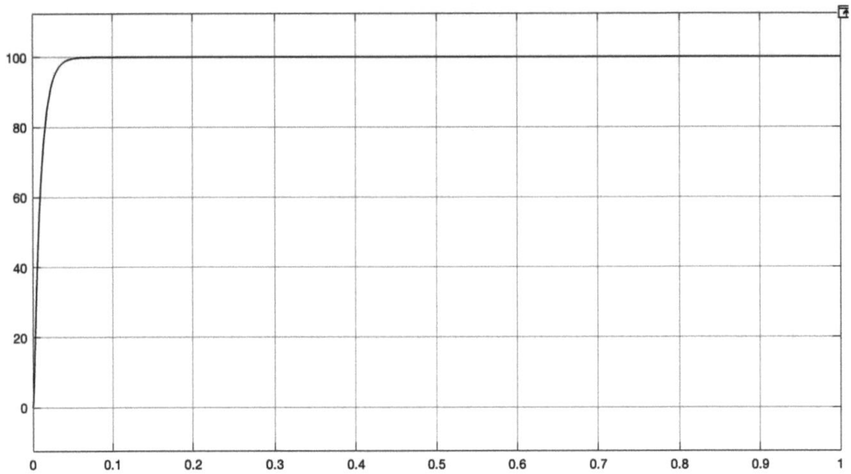

Fig. 2.22 Speed response waveform (oscilloscope block: wrm_rad/s)

By running the SIMULINK simulation, the motor speed simulation result as shown in Fig. 2.22 can be obtained.

It can be observed that the simulation results obtained using the transfer function model are consistent with those from the vector control model of the permanent magnet synchronous motor (PMSM). This consistency is due to the inclusion of current nonlinear decoupling compensation in the d- and q-axis current loops within the system shown in Fig. 2.21. As a result, the simulation results of the vector control system match those of the transfer function model.

2.2 Current Loop PI Controller Design with Delay Consideration

In the previous section, we provided a detailed introduction to the field-oriented control (FOC) strategies for both induction motors and permanent magnet synchronous motors (PMSMs), including feedforward compensation methods for nonlinear coupling terms in the current loops. The true purpose of applying field-oriented control to AC motors is to derive a linear control model, enabling independent control of torque and flux. Once a linear control model for the AC motor is obtained, linear control theory can then be applied to the design of the motor control system.

Therefore, in Sect. 2.1, we used the pole-zero cancellation method to design the PI controller parameters for both induction motors and PMSMs, aiming to meet the required bandwidth specifications. In practical applications, these control loops

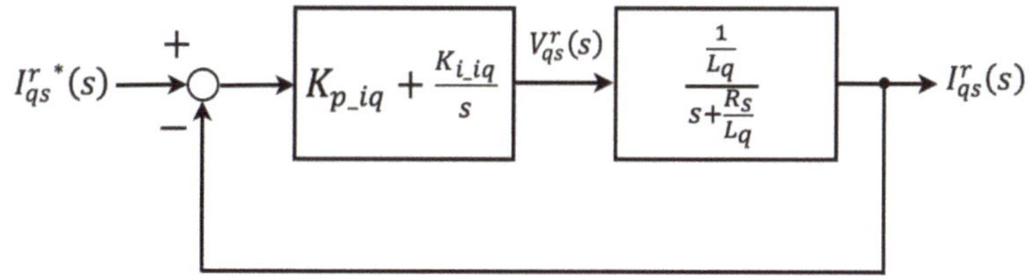

Fig. 2.23 PMSM q-axis current control block diagram with perfect nonlinear term compensation

are implemented using digital control systems. However, digital systems inevitably introduce various types of delays [4, 10]. The main sources of delay include:

- **Computation delay of the MCU**
- **PWM output delay of the MCU (Zero Order Hold delay, ZOH)**
- **Filtering delay in sensor signal feedback** (typically implemented using first- or second-order low-pass filters)
- **Sampling delay in sensor signal feedback**

- **Effects of Delay on the System**.

For example, for the **q-axis current loop of a Permanent Magnet Synchronous Motor (PMSM)**, the control architecture we designed in the previous section is shown in Fig. 2.23.

The current control loop shown in Fig. 2.23 does not include any delay effects. However, in practice, it will be affected by the four types of delays mentioned above [4, 10], as illustrated in Fig. 2.24.

Here, it is assumed that both the MCU's current sampling period and the PWM period operate at an interrupt execution rate of 8 kHz, that is:

$$\text{Sampling period } T_s = \frac{1}{8000} = 1.25\text{e} - 04 \text{ (s)} \tag{2.2.1}$$

Therefore, the computation delay T_c is equal to the sampling period T_s. Assuming that the voltage command calculated in the current control loop is output as a PWM

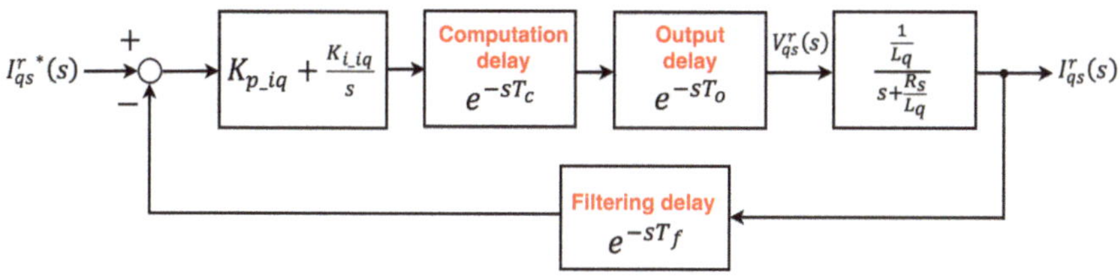

Fig. 2.24 PMSM q-axis current control block diagram considering delay effects

signal in the next current sampling period, the MCU's PWM output delay (ZOH delay) is half the sampling period, i.e., $T_o = \frac{T_s}{2}$.

(Note: For more details on the delay effects caused by different PWM sampling methods, please refer to Sect. 5.4*of this book.)*

Assuming the current signal is filtered through a first-order low-pass filter with a cutoff frequency $\omega_{fc} = 2000$ (rad/s), this first-order low-pass filter can be expressed as:

$$\text{The Feedback First-Order Low-Pass Filter } \frac{\omega_{fc}}{s + \omega_{fc}} = \frac{1}{T_f s + 1} \tag{2.2.2}$$

The time constant of the first-order low-pass filter is $T_f = 1/2000 = 5e - 04$, and the delay caused by the first-order low-pass filter can be approximated as e^{-sT_f}.

Finally, the filtered signal is sampled (Sample-and-Hold) before entering the MCU. The sampling delay introduced by this process can be approximated as half of the sampling period.

$$\text{The sampling delay } e^{-sT_{SH}} = e^{-s \times \frac{T_s}{2}} \tag{2.2.3}$$

If we take all the aforementioned delay effects into consideration and apply the q-axis PI controller parameters for the permanent magnet synchronous motor (PMSM) as designed in Sect. 2.1, we can use the example script m2_2_1 to plot the Bode plot of the closed-loop system. By comparing it with the Bode plot of the closed-loop system without delays, we obtain the comparative results shown in Fig. 2.25.

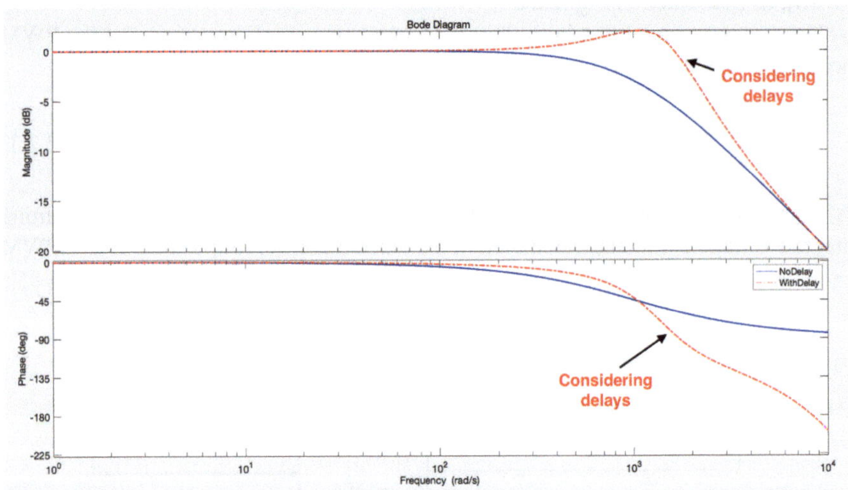

Fig. 2.25 Comparison of Bode plots for the current closed-loop system considering and neglecting delay effects

MATLAB Example Script: m2_2_1.m

```
Rs=1.2; Ld=0.0057; Lq=0.0125;
s = tf('s');
Ts = 1/8000;
lpf_T = 1/2000;
delay_computation = exp(-s*Ts);
delay_output = exp(-s*Ts/2);
tf_filter = tf(2000, [1 2000]);
delay_SH = exp(-s*Ts/2);
d_axis_plant = tf([1/Lq], [1 Rs/Lq]);
pi_controlller = tf([12.5 1200],[1 0]);
Go_no_delay = pi_controlller*d_axis_plant;
Gc_no_delay = Go_no_delay/(1+Go_no_delay);
Go_with_delay =
pi_controlller*delay_computation*delay_output*d_axis_plant;
Gc_with_delay = Go_with_delay/(1+Go_with_delay*tf_filter*delay_SH);
h=bodeoptions;
h.PhaseMatching='on';
h.Title.FontSize = 14;
h.XLabel.FontSize = 14;
h.YLabel.FontSize = 14;
h.TickLabel.FontSize = 14;
bodeplot(Gc_no_delay,'-b',Gc_with_delay,'-.r',{1,10000},h);
legend('NoDelay','WithDelay');
h = findobj(gcf,'type','line');
set(h,'linewidth',2);
```

The solid line in Fig. 2.25 represents the Bode plot of the ideal q-axis current closed-loop system without delay effects. From the magnitude and phase plots, we can observe that the closed-loop transfer function behaves as a first-order low-pass filter. The dashed line represents the Bode plot of the q-axis current closed-loop system with delay effects included. From the plots, we can see that before adding the delay effects, the system bandwidth was designed to be 1000 rad/s. However, after including the delay effects, the system bandwidth significantly increased to 2070 rad/s. At the same time, the damping characteristics and phase response of the closed-loop transfer function were also altered by the delay effects.

You can use the Example script m2_2_1b to plot the Bode diagram of the open-loop system for comparison, as shown in Fig. 2.26. Since delay elements erode the system's phase, it can be observed from the figure that after including delay effects, the phase margin of the q-axis current loop decreases from the original 90° to 52.5°. Phase margin is a key indicator of a control system's stability. As the operating frequency increases, phase delay also increases.

Figure 2.27 shows the step response of the system with delay included, compared to the response without delay. It can be observed that due to the reduced system stability caused by delay, the damping characteristics are also affected, resulting in an increased maximum overshoot in the time-domain response. If the delay continues to increase, it may lead to system oscillation or even instability.

In general, delays are unavoidable in digital control systems, and they inevitably reduce system stability. If delays are not taken into account during the control loop design phase, the designed system performance parameters—such as bandwidth—can deviate significantly from the actual system behavior, which is unacceptable.

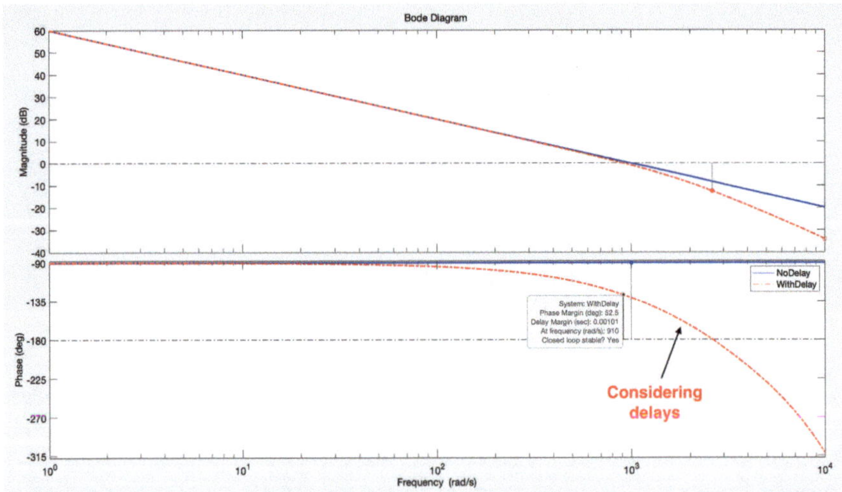

Fig. 2.26 Comparison of Bode plots for the current open-loop system considering and neglecting delay effects

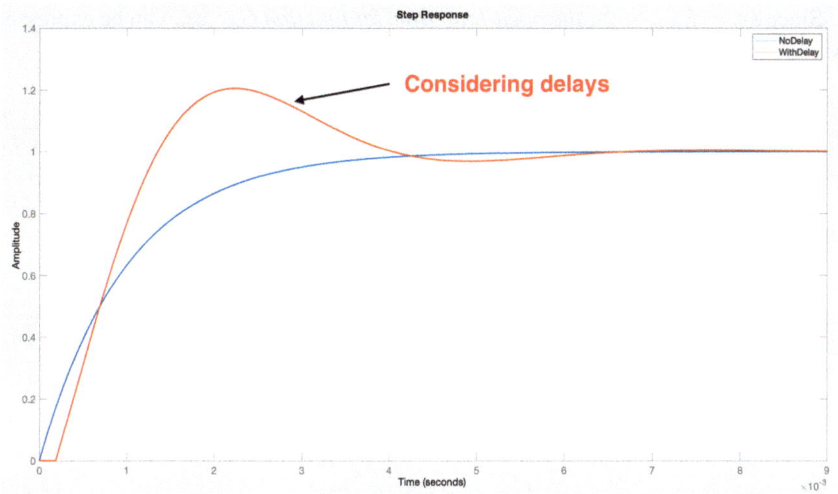

Fig. 2.27 Comparison of current step responses for the closed-loop system considering and neglecting delay effects

Therefore, delay effects must be considered from the beginning of the control loop design process. To ensure that the designed bandwidth specification matches the actual achieved bandwidth, we need to modify the method presented in Sect. 2.1 accordingly.

- **Current Loop Design Method Considering Delay Effects** [10]

First, all delay effects can be combined into a single total delay term e^{-sT_d}, where $T_d = T_c + T_o + T_f + T_{SH}$. Accordingly, the open-loop transfer function of the q-axis current loop considering delay effects, G_{q_open}, can be expressed as follows:

$$G_{qo\_delay} = \left(K_{p\_iq} + \frac{K_{i\_iq}}{s}\right) \times \frac{\frac{1}{L_q}}{s + \frac{R_s}{L_q}} \times e^{-sT_d} \tag{2.2.4}$$

Assuming that ω_q is the desired bandwidth specification for the q-axis current loop, we can use the **pole-zero cancellation method** to design the PI controller parameters as follows:

$$K_{p\_iq} = \frac{\omega_q}{N},\ K_{i\_iq} = \frac{D \times \omega_q}{N} \tag{2.2.5}$$

where $N = \frac{1}{L_q}$ and $D = \frac{R_s}{L_q}$, the **q-axis current loop open-loop transfer function** including delay, denoted as G_{qo_delay}, becomes:

$$G_{qo\_delay} = \frac{\omega_q}{s} \times e^{-sTd} \tag{2.2.6}$$

Since $\omega_q = K_{p\_iq} \times \frac{1}{L_q}$, *the open-loop transfer function* G_{qo_delay} can be expressed as:

$$G_{qo\_delay} = \frac{K_{p\_iq}}{sL_q} \times e^{-sT_d} \tag{2.2.7}$$

Multiplying both the numerator and denominator of G_{qo_delay} by T_d, and substituting s with $j\omega$, we obtain:

$$G_{qo\_delay} = \frac{K_{p\_iq} \times T_d}{j\omega L_q \times T_d} \times e^{-j\omega T_d} \tag{2.2.8}$$

Let $\alpha = \frac{K_{p\_iq} \times T_d}{L_q}$ and $\beta = \omega \times T_d$, then Eq. (2.2.8) can be expressed as:

$$G_{qo\_delay} = \alpha \times \frac{e^{-j\beta}}{j\beta} \tag{2.2.9}$$

The closed-loop transfer function of the q-axis current loop, G_{qc_delay}, can be expressed as:

$$G_{qc\_delay}(j\omega) = \frac{\alpha \times \frac{e^{-j\beta}}{j\beta}}{1 + \alpha \times \frac{e^{-j\beta}}{j\beta}} = \frac{\alpha}{\alpha + j\beta e^{j\beta}} \tag{2.2.10}$$

Using Euler's formula $e^{j\theta} = \cos\theta + j\sin\theta$, Eq. (2.2.10) can be rewritten as:

$$G_{qc\_delay}(j\omega) = \frac{\alpha}{\alpha - \beta\sin(\beta) + j\beta\cos(\beta)} \tag{2.2.11}$$

We can design the magnitude of the closed-loop transfer function G_{qc_delay} *to be* $\frac{1}{\sqrt{2}}$, *that is:*

$$\left|G_{qc\_delay}(j\omega)\right| = \left|\frac{\alpha}{\alpha - \beta\sin(\beta) + j\beta\cos(\beta)}\right| = \frac{1}{\sqrt{2}} \tag{2.2.12}$$

The frequency point that satisfies Eq. (2.2.12) is the system's bandwidth.

Next, we need to derive the relationship between α and β that satisfies Eq. (2.2.12). To do this, we compute the magnitude of both the numerator and the denominator of Eq. (2.2.12) separately.

$$\begin{aligned}\left|\frac{\alpha}{\alpha - \beta\sin(\beta) + j\beta\cos(\beta)}\right| &= \frac{\alpha}{\sqrt{\alpha^2 - 2\alpha\beta\sin(\beta) + [\beta\sin(\beta)]^2 + [\beta\cos(\beta)]^2}} \\ &= \frac{1}{\sqrt{2}}\end{aligned} \tag{2.2.13}$$

Taking the square of both sides of the equation, we obtain:

$$\frac{\alpha^2}{\alpha^2 - 2\alpha\beta\sin(\beta) + \beta^2\sin^2(\beta) + \beta^2\cos^2(\beta)} = \frac{1}{2} \tag{2.2.14}$$

Using the trigonometric identity: $\cos^2(\beta) = 1 - \sin^2(\beta)$, we can simplify Eq. (2.2.14) accordingly.

$$\frac{\alpha^2}{\alpha^2 - 2\alpha\beta\sin(\beta) + \beta^2\sin^2(\beta) + \beta^2\left[1 - \sin^2(\beta)\right]} = \frac{1}{2} \tag{2.2.15}$$

After simplifying Eq. (2.2.15), we obtain:

$$\alpha^2 + 2\alpha\beta\sin(\beta) - \beta^2 = 0 \tag{2.2.16}$$

Solving Eq. (2.2.16), we obtain:

$$\alpha = \frac{-2\beta\sin(\beta) + \sqrt{(2\beta\sin(\beta))^2 + 4\beta^2}}{2} \tag{2.2.17}$$

After simplification, we obtain:

$$\alpha = \beta\left(\sqrt{\sin^2(\beta) + 1} - \sin(\beta)\right) \tag{2.2.18}$$

Here we have completed the derivation. Therefore, to satisfy the magnitude condition of Eq. (2.2.12), the parameters α and β must satisfy the relationship given in Eq. (2.2.18).

Since it is known that $\alpha = \frac{K_{p\_iq} \times T_d}{L_q}$ and $\beta = \omega \times T_d$, then assuming the required control loop bandwidth is 1000 (rad/s) and the total delay time of the control loop is $T_d = T_c + T_o + T_f + T_{SH} = 7.5e - 4$,

we get:

$$\beta = \omega \times T_d = 1000 \times 7.5e - 4 = 0.75$$

After calculating β, we can use Eq. (2.2.18) to find $\alpha = 0.3964$.

Since $\alpha = \frac{K_{p\_iq} \times T_d}{L_q}$, then

$$K_{p\_iq} = \frac{\alpha \times L_q}{T_d} = \frac{0.3964 \times 0.0125}{7.5e - 04} = 6.6073 \tag{2.2.19}$$

And the integral gain K_{i_iq} can be calculated using the following equation:

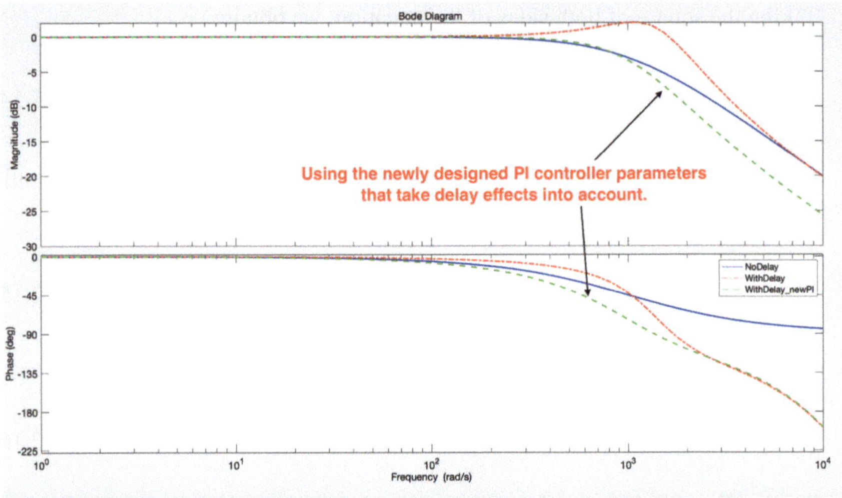

Fig. 2.28 Bode plot of the current closed-loop system using a PI controller considering delay effects

$$K_{i\_iq} = \frac{\alpha \times R_s}{T_d} = \frac{0.3964 \times 1.2}{7.5e - 04} = 634.297 \tag{2.2.20}$$

We can update the q-axis current loop PI controller with the newly obtained proportional gain K_{p_iq} and integral gain K_{i_iq}, which are designed with delay effects taken into account. You can use the Example script **m2_2_1c** to apply the updated parameters and plot the new closed-loop system Bode plot.

Figure 2.28 displays three sets of Bode plots:

- **Solid line**: Ideal closed-loop response without delay effects,
- **Short dashed line**: Closed-loop response with delay effects using the **original** PI controller,
- **Long dashed line**: Closed-loop response with delay effects using the **delay-compensated** PI controller.

The arrow in the figure points to the Bode plot corresponding to the system with both delay effects and delay-aware PI controller design. From the figure, we can observe that the closed-loop Bode plot using the delay-aware PI controller aligns much more closely with the ideal (no-delay) Bode plot.

This confirms that the new controller better meets the **desired bandwidth specification**, even in the presence of system delays.

The design of the d-axis current loop for a permanent magnet synchronous motor (PMSM) should also take delay effects into consideration. The method introduced in this section can be directly applied to the design of the d-axis current loop PI controller. Moreover, this design approach is also applicable to induction motors or other control loops with similar characteristics.

2.3 Speed Loop Controller Design with Delay Consideration

In Sect. 2.2, we redesigned the current loop PI controller by considering the delay effects introduced by digital control systems, ensuring that the loop bandwidth matched the specified design value. Since the motor speed control loop is also implemented using a digital system, similar delay effects will occur in the speed loop as well. Therefore, this section will focus on designing a motor speed control loop that accounts for delay effects.

In practical applications, the friction torque caused by the motor's viscous friction coefficient B is generally considered part of the motor's load torque T_L [1, 4, 10]. As a result, the motor's mechanical plant can be simplified to $\frac{1}{Js}$. A speed control loop for a permanent magnet synchronous motor (PMSM) that includes delay effects can be represented as shown in Fig. 2.29.

In general, since the speed control loop is also implemented using a microcontroller (MCU), its execution speed may be slower than that of the current loop. Therefore, the speed loop will also have its own computational delay, represented as e^{-sT_c}.

As for the speed feedback signal, a low-pass filter is commonly used to attenuate noise before the signal enters the MCU. This low-pass filter introduces an additional delay. Assuming a first-order low-pass filter is used for feedback, its transfer function is given by:

$$T_{filter}(s) = \frac{\omega_{fc}}{s + \omega_{fc}} \tag{2.3.1}$$

Equation (2.3.1) can be rearranged as:

$$T_{filter}(s) = \frac{1}{T_f s + 1} \tag{2.3.2}$$

where $T_f = \frac{1}{\omega_{fc}}$ is the time constant of the first-order low-pass filter, it can also be approximately treated as the filter's delay time. Therefore, the delay effect of the first-order low-pass filter can be equivalently represented as e^{-sT_f}.

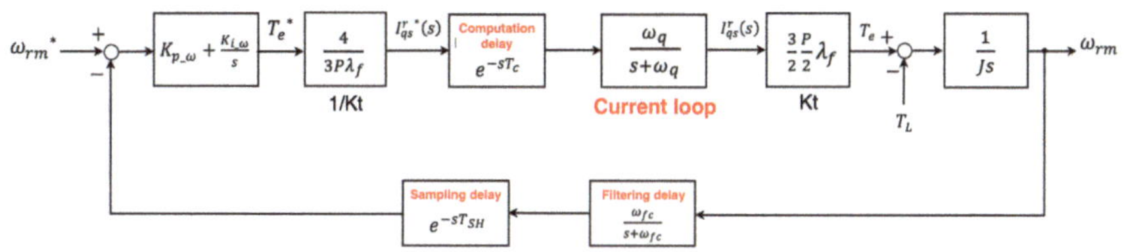

Fig. 2.29 Block diagram of the speed control of a permanent magnet synchronous motor considering delay effects

In Fig. 2.29, it is assumed that the current loop already includes the PWM output delay, and its cutoff frequency corresponds to the current loop bandwidth. Here, ω_q is used to represent the bandwidth of the q-axis current loop. This is because, in general, within a vector control system, the q-axis current loop serves as the inner loop of the speed control loop. Its PI controller must be designed using the method introduced in Sect. 2.2 in order to ensure the accuracy of the specified bandwidth.

The current loop will also introduce delay. Assuming the current loop can be approximated as a first-order low-pass filter, its transfer function is given by:

$$T_{curr\_loop}(s) = \frac{\omega_q}{s + \omega_q} \tag{2.3.3}$$

Equation (2.3.3) can be rewritten as:

$$T_{curr\_loop}(s) = \frac{1}{T_{curr}s + 1} \tag{2.3.4}$$

where $T_{curr} = \frac{1}{\omega_q}$ is the time constant of the current loop, it can also be approximated as the delay time of the current loop. Therefore, the delay effect of the current loop can be equivalently represented as $e^{-sT_{curr}}$.

Finally, the filtered signal will be sampled (Sample-and-Hold) before entering the MCU, and the sampling delay can be represented as:

$$\text{Sampling delay } e^{-sT_{SH}} = e^{-s \times \frac{T_s}{2}} \tag{2.3.5}$$

The signal entering the MCU through sampling (Sample-and-Hold) has a sampling delay that can be approximated as half of the sampling period.

2.3.1 Classical PI Speed Controller Design [1]

For the system in Fig. 2.29, assuming the delay effects are not yet considered, the open-loop transfer function of the speed control loop is:

$$G_{\omega\_open} = \left(K_{p\_\omega} + \frac{K_{i\_\omega}}{s}\right) \times \frac{\omega_q}{s + \omega_q} \times \frac{1}{Js} \tag{2.3.6}$$

In general, the current loop can be modeled as a unit gain within the frequency range below its bandwidth, i.e.,

$$\frac{\omega_q}{s + \omega_q} \cong 1 \tag{2.3.7}$$

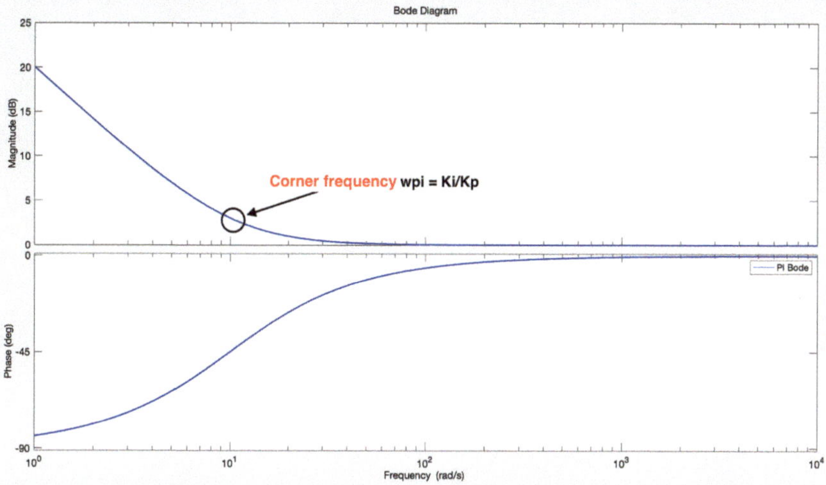

Fig. 2.30 Bode plot of the PI controller

The speed loop PI controller can be expressed in the following form:

$$K_{p\_\omega} + \frac{K_{i\_\omega}}{s} = K_{p\_\omega}\left(\frac{s + \frac{K_{i\_\omega}}{K_{p\_\omega}}}{s}\right) = K_{p\_\omega}\left(\frac{s + \omega_{PI}}{s}\right) \tag{2.3.8}$$

where $\frac{K_{i\_\omega}}{K_{p\_\omega}}$ is generally defined as the corner frequency ω_{PI} of the PI controller. Assuming $K_{i\_\omega} = 10$ and $K_{p\_\omega} = 1$, you can use the Example script m2_3_1 to plot the Bode plot of the PI controller, as shown in Fig. 2.30.

MATLAB Example Script: m2_3_1.m

```
kpw=1; kiw=10;
tf_pi=tf([kpw kiw],[1 0]);
h=bodeoptions;
h.PhaseMatching='on';
h.Title.FontSize = 14;
h.XLabel.FontSize = 14;
h.YLabel.FontSize = 14;
h.TickLabel.FontSize = 14;
bodeplot(tf_pi,'-b',{1,10000},h);
```

From Fig. 2.30, the corner frequency ω_{PI} is 10 (rad/s). Below this frequency, the PI controller behaves like an integrator, while above it, the PI controller behaves like a proportional controller.

Assuming $K_{i\_\omega} = 10$, $K_{p\_\omega} = 1$, and mechanical inertia $J = 0.00016$, substitute these values into Eq. (2.3.5) to plot the Bode diagram of the open-loop speed transfer function, as shown in Fig. 2.31.

Figure 2.31 shows that when the frequency is below the PI controller's crossover frequency of 10 rad/s, the slope of the Bode plot is − 40 dB/decade. When the frequency is above the crossover frequency, the slope becomes − 20 dB/decade. The gain crossover frequency ω_{sc} of the speed loop is approximately 6250 rad/s.

In speed loop design, the gain crossover frequency ω_{sc} of the speed control loop can be used as the target value for the speed bandwidth [1], and the corner frequency of the PI controller ω_{PI} is generally much lower than the loop's gain crossover frequency ω_{sc}.

When the frequency ω is near the gain crossover frequency ω_{sc}, i.e., $\omega \approx \omega_{sc}$, the PI controller can be approximated as a proportional (P) controller. Therefore,

$$G_{\omega\_open}(s) \approx K_{p\_\omega} \times \frac{1}{Js} \tag{2.3.9}$$

When $K_{p\_\omega} = J\omega_{sc}$, the speed open-loop transfer function becomes:

$$G_{\omega\_open}(s) \approx \frac{\omega_{sc}}{s} \tag{2.3.10}$$

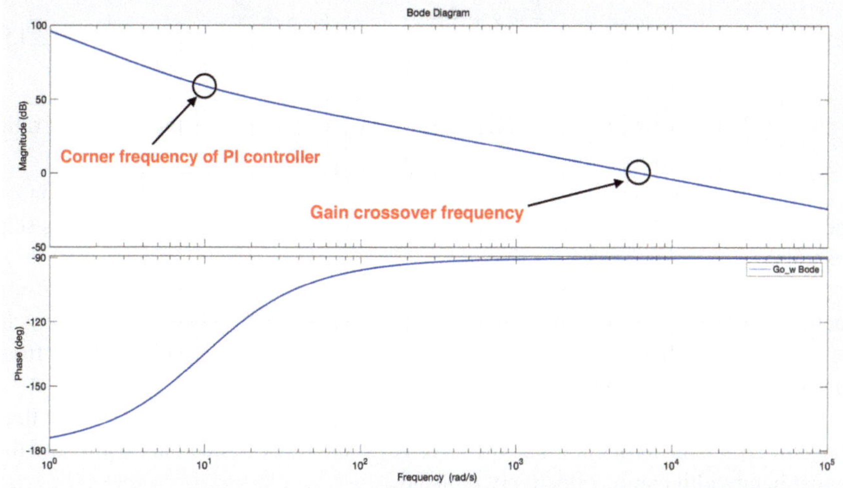

Fig. 2.31 Open-loop Bode plot of the speed control loop

Therefore, when $\omega \geq \omega_{sc}$, the speed closed-loop transfer function can be approximated as a first-order low-pass filter with a cutoff frequency of ω_{sc}, that is

$$G_{\omega\_close}(s) \approx \frac{\omega_{sc}}{s + \omega_{sc}} \tag{2.3.11}$$

At this point, the design of the proportional gain K_{p_ω} for the speed control loop is complete. Next, we proceed to design the integral gain K_{i_ω}.

It is known that the corner frequency ω_{PI} of the PI controller is typically much lower than the loop gain crossover frequency ω_{sc}. Here, we assume that the corner frequency ω_{PI} is at least one-fifth of the gain crossover frequency ω_{sc}, that is:

$$\omega_{PI} = \frac{K_{i\_\omega}}{K_{p\_\omega}} \leq \frac{\omega_{sc}}{5} \tag{2.3.12}$$

Therefore, we have:

$$K_{i\_\omega} \leq K_{p\_\omega} \times \frac{\omega_{sc}}{5} \tag{2.3.13}$$

Assuming the integral gain K_{i_ω} is set as follows:

$$K_{i\_\omega} = K_{p\_\omega} \times \frac{\omega_{sc}}{5} \tag{2.3.14}$$

Then, the closed-loop transfer function of the speed control loop is:

$$\frac{\omega_{rm}}{\omega_{rm}^*} = \frac{\frac{K_{p\_\omega}}{J}s + \frac{K_{i\_\omega}}{J}}{s^2 + \frac{K_{p\_\omega}}{J}s + \frac{K_{i\_\omega}}{J}} = \frac{\omega_{sc}s + \frac{\omega_{sc}^2}{5}}{s^2 + \omega_{sc}s + \frac{\omega_{sc}^2}{5}} \tag{2.3.15}$$

From the denominator of Eq. (2.3.15), we can see that the system damping ratio $\zeta = \frac{\sqrt{5}}{2}$, which corresponds to an overdamped response.

However, since the numerator includes a zero, it performs a differentiation on the input command, which can result in a maximum overshoot in the system's step response.

The above provides a complete introduction to the classical PI controller design method for motor speed loops. Since this method uses Bode plot approximations to design the controller, the designed bandwidth ω_{sc} will have some discrepancy from the actual speed loop bandwidth.

However, when the PI controller's corner frequency ω_{PI} is much smaller than the gain crossover frequency ω_{sc}, the error between the designed bandwidth and the actual bandwidth can be effectively reduced.

- **MATLAB/SIMULINK Simulation Verification**

Next, we use the actual motor parameters to design the speed control loop. Using a mechanical inertia of the motor $J = 0.00016$, and targeting a desired speed loop bandwidth of $\omega_{sc} = 100$ (rad/s), then according to the classical design method:

- The proportional gain is calculated as: $K_{p\_\omega} = J\omega_{sc} = 0.00016 \times 100 = 0.016$
- Using Eq. (2.3.14), the integral gain is: $K_{i\_\omega} = K_{p\_\omega} \times \frac{\omega_{sc}}{5} = 0.016 \times \frac{100}{5} = 0.32$.

If delay effects are not included, we can use the sample code m2_3_2 to plot the Bode plot of the open-loop transfer function of the speed loop, as shown in Fig. 2.32.

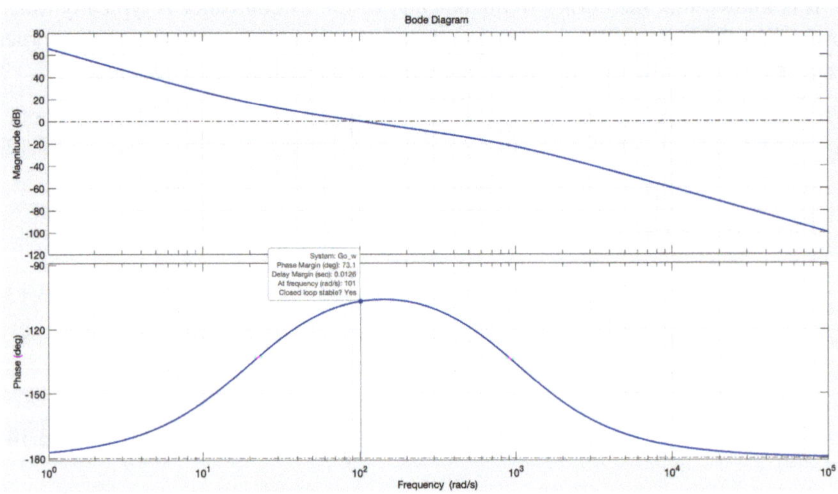

Fig. 2.32 Open-loop Bode plot of the speed control loop in the simulation system

MATLAB Example Script: m2_3_2.m

```
wsc=100; J=0.00016;
Kp_w=J*wsc; Ki_w=Kp_w*wsc/5;
tf_pi=tf([Kp_w Ki_w],[1 0]);
tf_currLoop = tf(1000,[1 1000]);
tf_plant=tf(1,[J 0]);
Go_w = tf_pi*tf_currLoop*tf_plant;
h=bodeoptions;
h.PhaseMatching='on';
h.Title.FontSize = 14;
h.XLabel.FontSize = 14;
h.YLabel.FontSize = 14;
h.TickLabel.FontSize = 14;
bodeplot(Go_w,'-b',{1,100000},h);
h = findobj(gcf,'type','line');
set(h,'linewidth',2);
grid on;
```

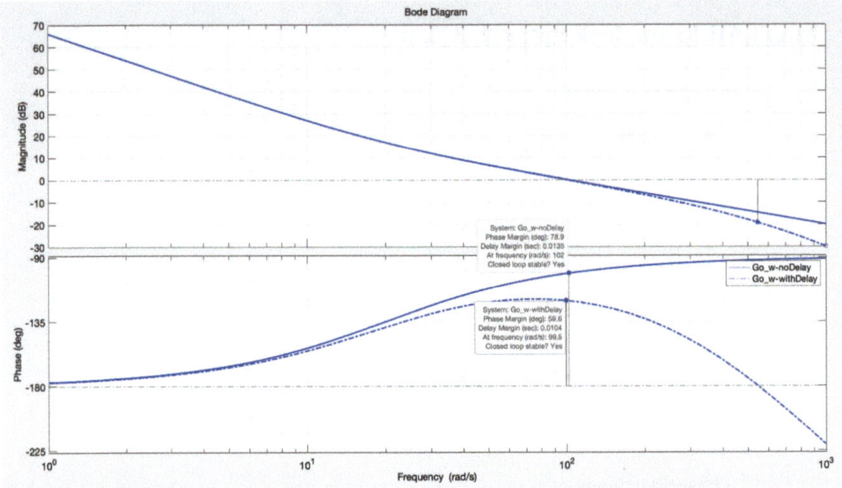

Fig. 2.33 Open-loop Bode plot of the speed control loop considering delay effects

From Fig. 2.32, it can be seen that the gain crossover frequency of the open-loop speed control system is 101 rad/s, which is very close to the design target $\omega_{sc} = 100$ (rad/s). Next, we incorporate the delay effects. Suppose the speed control loop is executed on an MCU with an execution frequency of 4 kHz, then the computation delay is $T_c = 1/4000 = 2.5\text{e} - 4$ (s). If the speed feedback signal passes through a first-order low-pass filter with a cutoff frequency of $\omega_{fc} = 500$ (rad/s), and the current loop can be approximated by a first-order low-pass filter with a cutoff frequency of $\omega_q = 1000$ (rad/s), then the sampling delay is $T_{SH} = \frac{T_s}{2} = 1.25e - 04$ (s).

We incorporate the above-mentioned delays into the open-loop speed gain. You may use the Example script **m2_3_3** to plot the Bode diagram with delay effects included, and compare it with the Bode diagram without delays, as shown in Fig. 2.33.

MATLAB Example Script: m2_3_3.m

```
s=tf('s'); Ts=1/4000;

wsc=100; J=0.00016;

Kp_w=J*wsc; Ki_w=Kp_w*wsc/5;

tf_pi=tf([Kp_w Ki_w],[1 0]);

tf_plant=tf(1,[J 0]);

tf_compuDelay = exp(-s*Ts);

tf_filter= tf(500, [1 500]);

tf_currLoop = tf(1000, [1 1000]);

tf_shDelay = exp(-s*Ts/2);

Go_w_noDelay = tf_pi*tf_plant;

Go_w_withDelay =
tf_pi*tf_plant*tf_compuDelay*tf_filter*tf_currLoop*tf_shDelay;

h=bodeoptions; h.PhaseMatching='on';

h.Title.FontSize = 14;

h.XLabel.FontSize = 14;

h.YLabel.FontSize = 14;

h.TickLabel.FontSize = 14;

bodeplot(Go_w_noDelay,'-b',Go_w_withDelay,'-.b',{1,1000},h);

legend('Go_w-noDelay','Go_w-withDelay');

h = findobj(gcf,'type','line');

set(h,'linewidth',2);

grid on;
```

From Fig. 2.33, it can be observed that adding delay effects does not affect the gain crossover frequency ω_{sc} of the speed loop. However, the phase is affected by the delay. As the frequency increases, the delay reduces the phase margin. Before adding the delay, the system's phase margin was **78.9°**; after introducing the delay, it decreased to **59.6°**, indicating a reduction in system stability.

You can use the example model **mdl_speedLoop_delay** to perform a SIMULINK simulation of both systems and observe the time-domain response. The SIMULINK block diagram is shown in Fig. 2.34, and the simulated speed response is presented in Fig. 2.35.

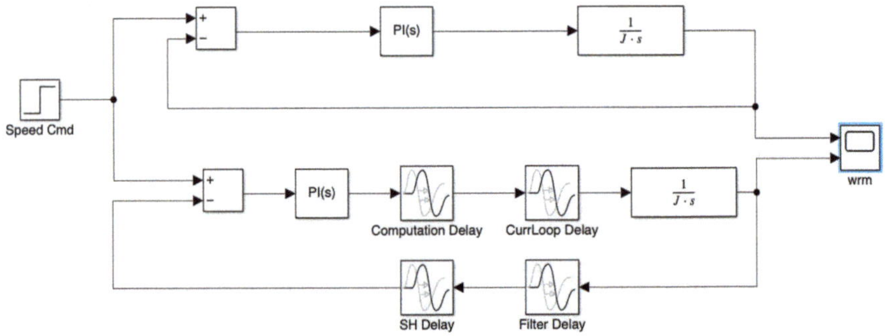

Fig. 2.34 SIMULINK simulation of the speed loop transfer function considering delay effects, Example model: mdl_speedLoop_delay.slx

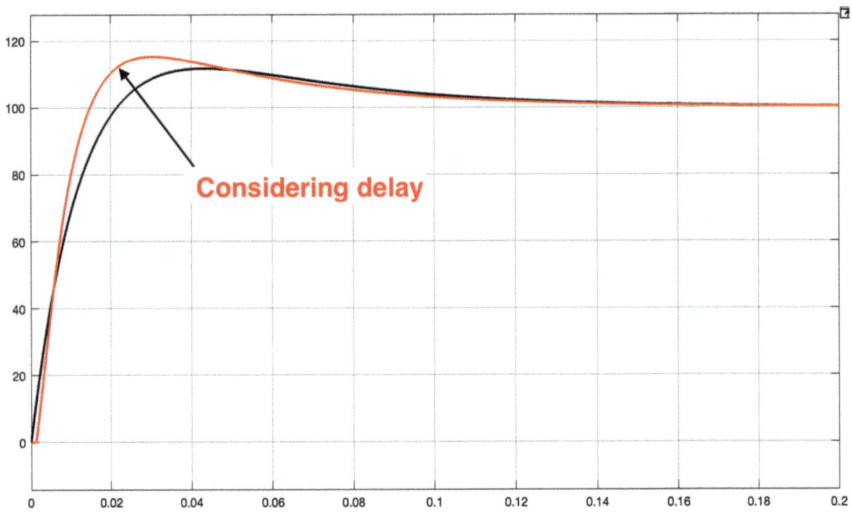

Fig. 2.35 Step response waveform of the speed loop considering delay effects

From the speed time-domain response waveform in Fig. 2.35, it can be seen that after introducing delay, the reduction in phase margin leads to a larger maximum overshoot in the system response.

The classical method for designing a speed-loop PI controller typically involves completing the PI controller design based on the desired bandwidth specification ω_{sc}, and then evaluating the system's actual phase margin after incorporating delay effects. If the resulting phase margin does not meet the design requirements, the PI controller may need to be redesigned accordingly. This method cannot incorporate the phase margin specification during the initial design phase, which is its major drawback.

In the next section, we will introduce a design technique called the Symmetrical Optimum Method. This method accounts for time-delay effects during the design stage of the speed-loop PI controller. It enables the designer to incorporate both the desired bandwidth ω_{sc} and the required phase margin as part of the design specifications. The speed control loop designed using the Symmetrical Optimum Method will precisely meet the specified bandwidth ω_{sc} and phase margin requirements.

2.3.2 *Delay-Aware PI Speed Controller Design*

In this section, we will introduce a speed loop design technique known as the Symmetrical Optimum Method [10, 11]. This method is a PI controller design approach for speed loops that takes time delay effects into account. Importantly, it allows both the desired bandwidth specification ω_{sc} and the phase margin to be incorporated during the design stage.

To begin, let us first revisit the block diagram of the speed control system with delay effects considered, as shown in Fig. 2.36.

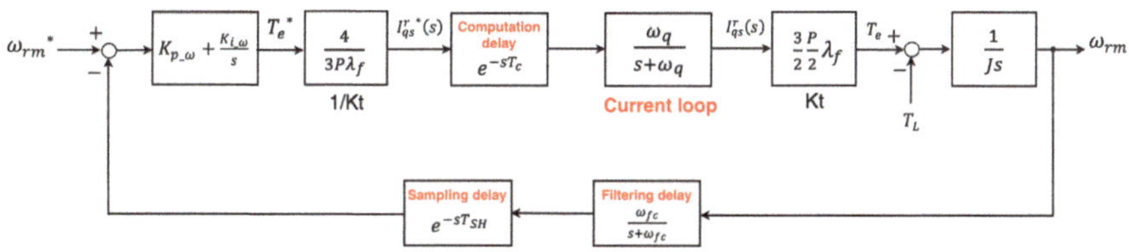

Fig. 2.36 Block diagram of the speed control of a permanent magnet synchronous motor considering delay effects

The open-loop transfer function of the speed loop, $G_{\omega\_open}(s)$, *can be expressed as:*

$$G_{\omega\_open}(s) = \left(K_{p\_\omega} + \frac{K_{i\_\omega}}{s}\right) \times e^{-sT_c} \times \frac{\omega_q}{s+\omega_q} \times \frac{1}{Js} \times \frac{\omega_{fc}}{s+\omega_{fc}} \times e^{-sT_{SH}} \tag{2.3.16}$$

We can combine the computation delay, current loop delay, feedback filter delay, and sampling delay into a total delay time T_{d_total}, that is:

$$T_{d\_total} = T_c + T_{curr} + T_{filter} + T_{SH}$$

where $T_{curr} = 1/\omega_q$ and $T_{filter} = 1/\omega_{fc}$.

Therefore, Eq. (2.3.16) can be expressed as:

$$G_{\omega\_open}(s) = \left(K_{p\_\omega} + \frac{K_{i\_\omega}}{s}\right) \times e^{-sT_{d\_total}} \times \frac{1}{Js} \tag{2.3.17}$$

The delay effect of T_{d_total} can be approximated using a first-order low-pass filter.

$$G_{\omega\_open}(s) = \left(K_{p\_\omega} + \frac{K_{i\_\omega}}{s}\right) \times \frac{1}{T_{d\_total}s+1} \times \frac{1}{Js} \tag{2.3.18}$$

If the control loop can be arranged into the form of Eq. (2.3.18), then according to the *Symmetrical Optimum method* design approach [10, 11], the PI controller parameters can be designed as follows:

$$K_{p\_\omega} = \frac{J}{\alpha T_{d\_total}} \cdot K_{i\_\omega} = \frac{K_{p\_\omega}}{\alpha^2 T_{d\_total}} \tag{2.3.19}$$

Then, the designed gain crossover frequency ω_{sc} can be expressed as:

$$\omega_{sc} = \frac{1}{\alpha T_{d\_total}} \tag{2.3.20}$$

The designed phase margin PM can be expressed as:

$$\mathrm{PM} = 2\tan^{-1}\alpha - 90^\circ \tag{2.3.21}$$

The following explains how to use the Symmetrical Optimum method. This design approach recommends selecting the parameter α in the range between $\sqrt{4}$ and $\sqrt{20}$, i.e., between 2 and approximately 4.47.

Assuming the target speed bandwidth $\omega_{sc} = 100$ (rad/s) and the total delay time $T_{d\_total} = 0.0034$ (s), we can calculate the corresponding α value using Eq. (2.3.20):

$$\alpha = \frac{1}{\omega_{sc} T_{d\_total}} = \frac{1}{100 \times 0.0034} = 2.9412 \tag{2.3.22}$$

At this point, according to Eq. (2.3.21), the **phase margin (PM)** at the given bandwidth can be calculated as follows:

$$\text{PM} = 2\tan^{-1}(2.9412) - 90° = 52.4442 \tag{2.3.23}$$

If this phase margin meets the design requirements, then the corresponding PI controller parameters can be calculated based on Eq. (2.3.19), as follows:

$$K_{p\_\omega} = \frac{J}{\alpha T_{d\_total}} = 0.016 \cdot K_{i\_\omega} = \frac{K_{p\_\omega}}{\alpha^2 T_{d\_total}} = 0.54 \tag{2.3.24}$$

We can use the Example script **m2_3_4** for verification. After running the script, the open-loop Bode plot of the system can be generated, as shown in Fig. 2.37.

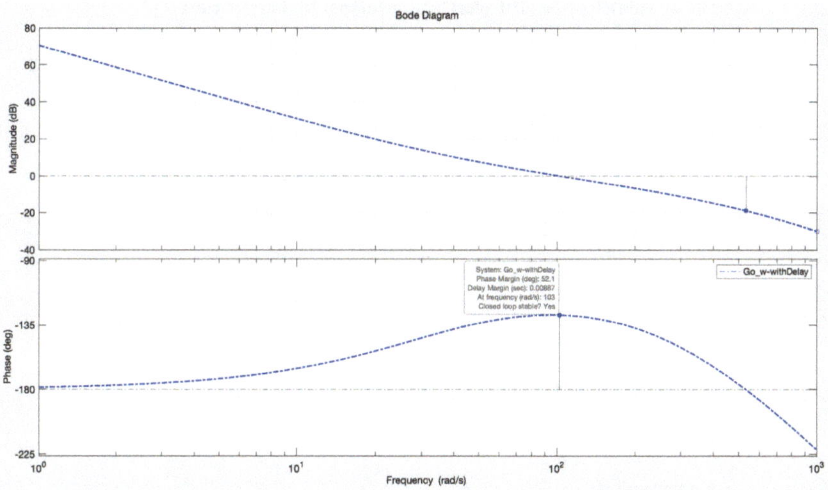

Fig. 2.37 Open-loop Bode plot of the speed loop

MATLAB Example Script: m2_3_4.m

```
s=tf('s'); Ts=1/4000;

wsc=100; J=0.00016;

Td_total=Ts+0.002+0.001+Ts/2;

alpha=1/(wsc*Td_total);

Kp_w=J/(alpha*Td_total); Ki_w=Kp_w/(alpha^2*Td_total);

tf_pi=tf([Kp_w Ki_w],[1 0]);

tf_plant=tf(1,[J 0]);

tf_compuDelay = exp(-s*Ts);

tf_filter = tf(500, [1 500]);

tf_currLoop = tf(1000, [1 1000]);

tf_shDelay = exp(-s*Ts/2);

Go_w_withDelay = tf_pi*tf_plant*tf_compuDelay*tf_filter*tf_currLoop*
tf_shDelay;

Gc_w_withDelay = Go_w_withDelay/(1+Go_w_withDelay);

h=bodeoptions; h.PhaseMatching='on';

h.Title.FontSize = 14;

h.XLabel.FontSize = 14;

h.YLabel.FontSize = 14;

h.TickLabel.FontSize = 14;

bodeplot(Go_w_withDelay,'-.b',{1,1000},h);

legend('Go_w-withDelay');

h = findobj(gcf,'type','line');

set(h,'linewidth',2);

grid on;
```

From Fig. 2.37, we can observe that the open-loop gain crossover frequency is 103 (rad/s), which is very close to the design target of 100 (rad/s). Meanwhile, the system's phase margin is 52.1°, also very close to the design specification of 52.4°.

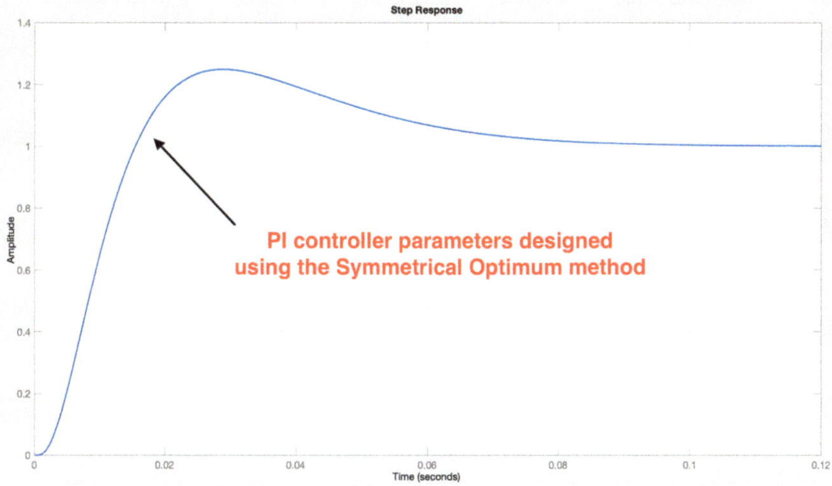

Fig. 2.38 Step response waveform of the speed loop

Although the Symmetrical Optimum method involves certain approximation techniques, the minor discrepancy between the actual and designed values is generally acceptable and sufficient for most application requirements.

If you have already run the Example script m2_3_4, you can use the following command in the MATLAB command window to plot the step response, as shown in Fig. 2.38.

step(Gc_ w_ withDelay)

For certain applications where phase margin is a more critical design specification, one can first use Eq. (2.3.20) during the design stage to calculate the α value that meets the required phase margin. Then, Eq. (2.3.19) can be used to determine the corresponding gain crossover frequency ω_{sc}. If the result meets the design requirements, the corresponding PI controller parameters can be calculated using Eq. (2.3.18). This completes the design of the speed loop PI controller considering delay effects.

2.3.3 IP Controller Design for Speed Loops [1, 4, 5]

In the previous two sections (Sects. 2.3.1 and 2.3.2), we introduced two methods for designing the motor speed loop PI controller. Regardless of the method used, the closed-loop transfer function of the speed loop with a PI controller always includes a zero in the numerator. This zero performs a differentiation operation on the speed command, which leads to a maximum overshoot in the speed response. Here, let us revisit the typical permanent magnet synchronous motor (PMSM) speed control loop architecture using a PI controller, as shown in Fig. 2.39 (for simplification, delay effects are omitted in this derivation).

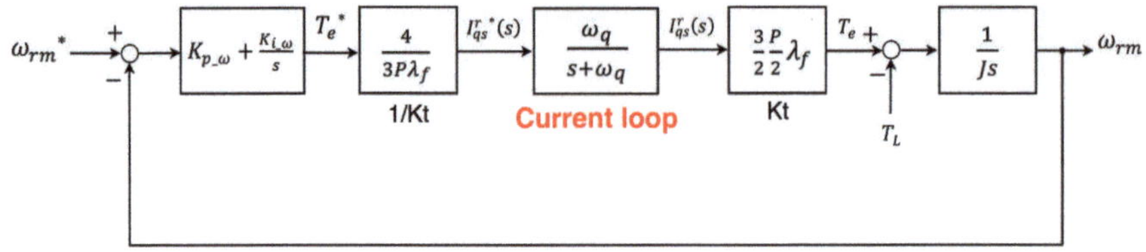

Fig. 2.39 PMSM speed control block diagram with PI controller

Since the speed loop bandwidth is much lower than the current loop bandwidth, the current loop transfer function can be approximated as a unit gain, that is

$$\frac{\omega_q}{s+\omega_q} \cong 1 \tag{2.3.25}$$

In Fig. 2.39, the gain $\frac{4}{3P\lambda_f}$ represents the conversion gain from torque command to current command for a permanent magnet synchronous motor (i.e., the inverse of the torque constant K_T), while the gain $\frac{3P\lambda_f}{4}$ represents the conversion gain from current to torque (i.e., the torque constant K_T). The product of these two gains is exactly 1.

Through derivation, the transfer function between the output speed ω_{rm} and the speed command ω_{rm}^* can be expressed as:

$$\frac{\omega_{rm}}{\omega_{rm}^*} = \frac{\frac{K_{p\_\omega}}{J}s + \frac{K_{i\_\omega}}{J}}{s^2 + \frac{K_{p\_\omega}}{J}s + \frac{K_{i\_\omega}}{J}} \tag{2.3.26}$$

As shown in Eq. (2.3.26), the numerator of the speed closed-loop transfer function contains a zero. This zero performs a differentiation on the input command, which causes the output response to exhibit a maximum overshoot. Therefore, if the zero in the numerator of the speed closed-loop transfer function can be eliminated, the maximum overshoot can be effectively reduced. To achieve this, we can use an IP (Integral-Proportional) controller structure, as illustrated in Fig. 2.40.

In the control structure shown in Fig. 2.40, the section enclosed by the dashed line represents the IP controller. Using the equivalent transformation process illustrated in Fig. 2.41, we can easily derive the transfer function between the output speed ω_{rm}

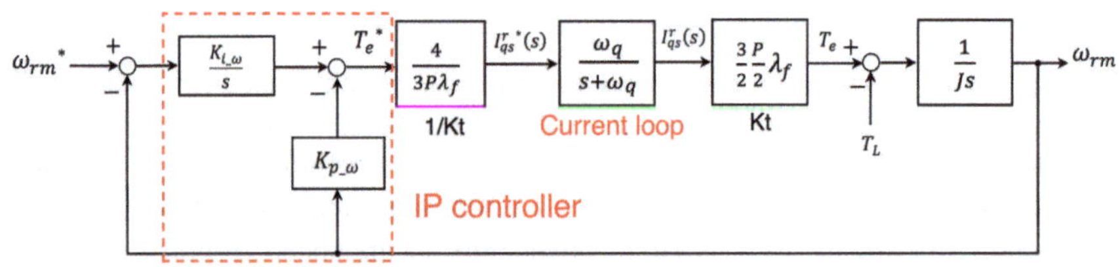

Fig. 2.40 PMSM speed control block diagram with IP controller

and the speed command ω_{rm}^* under the IP controller structure (assuming $\frac{\omega_q}{s+\omega_q} \cong 1$).

$$\frac{\omega_{rm}}{\omega_{rm}^*} = \frac{\frac{K_{i\_\omega}}{J}}{s^2 + \frac{K_{p\_\omega}}{J}s + \frac{K_{i\_\omega}}{J}} \tag{2.3.27}$$

From Eq. (2.3.27), it can be observed that the zero in the numerator has been eliminated. This is because the pole of the low-pass filter applied to the command signal cancels out the zero in the closed-loop transfer function. Therefore, compared to the PI controller, the IP controller results in less oscillation in the speed control loop output.

As shown in Fig. 2.41, the IP controller essentially consists of a PI controller combined with a low-pass filter applied to the command signal. From a signal processing perspective, the reason the IP controller produces less output oscillation than the PI controller is that it first applies low-pass filtering to the speed command before it enters the control loop.

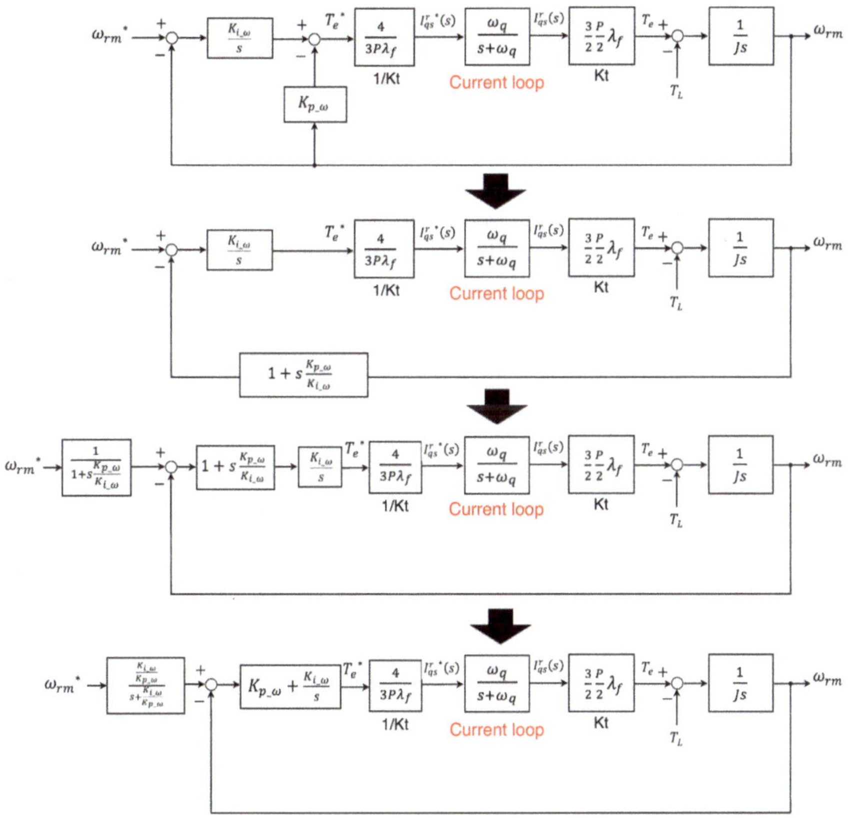

Fig. 2.41 Simplified PMSM speed control block diagram with IP controller

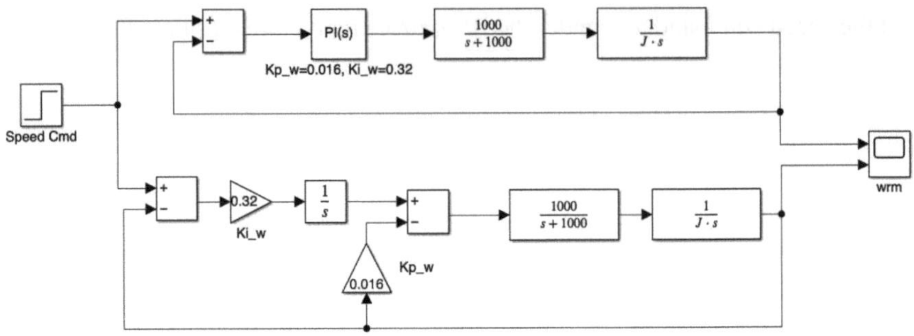

Fig. 2.42 SIMULINK simulation of speed loop with PI and IP controllers, Example model: mdl_speedLoop_PI_IP.slx

- **MATLAB/SIMULINK Simulation Verification**

The PI controller parameters designed using the classical PI controller design method from Sect. 2.3.1 are as follows:

$$K_{p\_\omega} = 0.016$$
$$K_{i\_\omega} = 0.32$$

Enter the parameters into the SIMULINK blocks. Here, we assume that the current control loop can be approximated as a first-order low-pass filter with a cutoff frequency of 1000 (rad/s). At the same time, construct the IP controller loop. The completed SIMULINK control system block diagram is shown in Fig. 2.42.

After running the SIMULINK simulation of Fig. 2.42, the comparison of the speed responses can be obtained, as shown in Fig. 2.43.

From the speed response comparison waveform in Fig. 2.43, it can be observed that the IP controller completely suppresses the maximum overshoot caused by the PI controller.

Returning to the final form of the equivalent transformation of the IP controller in Fig. 2.41, if the IP controller is considered as a combination of a command low-pass filter and a PI controller, then the speed command low-pass filter must be designed as shown in Eq. (2.3.28).

$$\text{Speed command lowpass filter} = \frac{\frac{K_{i\_\omega}}{K_{p\_\omega}}}{s + \frac{K_{i\_\omega}}{K_{p\_\omega}}} \tag{2.3.28}$$

Here, $\frac{K_{i\_\omega}}{K_{p\_\omega}}$ is the cutoff frequency of the low-pass filter. The pole in the denominator is designed to cancel out the zero in the numerator of the closed-loop transfer function in Eq. (2.3.26), resulting in the simplified form shown in Eq. (2.3.27).

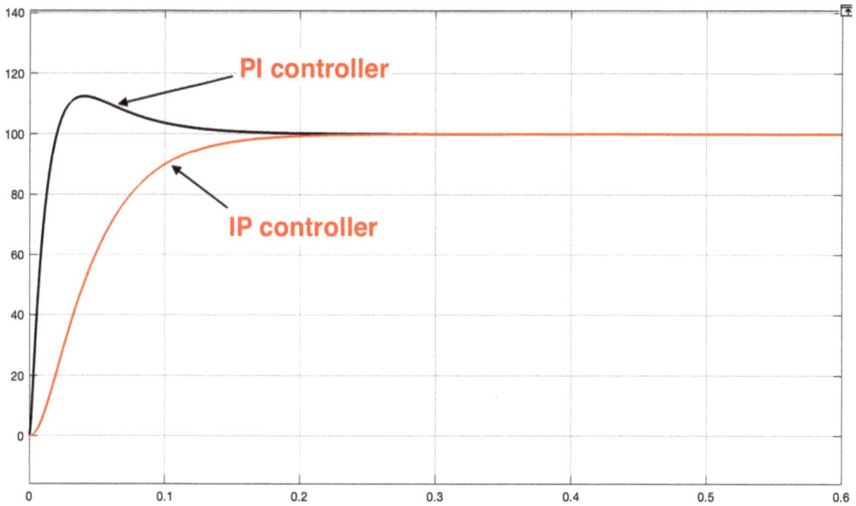

Fig. 2.43 Step response waveforms of the speed loop comparing PI and IP controllers

If the gain K_{i_ω} of the IP controller in Fig. 2.41 is increased independently, the bandwidth of the command low-pass filter in the IP controller will increase. This can improve the response speed of the IP controller. However, the trade-off is that it may introduce overshoot in the response. This occurs because the pole of the command low-pass filter can no longer perfectly cancel the zero in the numerator of Eq. (2.3.26).

Using the classical PI controller design method, we obtained the PI controller parameters $K_{p\_\omega} = 0.016$ and $K_{i\_\omega} = 0.32$. This set of parameters was selected to provide a damping ratio $\frac{\sqrt{5}}{2}$, which corresponds to an **overdamped** response. Therefore, when these parameters are used in an IP controller, the main purpose is simply to **cancel out the zero** in the numerator of the transfer function. As a result, the system's closed-loop transfer function becomes equivalent to that of a **standard second-order system**.

$$\frac{\omega_{rm}}{\omega_{rm}^*} = \frac{\frac{K_{i\_\omega}}{J}}{s^2 + \frac{K_{p\_\omega}}{J}s + \frac{K_{i\_\omega}}{J}} \tag{2.3.29}$$

When the system is overdamped, the roots of the characteristic equation in (2.3.29) are two negative real roots, so the system will not oscillate. However, a correct understanding of the IP controller should be: "It is not that the IP controller inherently prevents oscillation—whether oscillation occurs is determined by the damping ratio of the characteristic equation. The greatest contribution of the IP controller is that it eliminates the zero in the transfer function that would otherwise cause oscillation."

Although the IP controller produces less oscillation than the PI controller, it applies a low-pass filter to the input command, which results in a longer settling time. From a frequency-domain perspective, this means that the bandwidth of the IP control loop is lower than that of the PI control loop. You can use the example script **m2_3_5** to plot the closed-loop transfer function of the IP control loop and compare it with the PI control loop, as shown in Fig. 2.43.

MATLAB Example Script: m2_3_5.m

```
s=tf('s'); Ts=1/4000;

wsc=100; J=0.00016;

Kp_w=J*wsc; Ki_w=Kp_w*wsc/5;

tf_pi=tf([Kp_w Ki_w],[1 0]);

tf_plant=tf(1,[J 0]);

tf_currLoop = tf(1000, [1 1000]);

tf_cmdFilter = tf(Ki_w/Kp_w, [1 Ki_w/Kp_w]);

Go_w_PI = tf_pi*tf_currLoop*tf_plant;

Gc_w_PI = Go_w_PI/(1+Go_w_PI);

Gc_w_IP = tf_cmdFilter*Go_w_PI/(1+Go_w_PI);

h=bodeoptions; h.PhaseMatching='on';

h.Title.FontSize = 14;

h.XLabel.FontSize = 14;

h.YLabel.FontSize = 14;

h.TickLabel.FontSize = 14;

bodeplot(Gc_w_PI,'-b',Gc_w_IP,'-.b',{1,1000},h);

legend('Go_w-PI','Gc_w-IP');

h = findobj(gcf,'type','line');

set(h,'linewidth',2);

grid on;
```

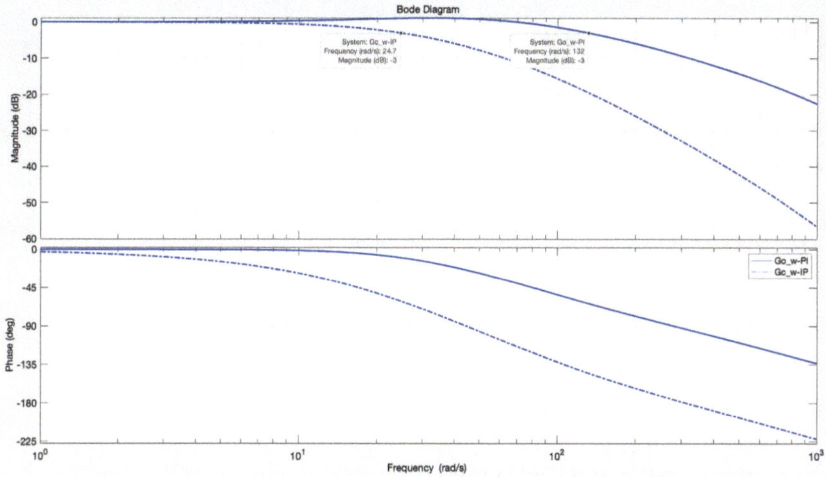

Fig. 2.44 Speed loop closed-loop Bode plot comparing PI and IP controllers

In Fig. 2.44, the dashed line represents the Bode plot of the closed-loop transfer function for the IP control loop, while the solid line represents that of the PI control loop. As shown in the figure, the bandwidth of the PI control loop is 132 (rad/s). (Note: The target bandwidth set using the classical design method was 100 (rad/s). However, since the method involves approximations, there is a certain degree of deviation between the actual and designed bandwidth.)

On the other hand, the bandwidth of the IP control loop is only 24.7 (rad/s). This clearly indicates that the command low-pass filter included in the IP controller has a suppressive effect on the bandwidth.

Therefore, to effectively increase the bandwidth of the IP control loop, it is necessary to co-design both the command low-pass filter and the PI controller parameters. You can use the example script **m2_3_6** to have the computer automatically search for the IP controller parameters that meet the desired bandwidth specifications.

MATLAB Example script: m2_3_6.m

```
J=0.00016;
target_BW=200; % rad/s
for wsc=target_BW/3:0.1:target_BW*5
    Kp_w = J*wsc;
    Ki_w = Kp_w*(wsc/5);
    tf_cmdFilter = tf([Ki_w/Kp_w], [1 Ki_w/Kp_w]);
    tf_PI = tf([Kp_w Ki_w],[1 0]);
    tf_currLoop = tf(1000, [1 1000]);
    tf_plant = tf(1,[J 0]);
    Go_PI = tf_PI*tf_currLoop*tf_plant;
    Gc_PI = Go_PI/(1+Go_PI);
    G_IP = tf_cmdFilter*Gc_PI;
    IP_BW = bandwidth(G_IP);
        if (abs(IP_BW - target_BW)<1)
            break;
        end
    end
    Kp_w
    Ki_w
```

Running the example script **m2_3_6** allows the computer to automatically determine the IP controller parameters that meet the target speed bandwidth of 200 rad/s. The obtained parameters are as follows:

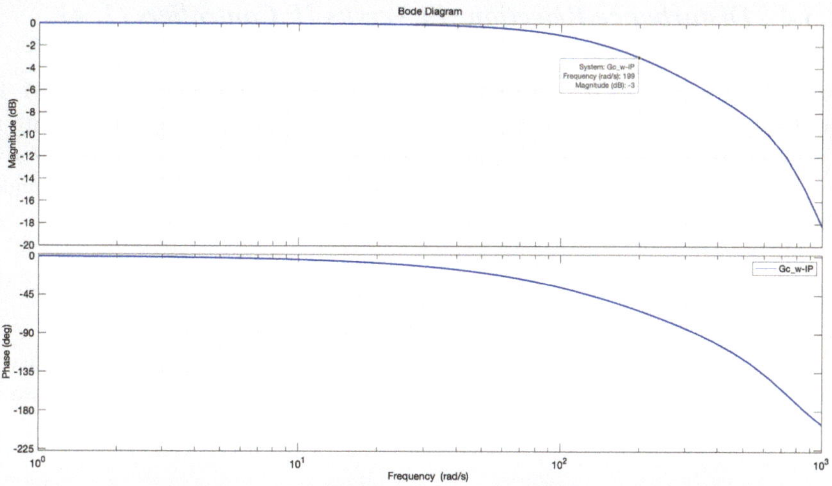

Fig. 2.45 Speed loop closed-loop Bode plot using the IP controller

$$K_{p\_\omega} = 0.1202$$
$$K_{i\_\omega} = 18.0657$$

Using the example script **m2_3_6b**, the Bode plot of the IP control loop with this set of parameters can be generated, as shown in Fig. 2.45. From the Bode plot, it can be seen that the − 3 dB frequency of the IP control loop is **199 rad/s**, which is very close to the target bandwidth specified for the control loop.

Please note in example script **m2_3_6** that the relationship between K_{p_w}, K_{i_w}, and wsc follows:

$$K_{p\_w} = J^{*}wsc, \;\; K_{i\_w} = K_{p\_w} * wsc/5$$

This is identical to the classical PI controller design method presented in Sect. 2.3.1, which sets the system damping ratio ζ to $\frac{\sqrt{5}}{2}$, resulting in an **over-damped** response. Based on this principle, the IP controller parameters found to meet the desired bandwidth will naturally yield an **overdamped** system response as well.

Moreover, a PI controller designed using the **Symmetrical Optimum method** described in Sect. 2.3.2 can also benefit from the IP controller concept: by placing a low-pass filter corresponding to the PI controller at the command input, the oscillation effect caused by the numerator zero can be effectively suppressed.

2.3.4 Disturbance Rejection: PI Versus IP Controllers [1, 4]

Next, let us analyze the **disturbance rejection performance** of **PI and IP controllers** when facing external disturbances such as **load torque**. Referring to Fig. 2.39, and assuming that the input speed command $\omega_{rm}^* = 0$, we can easily derive the transfer function between the **output speed** ω_{rm} and the **load torque** T_L under the **PI controller** structure as follows:

$$\frac{\omega_{rm}}{T_L} = \frac{s}{Js^2 + K_{p\_\omega}s + K_{i\_\omega}} \tag{2.3.30}$$

From Figs. 2.40 and 2.41, we can also easily derive the **transfer function** between the **output speed** ω_{rm} and the **load torque** T_L under the **IP controller** structure as follows:

$$\frac{\omega_{rm}}{T_L} = \frac{s}{Js^2 + K_{p\_\omega}s + K_{i\_\omega}} \tag{2.3.31}$$

As you can clearly see, the disturbance rejection transfer functions of both the **PI** and **IP** control structures are identical. This implies that both controllers have the **same disturbance rejection capability**. To verify this, we can slightly modify the system block diagram from Fig. 2.41. The modified system block diagram is shown in Fig. 2.46.

By simulating the system in Fig. 2.46 using **SIMULINK**, we can obtain the speed response waveform shown in Fig. 2.47. From the waveform, it is evident that after applying a 0.5 Nm impulse load at **0.5 s**, the disturbance responses of both the **PI** and **IP** controllers are identical. Therefore, it can be concluded that the **disturbance rejection capabilities** of the PI and IP controllers are the **same**.

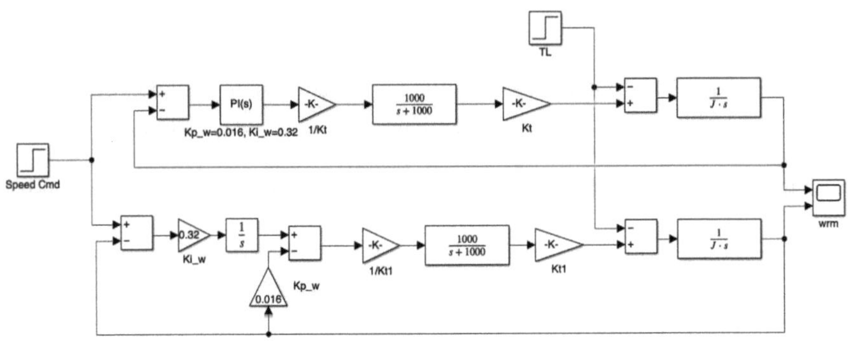

Fig. 2.46 SIMULINK simulation of speed loop disturbance rejection performance: PI versus IP controllers, Example model: mdl_speedLoop_PI_IP_disturbance.slx

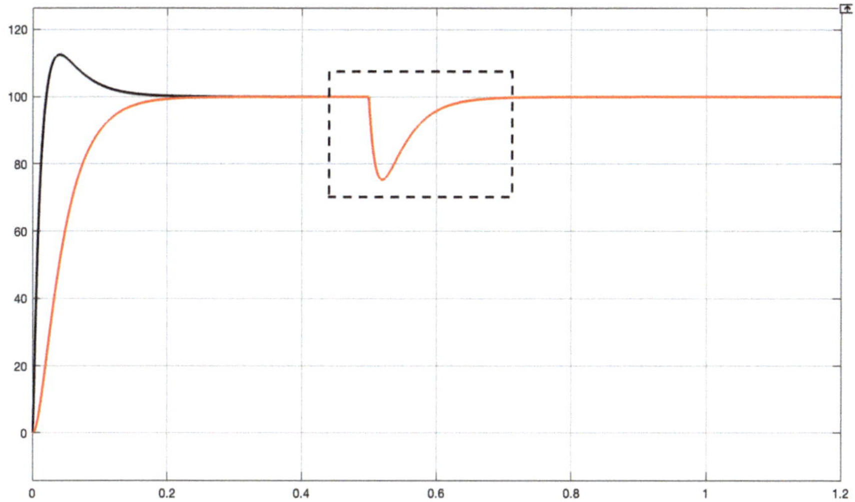

Fig. 2.47 Step response of speed loop disturbance rejection performance: PI versus IP controllers

- **Disturbance Transfer Function Analysis**

For the transfer function in Eq. (2.3.30), the load disturbance is treated as the input. When a disturbance occurs, we hope its effect on the output is minimized. Therefore, we typically want the **transfer function from disturbance to output**—such as that in Eq. (2.3.30)—to be **as small as possible**.

Now let's analyze the behavior of Eq. (2.3.30) across different frequency bands by substituting $s = j\omega$, and examining the **approximate behavior** of the disturbance transfer function at various frequencies:

- **Low-Frequency Region**: At low frequencies, since the frequency ω is very small, s is small, and s^2 is even smaller. Therefore, the denominator is dominated by the K_{i_ω} term. The transfer function $\frac{\omega_{rm}}{T_L}$ can be approximated as $\frac{s}{K_{i\_\omega}}$, i.e., $\frac{\omega_{rm}}{T_L} \cong \frac{s}{K_{i\_\omega}}$. As the frequency decreases, $\frac{s}{K_{i\_\omega}}$ becomes smaller.
- **High-Frequency Region**: At high frequencies, since the frequency ω is large, s is large, and s^2 is even larger. Therefore, the denominator is dominated by the Js^2 term. The transfer function $\frac{\omega_{rm}}{T_L}$ can be approximated as $\frac{s}{Js^2}$, that is: $\frac{\omega_{rm}}{T_L} \cong \frac{s}{Js^2} = \frac{1}{Js}$. As the frequency increases, $\frac{1}{Js}$ becomes smaller.
- **Mid-Frequency Region**: At mid frequencies, the denominator is dominated by the $K_{p\_\omega}s$ term. Therefore, the transfer function $\frac{\omega_{rm}}{T_L}$ can be approximated as: $\frac{\omega_{rm}}{T_L} \cong \frac{s}{K_{p\_\omega}s} = \frac{1}{K_{p\_\omega}}$. This implies that in the mid-frequency range, the system's sensitivity to load disturbances is approximately constant and inversely proportional to the proportional gain K_{p_ω}.

MATLAB Example Script: m2_3_7.m

```
wsc=100; J=0.00016;
Kp_w=J*wsc; Ki_w=Kp_w*wsc/5;
tf_disturbance = tf([1 0],[J Kp_w Ki_w]);
h=bodeoptions;
h.PhaseMatching='on';
h.Title.FontSize = 14;
h.XLabel.FontSize = 14;
h.YLabel.FontSize = 14;
h.TickLabel.FontSize = 14;
bodeplot(tf_disturbance,'-b',{1,10000},h);
legend('TF-disturbance');
h = findobj(gcf,'type','line');
set(h,'linewidth',2);
grid on;
```

You can use the example script m2_3_7 to plot the Bode diagram of the transfer function in Eq. (2.3.29), as shown in Fig. 2.48. The above approximate analysis of the disturbance-to-output transfer function is validated by the Bode plot in Fig. 2.48. In different frequency ranges, the transfer function can be approximated by different forms: in both the low- and high-frequency ranges, the magnitude decreases as frequency increases; while in the mid-frequency range, the magnitude of the transfer function can be approximated as $\frac{1}{K_{p\_\omega}}$. In this example, $K_{p\_\omega} = 0.016$, which can be converted to a dB value as follows:

$$20\log\left(\frac{1}{K_{p_\omega}}\right) = 20\log\left(\frac{1}{0.016}\right) = 35.91\ (\text{dB})$$

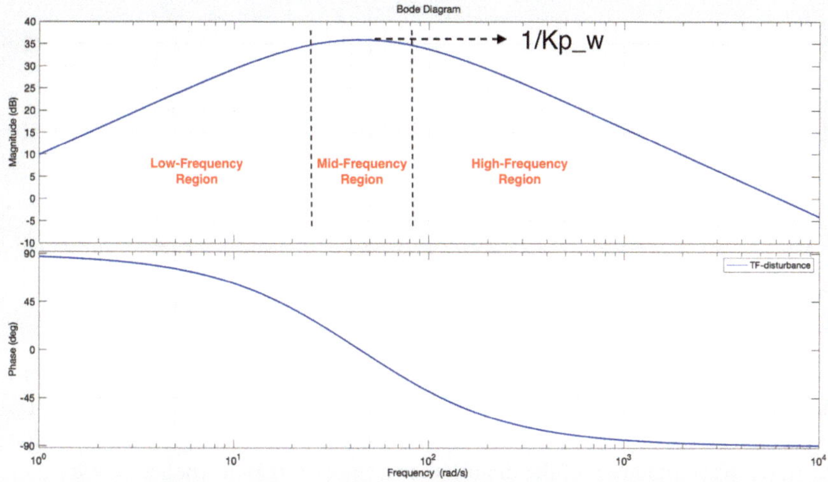

Fig. 2.48 Bode plot of disturbance-to-output transfer function

The magnitude in the mid-frequency range can be validated by the Bode plot in Fig. 2.48. As the control loop bandwidth increases, it indicates that the loop gain is higher—meaning K_{p_ω} is larger. A larger K_{p_ω} results in a smaller $\frac{1}{K_{p\_\omega}}$, which means the magnitude of the disturbance transfer function can shift further downward in the Bode plot, reducing the disturbance's effect on the output. Therefore, the higher the control loop bandwidth, the better the disturbance rejection capability.

Figure 2.49 shows the change in the magnitude of the disturbance transfer function's Bode plot when K_{p_ω} is increased from the original value of 0.016 to twice that, i.e., 0.032. From the figure, we can see that increasing the proportional gain shifts the Bode magnitude curve downward, further reducing the impact of disturbances on the output. This demonstrates that the higher the controller gain, the better the disturbance rejection performance of the control loop.

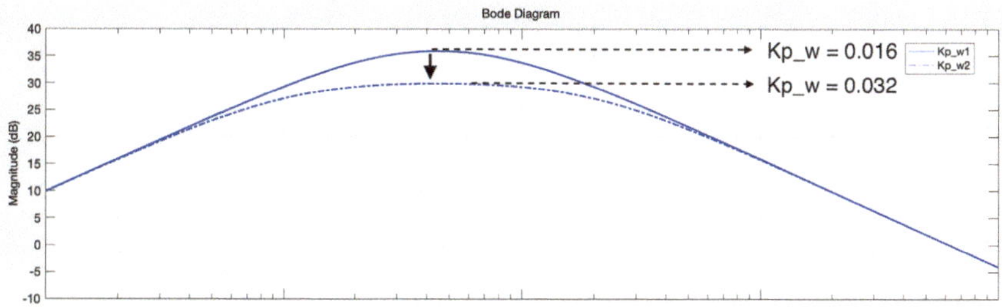

Fig. 2.49 Bode plot showing the effect of doubling the proportional gain, Example script: m2_3_7b.m

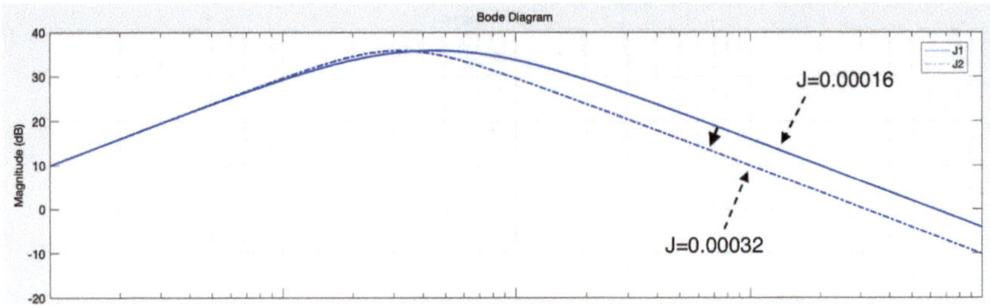

Fig. 2.50 Bode plot showing the effect of doubling the inertia, Example script: m2_3_7c.m

If the parameters of the controlled plant change—for example, when the motor's moment of inertia J increases—the magnitude of the disturbance transfer function in the high-frequency range can be further reduced. Figure 2.50 shows the change in the Bode magnitude plot of the disturbance transfer function when the motor inertia J is increased from the original value of 0.00016 to twice that, 0.00032. From the figure, we can see that as the system's overall inertia increases, the magnitude of the disturbance transfer function in the high-frequency range shifts downward, making the output less affected by disturbances. This implies that the larger the mechanical inertia, the less susceptible the speed is to high-frequency disturbances.

If the integral gain of the PI controller is increased alone, the magnitude of the disturbance transfer function in the low-frequency range can be further reduced. Figure 2.51 shows the change in the Bode magnitude plot of the disturbance transfer function when the integral gain K_{i_ω} of the PI controller is increased from the original value of 0.32 to twice that, 0.64. From the figure, it can be seen that as the integral gain increases, the Bode magnitude in the low-frequency region shifts downward, reducing the impact of disturbances on the output. This implies that the greater the integral gain, the less susceptible the motor speed is to low-frequency disturbances.

- **Analysis of Disturbance Rejection Capability for Systems with Different Moments of Inertia**

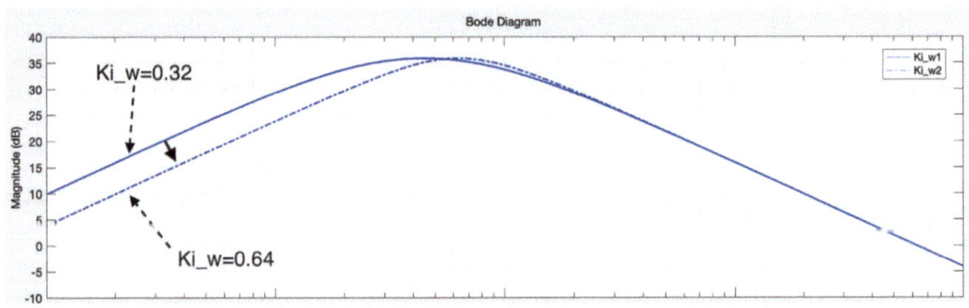

Fig. 2.51 Bode plot showing the effect of doubling the integral gain, Example script: m2_3_7d.m

Based on the above analysis, when the motor's moment of inertia J is doubled, the system's high-frequency disturbance rejection capability is directly improved. This increase in inertia may also indirectly enhance the PI controller gains. Specifically:

- **To maintain the same control loop bandwidth**, when J is doubled, the **proportional gain** K_{p_ω} of the PI controller must also be doubled (since $K_{p\_\omega} = J\omega_{sc}$).
- To preserve the same **damping ratio**, the **integral gain** K_{i_ω} must also be doubled accordingly (since $K_{i\_\omega} = K_{p\_\omega} \times \frac{\omega_{sc}}{5}$).

As shown in the Bode plot of Fig. 2.52, when all three—**moment of inertia**, **proportional gain**, and **integral gain**—are doubled, the magnitude of the disturbance transfer function uniformly shifts downward. This means that even though both systems have the **same bandwidth**, the system with the **larger inertia** demonstrates **better disturbance rejection**.

To verify this, a SIMULINK model can be constructed (see Fig. 2.53), and the simulation results are shown in Fig. 2.54. From the speed response:

- The **transient responses are identical**, confirming the same closed-loop bandwidth.

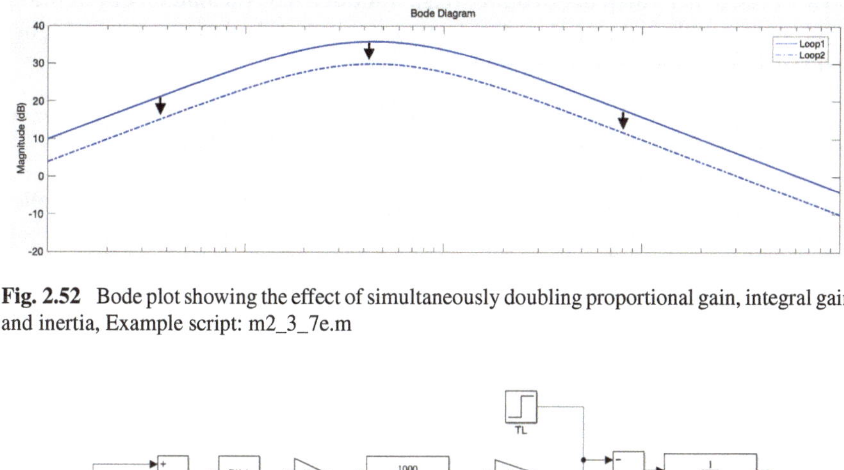

Fig. 2.52 Bode plot showing the effect of simultaneously doubling proportional gain, integral gain, and inertia, Example script: m2_3_7e.m

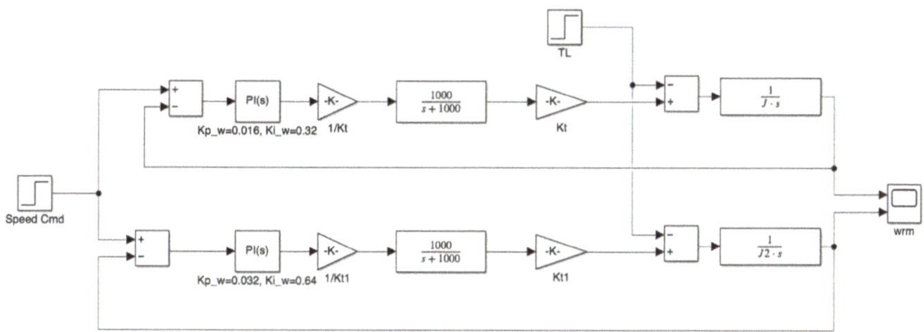

Fig. 2.53 SIMULINK simulation showing the effect of simultaneously doubling proportional gain, integral gain, and inertia, Example model: mdl_speedLoop_compare_J.slx

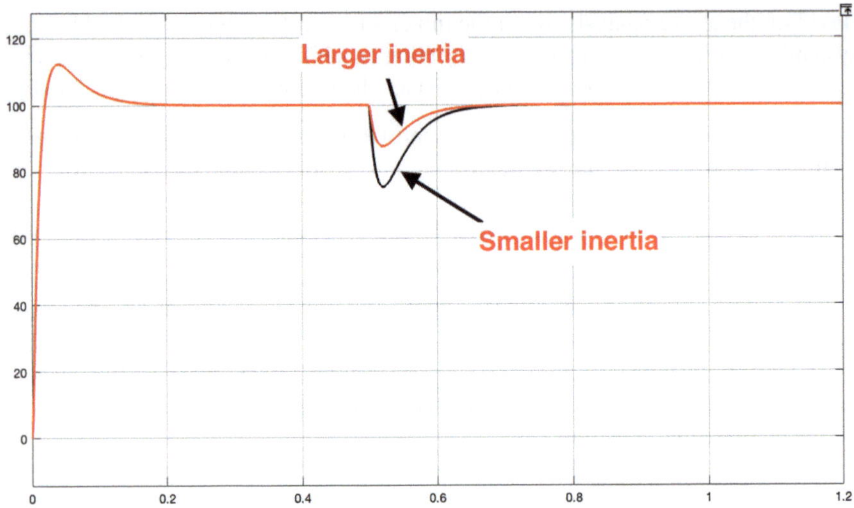

Fig. 2.54 Speed response waveform when proportional gain, integral gain, and inertia are all doubled

- However, under a **step disturbance (e.g., a sudden 0.5 Nm load torque at 0.5 s)**, the system with the **larger inertia** clearly exhibits **less deviation**, proving **higher disturbance rejection performance**.

This confirms that **increasing system inertia**, while adjusting PI controller gains accordingly, enhances the system's robustness to external load disturbances.

- **Disturbance Rejection Capability Analysis with Different PI Controller Gains**

Next, we will analyze the disturbance rejection capability of control loops using different PI controller gains. Both control loops have identical plant inertia. We prepare two sets of PI controller parameters:

The first set corresponds to a PI controller designed for a speed loop bandwidth of 100 rad/s (the same parameters used in example program **m2_3_7**):

$$K_{p\_\omega} = 0.016$$
$$K_{i\_\omega} = 0.32$$

The second set corresponds to a PI controller designed for a speed loop bandwidth of **200 rad/s**:

$$K_{p\_\omega} = 0.032$$
$$K_{i\_\omega} = 1.28$$

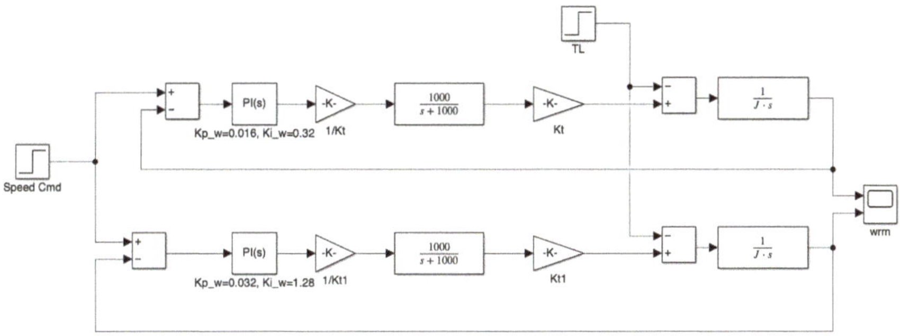

Fig. 2.55 SIMULINK simulation of disturbance rejection performance of the speed loop using different PI controller gains, Example model: mdl_speedLoop_compare_PI.slx

Input both sets of parameters into the SIMULINK system. The completed system block diagram is shown in Fig. 2.55.

Run the SIMULINK simulation for the system shown in Fig. 2.55, and the resulting speed response waveform is shown in Fig. 2.56. From the waveform, we observe that after the impact load is applied, the output speed of the PI control loop with a bandwidth of 100 rad/s drops by approximately 24 rad/s, while the output speed of the PI control loop with a bandwidth of 200 rad/s drops by only about 12 rad/s. Therefore, in this example, the PI control loop with a 200 rad/s bandwidth exhibits **twice the disturbance rejection capability** compared to the one with a 100 rad/s bandwidth.

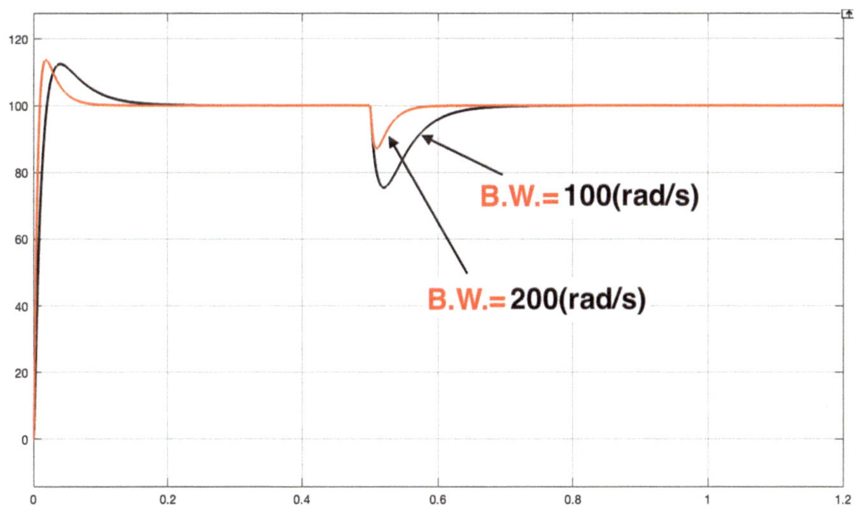

Fig. 2.56 Speed response waveforms using different PI controller gains

2.4 Feedforward Compensation Techniques

This section introduces **feedforward compensation techniques**. We begin by examining a typical control system block diagram that utilizes feedforward compensation [4], as shown in Fig. 2.57.

In the diagram:

- $R(s)$ is the input signal,
- $Y(s)$ is the output signal,
- $G_C(s)$ is the transfer function of the controller,
- $G_{PC}(s)$ is the transfer function of the power converter,
- $G_P(s)$ is the transfer function of the plant (controlled system),
- $G_F(s)$ is the transfer function of the feedforward path,
- K_F is the feedforward gain.

The dashed block in Fig. 2.57 represents the **feedforward path**. By applying **Mason's Gain Formula**, we can derive the transfer function between the output $Y(s)$ and the input $R(s)$ as follows:

$$\frac{Y(s)}{R(s)} = \frac{[K_F G_F(s) + G_C(s)]G_{PC}(s)G_P(s)}{1 + G_C(s)G_P(s)G_{PC}(s)} \tag{2.4.1}$$

Without loss of generality, we can assume the power converter transfer function $G_{PC}(s) \cong 1$. Under this assumption, Eq. (2.4.1) can be simplified as:

$$\frac{Y(s)}{R(s)} = \frac{[K_F G_F(s) + G_C(s)]G_P(s)}{1 + G_C(s)G_P(s)} \tag{2.4.2}$$

If the feedforward transfer function $G_F(s) = G_P^{-1}(s)$ and the feedforward gain K_F is 1, then Eq. (2.4.2) becomes:

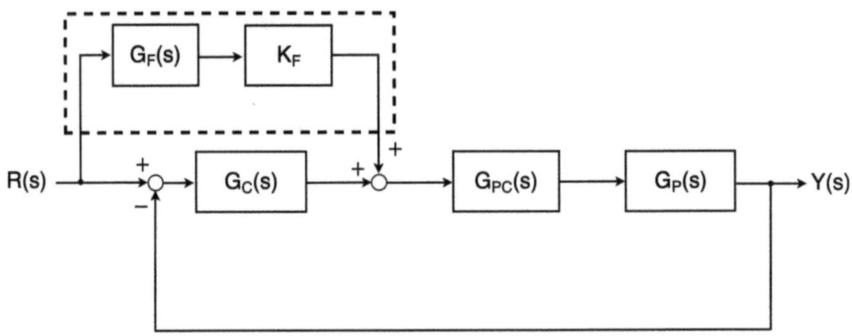

Fig. 2.57 Block diagram of a closed-loop system with feedforward control

$$\frac{Y(s)}{R(s)} = \frac{[K_F G_F(s) + G_C(s)]G_P(s)}{1 + G_C(s)G_P(s)}$$
$$= \frac{\left[1 \times G_P^{-1}(s) + G_C(s)\right]G_P(s)}{1 + G_C(s)G_P(s)} = \frac{1 + G_C(s)G_P(s)}{1 + G_C(s)G_P(s)} = 1 \quad (2.4.3)$$

It can be seen that using feedforward control allows the transfer function between the output $Y(s)$ and the input $R(s)$ to become unity gain. In control engineering, this represents a nearly perfect and ideal condition, meaning that the system output can perfectly follow the input command. However, this ideal state described by Eq. (2.4.3) can only be achieved if the feedforward transfer function $G_F(s)$ is the exact inverse of the plant transfer function $G_P(s)$, and the precise, error-free parameters of the plant must be known at all times.

In practice, however, the plant parameters are usually time-varying, and the parameters obtained through estimation techniques often contain errors. Moreover, the power converter transfer function $G_{PC}(s)$ is not actually equal to 1. Therefore, the ideal condition described by Eq. (2.4.3) is difficult to achieve. Nonetheless, feedforward control still holds significant value, as it can substantially improve the response speed to command inputs.

By observing the system block diagram in Fig. 2.57, if we consider only the effect of the **feedforward path** on the output $Y(s)$, we can calculate the gain from the input $R(s)$ through the feedforward path to the output $Y(s)$, denoted as G_{ff_path}, *under the assumption that the power converter transfer function* $G_{PC}(s) \cong 1$:

$$G_{ff\_path} = K_F G_F(s) G_{PC}(s) G_P(s) \cong K_F G_F(s) G_P(s) \quad (2.4.4)$$

By substituting the feedforward gain $K_F = 1$ and the feedforward transfer function $G_F(s) = G_P^{-1}(s)$ into Eq. (2.4.4), we obtain:

$$G_{ff\_path} = 1 \times G_P^{-1}(s) G_P(s) = 1 \quad (2.4.5)$$

Equation (2.4.5) indicates that the feedforward path alone can provide most of the control effort required by the system, allowing the output to equal the input. The feedforward path essentially predicts how much signal needs to be sent to the plant to generate the desired response. Therefore, feedforward can significantly reduce the burden on the feedback control loop. Most of the control effort required by the power converter can be supplied by the feedforward path, while the controller only needs to make corrections when the output deviates from the reference.

- **Feedforward Compensation in PMSM Speed Control Loop**

To verify the effect of feedforward compensation, we substitute the control block diagram in Fig. 2.57 with the Permanent Magnet Synchronous Motor (PMSM) speed control loop described in Sect. 2.3.1. The equivalent substitutions for the control blocks are as follows:

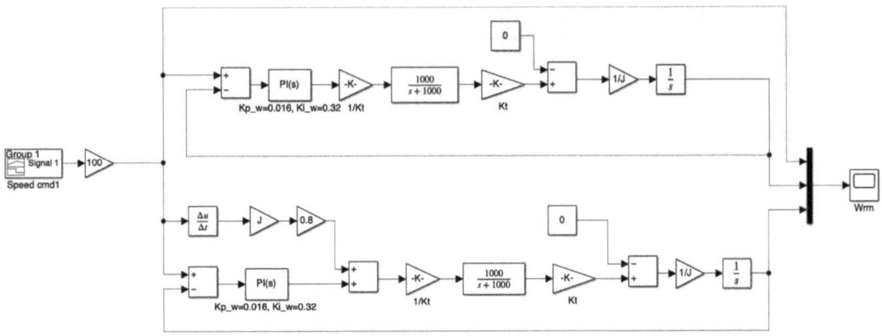

Fig. 2.58 SIMULINK simulation of speed loop with and without feedforward control, Example model: mdl_speedLoop_ff1.slx

$$G_P(s) = \frac{1}{Js}$$
$$G_F(s) = Js$$
$$G_{PC}(s) = \frac{1000}{s + 1000}$$
$$G_C(s) = K_{p\_\omega} + \frac{K_{i\_\omega}}{s}$$
$$K_F = 0.8$$

Given the parameters $J = 0.00016$, $K_{p\_\omega} = 0.016$, and $K_{i\_\omega} = 0.32$, Fig. 2.58 shows the constructed speed control loop with feedforward compensation, alongside a comparison with a control loop without feedforward compensation. To better compare the difference in response waveforms, the feedforward gain K_F is initially set to 0.8. In practice, the feedforward gain is typically less than 1, with a common setting range between 0.4 and 0.8.

After running the SIMULINK simulation of Fig. 2.58, the resulting response waveform is shown in Fig. 2.59.

In the example program **mdl_speedLoop_ff1**, a trapezoidal wave is used as the speed command. From the waveform in Fig. 2.59, it is evident that the speed response with feedforward is significantly better than the response without feedforward. The feedforward-enhanced response shows **superior command tracking**, as feedforward can **increase the command-following speed by several times**, allowing the output response speed to **no longer rely solely on the control loop bandwidth.**

You can use the example program **m2_4_1** to plot the **Bode diagram of the speed control loop with feedforward**, and compare it with the Bode diagram of the loop **without feedforward**. Executing **m2_4_1** will generate the Bode plot shown in Fig. 2.60.

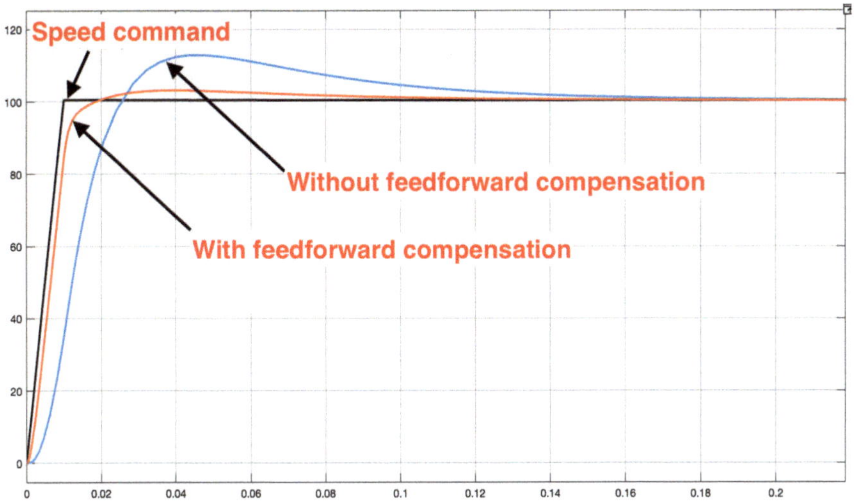

Fig. 2.59 Speed response waveforms with and without feedforward control

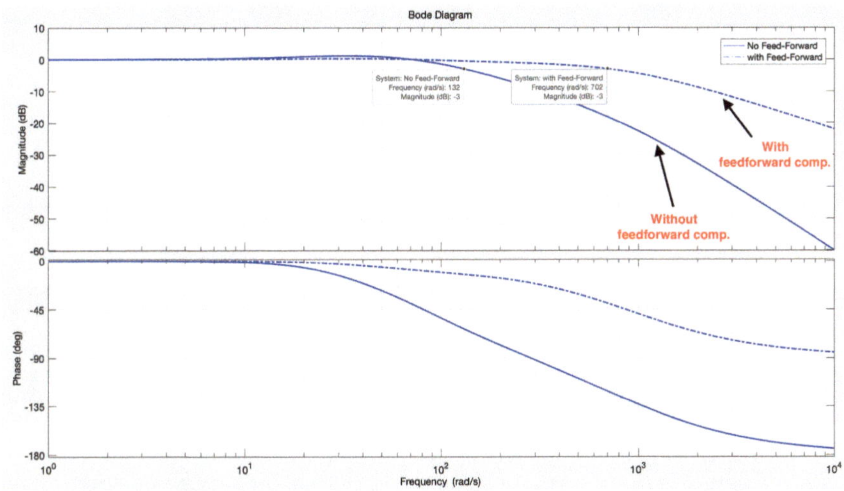

Fig. 2.60 Speed loop closed-loop Bode plots with and without feedforward control

MATLAB Example Script: m2_4_1.m

```
wsc=100; J=0.00016;
s = tf('s');
Kp_w=J*wsc; Ki_w=Kp_w*wsc/5;
tf_pi = tf([Kp_w Ki_w],[1 0]);
tf_pc = tf([1000], [1 1000]);
tf_plant = tf([1], [J 0]);
Gw_open = tf_pi*tf_pc*tf_plant;
Gw_close = Gw_open/(1+Gw_open);
Gf = s*J;
kf = 0.8;
Gw_ff = (kf*Gf+tf_pi)*tf_pc*tf_plant/(1+tf_pi*tf_pc*tf_plant);
h=bodeoptions; h.PhaseMatching='on';
h.Title.FontSize = 14; h.XLabel.FontSize = 14;
h.YLabel.FontSize = 14; h.TickLabel.FontSize = 14;
bodeplot(Gw_close,'-b',Gw_ff,'-.b',{1,10000},h);
legend('No Feed-Forward', 'with Feed-Forward');
h = findobj(gcf,'type','line');
set(h,'linewidth',2);
grid on;
```

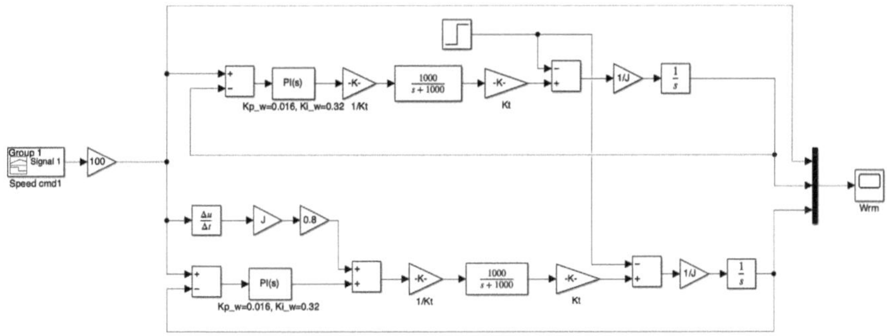

Fig. 2.61 SIMULINK simulation of the speed loop disturbance rejection performance: comparison with and without feedforward control, Example model: mdl_speedLoop_ff1_dist

From the Bode magnitude plot in Fig. 2.60, it is clearly observed that the speed loop bandwidth without feedforward is 132 rad/s, whereas the bandwidth with feedforward exceeds five times that value, reaching 702 rad/s. Additionally, feedforward improves the peaking effect in the Bode plot and reduces phase delay in the closed-loop system.

Although feedforward greatly enhances the speed of command response, it does not form a feedback loop, and therefore does not affect the system's stability. Moreover, since the feedforward path does not respond to disturbances, it cannot improve disturbance rejection.

You can verify this by running the example program mdl_speedLoop_ff1_dist. The corresponding SIMULINK block diagram is shown in Fig. 2.61.

The controller parameters used in the system of Fig. 2.61 are the same as those in Fig. 2.58, so the two control loops in Fig. 2.61 share the same bandwidth. To test disturbance rejection capability, a step load torque of 0.2 Nm is applied at 0.5 s as an impulse disturbance. After running the SIMULINK simulation, the resulting speed response waveform is shown in Fig. 2.62. As illustrated, regardless of whether feedforward is included, systems with the same loop bandwidth exhibit the same disturbance rejection capability.

- **Feedforward Compensation Considering the Power Converter**

In the previous derivation of the feedforward path, we assumed that the power converter transfer function $G_{PC}(s) = 1$. Based on this assumption, the feedforward transfer function $G_F(s)$ could be designed as the inverse of the plant transfer function, i.e., $G_F(s) = G_P^{-1}(s)$.

However, in reality, the power converter transfer function $G_{PC}(s)$ is not equal to unity. In the context of motor speed control loops, $G_{PC}(s)$ typically includes not only the dynamics of the power converter itself but also the transfer function of the current control loop.

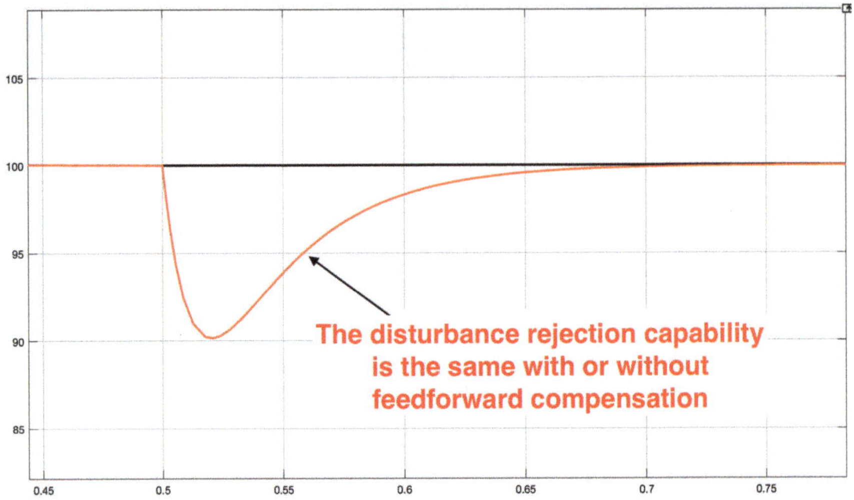

Fig. 2.62 Disturbance rejection response waveforms with and without feedforward control

In practical applications, $G_{PC}(s)$ can often be modeled using either a second-order or a first-order low-pass filter. Here, we assume that the power converter transfer function $G_{PC}(s)$ can be approximated as a first-order low-pass filter with a cutoff frequency of 1000 rad/s, expressed as:

$$G_{PC}(s) = \frac{1000}{s + 1000} \tag{2.4.6}$$

Since $G_{PC}(s) \neq 1$, we need to rewrite the transfer function between the output $Y(s)$ and the input $R(s)$ in the feedforward control structure shown in Fig. 2.57.

$$\frac{Y(s)}{R(s)} = \frac{[K_F G_F(s) + G_C(s)]G_{PC}(s)G_P(s)}{1 + G_C(s)G_P(s)G_{PC}(s)} \tag{2.4.7}$$

If the feedforward transfer function is set as $G_F(s) = G_P^{-1}(s)G_{PC}^{-1}(s)$ and the feedforward gain $K_F = 1$, then Eq. (2.4.7) becomes:

$$\begin{aligned}\frac{Y(s)}{R(s)} &= \frac{\left[G_P^{-1}(s)G_{PC}^{-1}(s) + G_C(s)\right]G_{PC}(s)G_P(s)}{1 + G_C(s)G_P(s)G_{PC}(s)} \\ &= \frac{1 + G_C(s)G_P(s)G_{PC}(s)}{1 + G_C(s)G_P(s)G_{PC}(s)} = 1\end{aligned} \tag{2.4.8}$$

Therefore, if the power converter transfer function $G_{PC}(s)$ is to be taken into account, the feedforward transfer function $G_F(s)$ must be designed as:

$$G_F(s) = G_P^{-1}(s)G_{PC}^{-1}(s) = Js \cdot \frac{s + 1000}{1000} \tag{2.4.9}$$

The numerator of Eq. (2.4.9) contains a second-order derivative, which may lead to significant high-frequency effects. Therefore, in practical applications, the feedforward transfer function $G_F(s)$ can be cascaded with a low-pass filter $G_{LPF}(s)$ to attenuate the high-frequency effects caused by the differentiation.

$$G_F(s) = G_P^{-1}(s) \cdot G_{PC}^{-1}(s) \cdot G_{LPF}(s) = Js \cdot \frac{s + 1000}{1000} \cdot \frac{\omega_c}{s + \omega_c} \tag{2.4.10}$$

Here, the low-pass filter $G_{LPF}(s)$ is chosen with a cutoff frequency $\omega_c = 2000$ (rad/s), so Eq. (2.4.10) can be expressed as:

$$G_F(s) = Js \cdot \frac{s + 1000}{1000} \cdot \frac{2000}{s + 2000} = Js \cdot \frac{2000s + 2e06}{1000s + 2e06} \tag{2.4.11}$$

Next, we can incorporate the feedforward path that takes the power converter transfer function into account into the SIMULINK simulation model. The completed SIMULINK model is shown in Fig. 2.63. After running the simulation, the results are shown in Fig. 2.64.

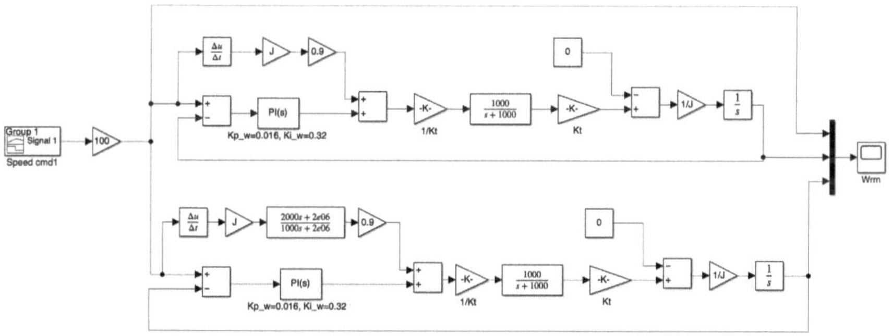

Fig. 2.63 SIMULINK simulation of feedforward compensation considering the power converter, Example model: mdl_speedLoop_ff2.slx

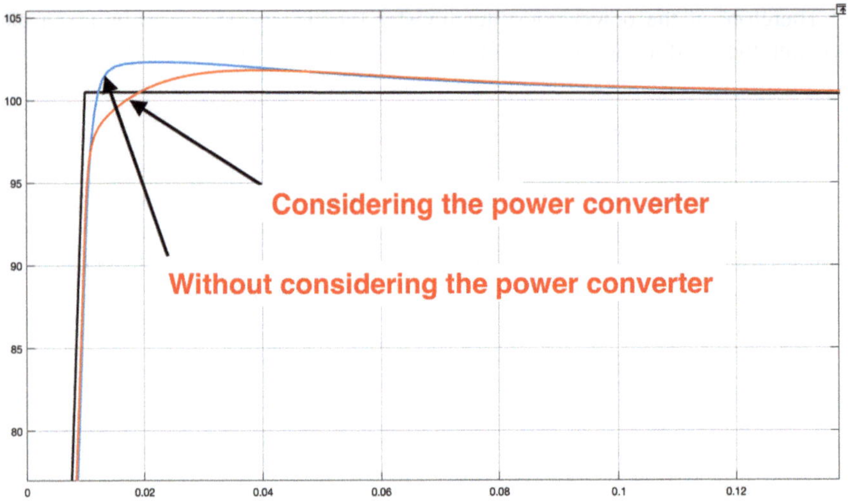

Fig. 2.64 Speed response waveforms with feedforward compensations with and without considering the power converter

In Fig. 2.63, the upper control loop represents the feedforward control system without considering the power converter, while the lower control loop represents the feedforward control system with the power converter taken into account. For easier waveform comparison, the feedforward gain K_F is set to 0.9.

From the simulation results shown in Fig. 2.64, it can be observed that the feedforward control system considering the power converter provides better command tracking performance and a smaller overshoot compared to the system that does not take the power converter into account.

MATLAB Example Script: m2_4_2.m

```
wsc=100; J=0.00016;
s = tf('s');
Kp_w=J*wsc; Ki_w=Kp_w*wsc/5;
tf_pi = tf([Kp_w Ki_w],[1 0]);
tf_pc = tf([1000], [1 1000]);
tf_plant = tf([1], [J 0]);
Gw_open = tf_pi*tf_pc*tf_plant;
Gw_close = Gw_open/(1+Gw_open);
Gf1 = s*J;
Gf2 = (s*J)*(2000*s+2e06)/(1000*s+2e06);
kf = 0.9;
Gw_ff1 = (kf*Gf1+tf_pi)*tf_pc*tf_plant/(1+tf_pi*tf_pc*tf_plant);
Gw_ff2 = (kf*Gf2+tf_pi)*tf_pc*tf_plant/(1+tf_pi*tf_pc*tf_plant);
h=bodeoptions; h.PhaseMatching='on';
h.Title.FontSize = 14;
h.XLabel.FontSize = 14;
h.YLabel.FontSize = 14;
h.TickLabel.FontSize = 14;
bodeplot(Gw_ff1,'-b',Gw_ff2,'-.b',{1,10000},h);
legend('Feed-Forward-No-Gpc', 'with Feed-Forward-With-Gpc');
h = findobj(gcf,'type','line');
set(h,'linewidth',2);
grid on;
```

You can use the sample program m2_4_2 to plot the Bode diagram of the feed-forward control system that considers the power converter, and compare it with the

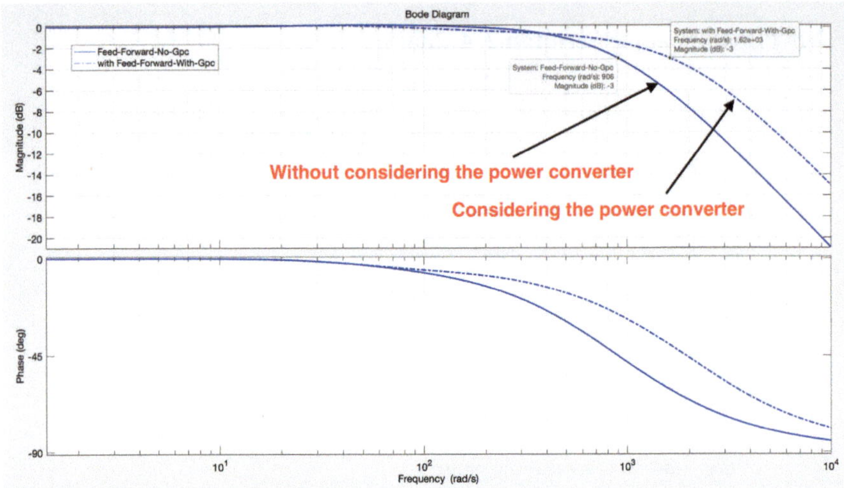

Fig. 2.65 Closed-loop Bode plots of the speed loop with and without considering the power converter

Bode diagram of the feedforward control system that does not consider the power converter.

By running the sample program m2_4_2, you can obtain the Bode plot shown in Fig. 2.65. From Fig. 2.65, it is clearly observed that:

- The bandwidth of the feedforward control system without considering the power converter is 906 (rad/s),
- While the bandwidth of the system with the power converter taken into account is 1620 (rad/s).

Therefore, if the influence of the power converter is incorporated into the feedforward path, the system can achieve a larger bandwidth, resulting in faster command response.

- **Feedforward Compensation Considering Feedback Delay**

The control loop shown in Fig. 2.57 does not take feedback delay into account. In practice, however, feedback delay is present and primarily arises from sampling delay and filtering delay. If feedback delay is considered, then Fig. 2.57 can be modified into Fig. 2.66.

Assuming the sampling period of the speed control loop is $T_s = 1$ (ms), the feedback delay effect due to sampling can be equivalently modeled as $T_s/2$. Additionally, considering the possible filtering delay, we assume the total delay time is equal to one sampling period T_s. Therefore, the feedback delay effect $G_{Delay}(s)$ can be expressed as:

$$G_{Delay}(s) = e^{-sT_s} \tag{2.4.12}$$

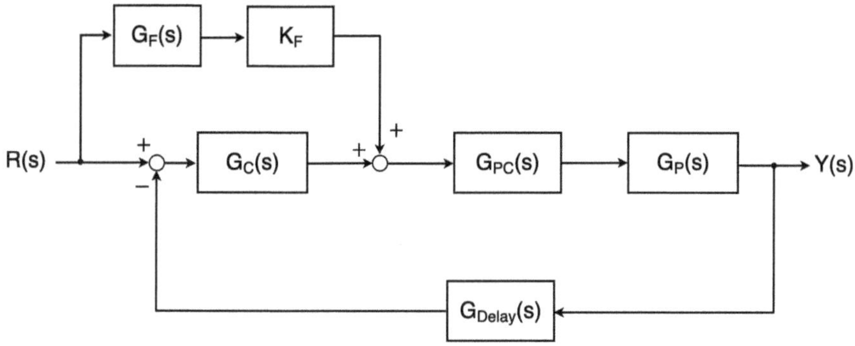

Fig. 2.66 Closed-loop control block diagram with feedforward compensation considering feedback delay

We can first use SIMULINK to simulate the effect of feedback delay on system response and compare it with the system response without feedback delay. The completed SIMULINK block diagram is shown in Fig. 2.67. In this diagram, the lower control loop includes the feedback delay, while the upper control loop does not. After running the simulation, the speed response waveform shown in Fig. 2.68 can be obtained.

From the speed response waveform in Fig. 2.68, it can be observed that after introducing feedback delay, the transient peak overshoot of the speed increases significantly. To address this issue, a command delay element $G_{R\_Delay}(s)$ can be added in Fig. 2.67, as shown in Fig. 2.69.

We can re-derive the transfer function between the output $Y(s)$ and the input $R(s)$.

$$\frac{Y(s)}{R(s)} = \frac{\left[K_F G_F(s) + G_{R\_Delay}(s) G_C(s)\right] G_{PC}(s) G_P(s)}{1 + G_C(s) G_P(s) G_{PC}(s) G_{Delay}(s)} \tag{2.4.13}$$

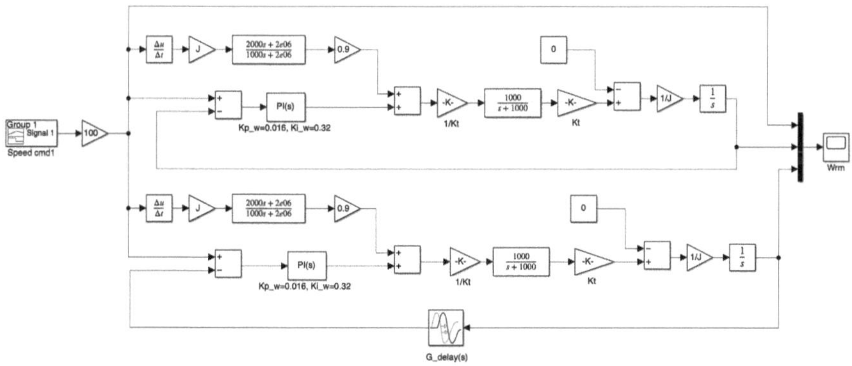

Fig. 2.67 SIMULINK simulation of speed closed-loop control with feedforward compensations with and without considering feedback delay, Example model: mdl_speedLoop_ff3.slx

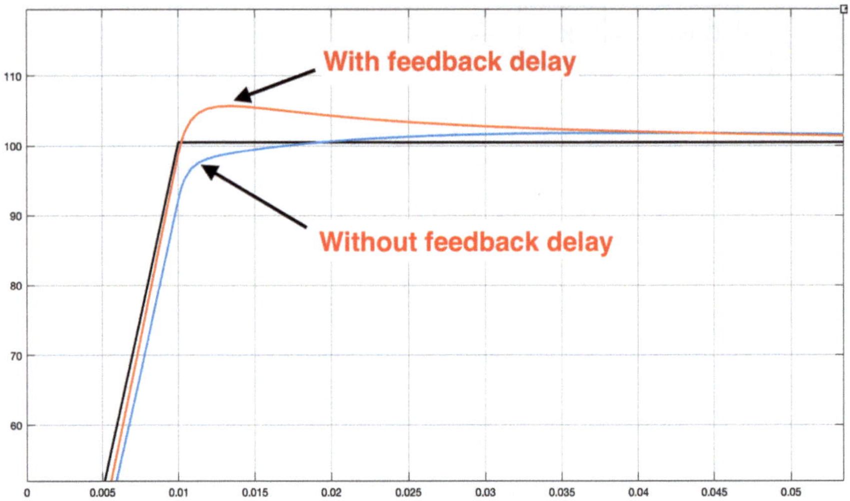

Fig. 2.68 Speed response waveforms with feedforward compensations with and without considering feedback delay

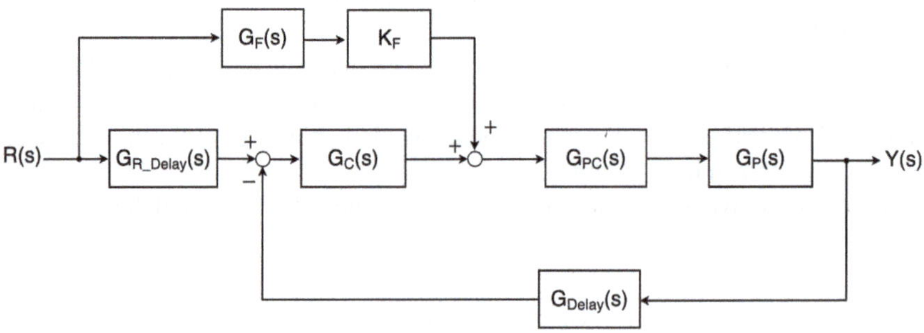

Fig. 2.69 Closed-loop control block diagram with feedforward compensation considering command delay

When the feedforward gain $K_F = 1$, and the feedforward transfer function is designed as $G_F(s) = G_P^{-1}(s)G_{PC}^{-1}(s)$, and the command delay element is set equal to the feedback delay element, i.e., $G_{R\_Delay}(s) = G_{Delay}(s)$, then Eq. (2.4.13) becomes:

$$\begin{aligned} \frac{Y(s)}{R(s)} &= \frac{\left[G_P^{-1}(s)G_{PC}^{-1}(s) + G_{Delay}(s)G_C(s)\right]G_{PC}(s)G_P(s)}{1 + G_C(s)G_P(s)G_{PC}(s)G_{Delay}(s)} \\ &= \frac{1 + G_C(s)G_P(s)G_{PC}(s)G_{Delay}(s)}{1 + G_C(s)G_P(s)G_{PC}(s)G_{Delay}(s)} = 1 \end{aligned} \tag{2.4.14}$$

Therefore, when the command delay element is designed to match the feedback delay element, the feedback delay effect can be canceled, making Eq. (2.4.14) valid.

You can run the simulation using the SIMULINK model mdl_speedLoop_ff4.slx. Figure 2.70 shows the completed SIMULINK system, where the upper control loop represents the feedforward system **without** the command delay element, and the lower control loop represents the feedforward system **with** the command delay element. Running the SIMULINK simulation yields the speed response waveform shown in Fig. 2.71.

From the response waveform in Fig. 2.71, it can be observed that the speed response with the command delay element is better than that without it. The reason

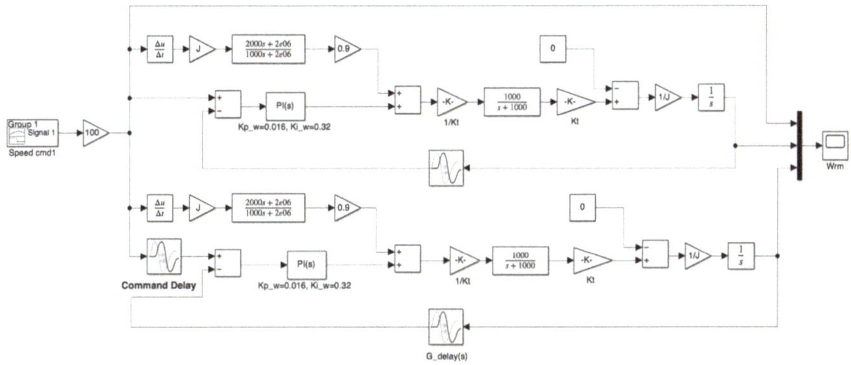

Fig. 2.70 SIMULINK simulation of speed closed-loop control with feedforward compensation with and without considering command delay, Example model: mdl_speedLoop_ff4.slx

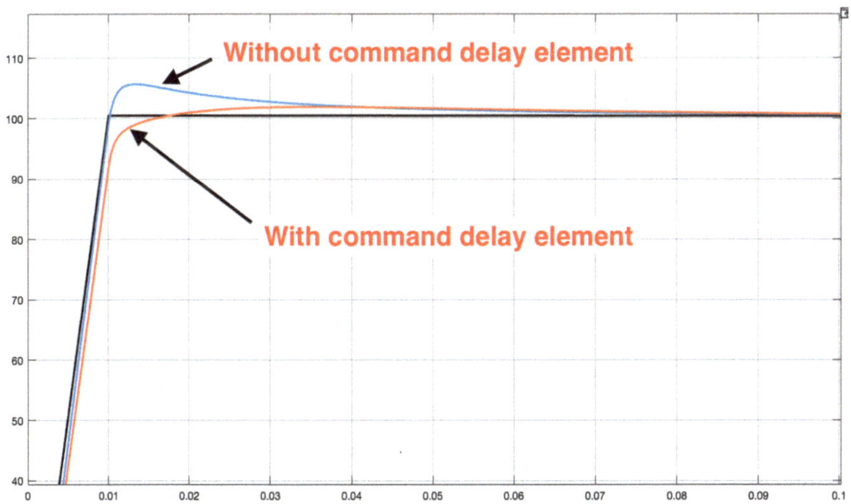

Fig. 2.71 Speed response waveforms with feedforward compensations with and without considering feedback delay

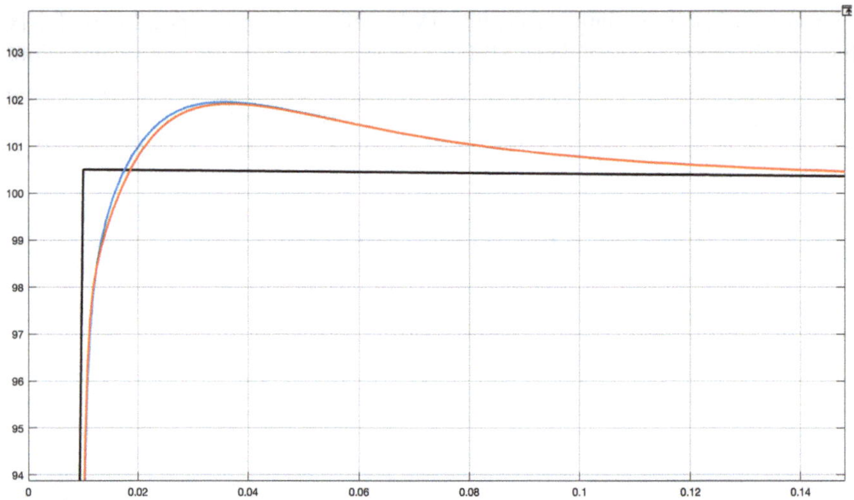

Fig. 2.72 Effect of approximating command delay using a first-order low-pass filter, Example model: mdl_speedLoop_ff5.slx

is that the command delay element can cancel out the delay caused by feedback. However, this cancellation is only effective if both delay times are identical.

In Fig. 2.70, the command delay element we used is e^{-sT_s}. In practice, this can be approximated using a first-order low-pass filter.

$$e^{-sT_s} \cong \frac{1}{T_s s + 1} \tag{2.4.15}$$

We use Eq. (2.4.15) to replace the command delay element in Fig. 2.71. Figure 2.72 shows the simulation result using the command delay element defined by Eq. (2.4.15), along with a comparison to the ideal delay element e^{-sT_s}. As shown in the figure, although there is a slight discrepancy between the two, approximating the ideal command delay element using the first-order low-pass filter of Eq. (2.4.15) is sufficiently accurate to meet practical industrial requirements.

2.5 Position Loop Controller Design

Next, we will introduce the design of the position control loop [4]. Generally, the position loop serves as the outer loop of the speed loop. In most cases, a proportional controller is used for the position loop. However, to avoid oscillations caused by the zero in the numerator, this section adopts the IP controller introduced in Sect. 2.3.3 as the controller for the speed loop. The completed position control loop for the permanent magnet synchronous motor is shown in Fig. 2.73.

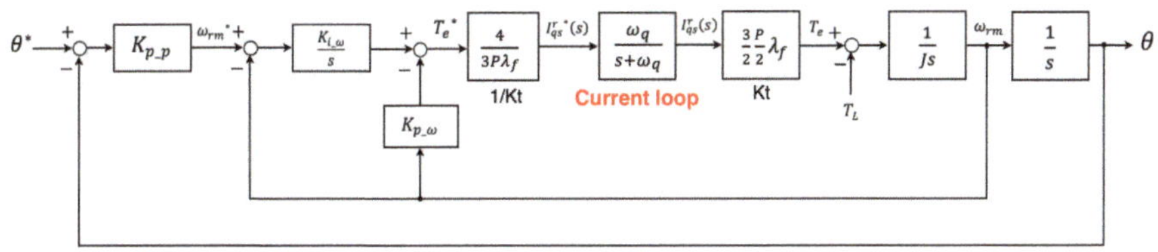

Fig. 2.73 Block diagram of PMSM position control system

Here, the current loop can be approximated as a unity gain, i.e.,

$$T_{curr\_loop}(s) = \frac{\omega_q}{s + \omega_q} \cong 1 \tag{2.5.1}$$

The closed-loop transfer function of the position control loop can be derived.

$$\frac{\theta}{\theta *} = \frac{K_{p\_p} K_{i\_\omega}}{Js^3 + K_{p\_\omega} s^2 + K_{i\_\omega} s + K_{p\_p} K_{i\_\omega}} \tag{2.5.2}$$

Next, we will use SIMULINK to perform a computer simulation. Since a trapezoidal position command is used with a slope as high as 10,000 (rad/s), we need to simultaneously increase the bandwidths of both the current loop and the speed loop in order to track such a high-speed position command. In the following simulation, the current loop bandwidth will be increased to 5000 (rad/s), that is:

$$T_{curr\_loop}(s) = \frac{5000}{s + 5000} \tag{2.5.3}$$

Next, the speed loop bandwidth of the IP controller will be designed to be 628 (rad/s). You can use the example program **m2_3_6** to find the IP controller parameters that meet this bandwidth requirement. The designed IP controller parameters are as follows:

$$K_{p\_\omega} = 0.3905, \ K_{i\_\omega} = 190.63$$

Enter the above parameters into SIMULINK and construct a position control loop using proportional gain, as shown in Fig. 2.74.

The method for tuning the position proportional controller gain K_{p_p} is as follows:

After the current and speed loops have been properly designed, gradually increase the position proportional gain K_{p_p} until it reaches a value just below the point where overshoot begins to occur.

In this example, the position proportional controller gain K_{p_p} is set to **150**.

Run the simulation to obtain:

- Figure 2.75: the position response waveform
- Figure 2.76: the position error waveform.

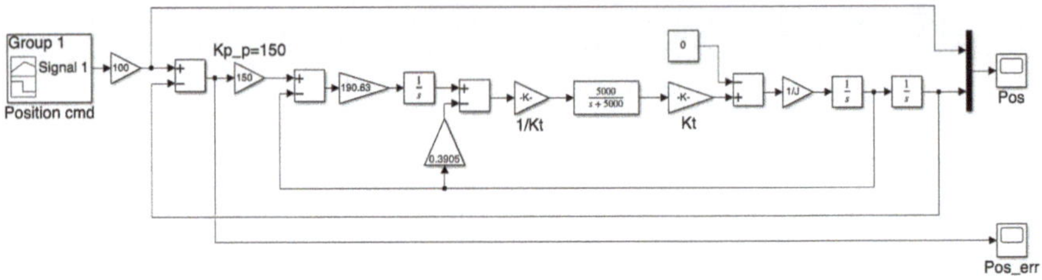

Fig. 2.74 SIMULINK simulation of PMSM position control system, Example model: mdl_positionLoop_1.slx

From the position response waveform in Fig. 2.75, we can observe that the tuned position proportional controller gain K_{p_p} in this example does not cause any overshoot. If the criterion is to avoid overshoot in the position response, then the current value of K_{p_p} appears to be close to the maximum allowable for achieving the fastest position response without overshoot.

However, from the position error waveform in Fig. 2.76, we can see that there is still a significant transient position error, with a maximum value of approximately 60 rad.

To further analyze this, we can use example program m2_5_1 to plot the Bode diagram of the position closed-loop system, and compare it with the Bode diagram of the speed closed-loop system, as shown in Fig. 2.77.

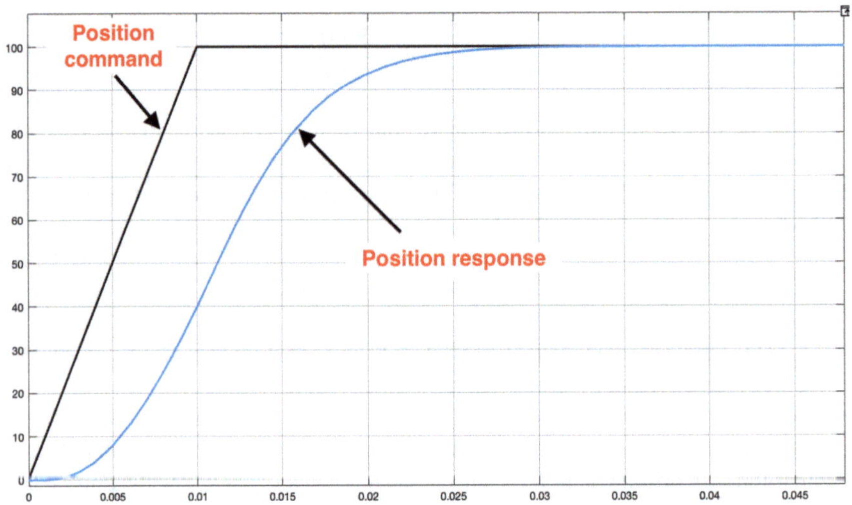

Fig. 2.75 Position response waveform

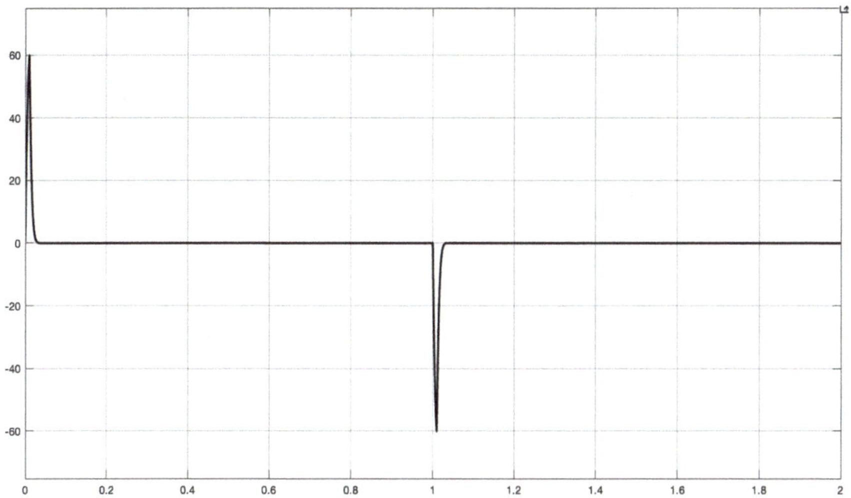

Fig. 2.76 Position error waveform

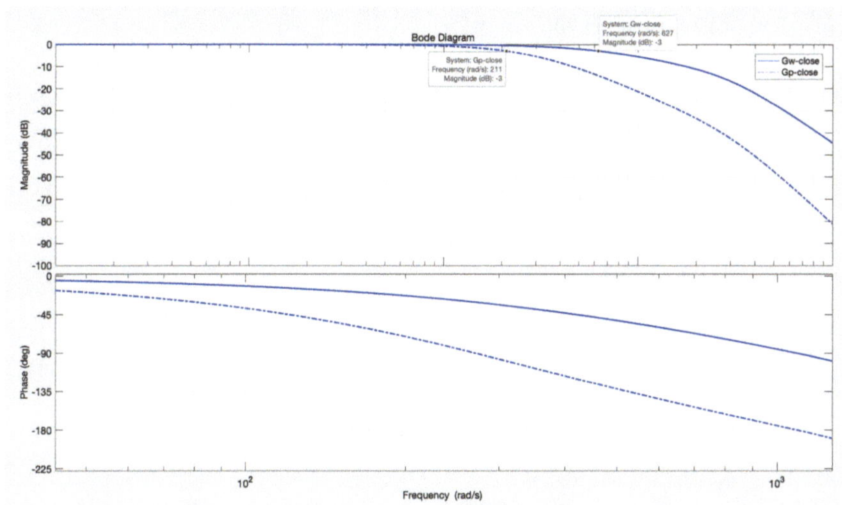

Fig. 2.77 Closed-loop Bode plots of the speed and position loops

In this case:

- The speed loop bandwidth, using the IP controller, is 627 rad/s.
- When the position proportional controller gain $K_{p\_p} = 150$, the position loop bandwidth is 211 rad/s.

MATLAB Example Script: m2_5_1.m

```
J=0.00016;
s = tf('s');
Kp_w=0.3905; Ki_w=190.6307;
tf_cmdFilter = tf(Ki_w/Kp_w, [1 Ki_w/Kp_w]);
tf_pi = tf([Kp_w Ki_w],[1 0]);
tf_pc = tf([5000], [1 5000]);
tf_plant = tf([1], [J 0]);
Gw_open = tf_pi*tf_pc*tf_plant;
Gw_close = tf_cmdFilter*Gw_open/(1+Gw_open);
Kp_p = 150;
tf_plant_p = tf(1,[1 0]);
Gp_open = Kp_p*Gw_close*tf_plant_p;
Gp_close = Gp_open/(1+Gp_open);
h=bodeoptions; h.PhaseMatching='on';
h.Title.FontSize = 14;
h.XLabel.FontSize = 14;
h.YLabel.FontSize = 14;
h.TickLabel.FontSize = 14;
bodeplot(Gw_close,'-b',Gp_close,'-.b',{1,10000},h);
legend('Gw-close', 'Gp-close');
h = findobj(gcf,'type','line');
set(h,'linewidth',2);
grid on;
```

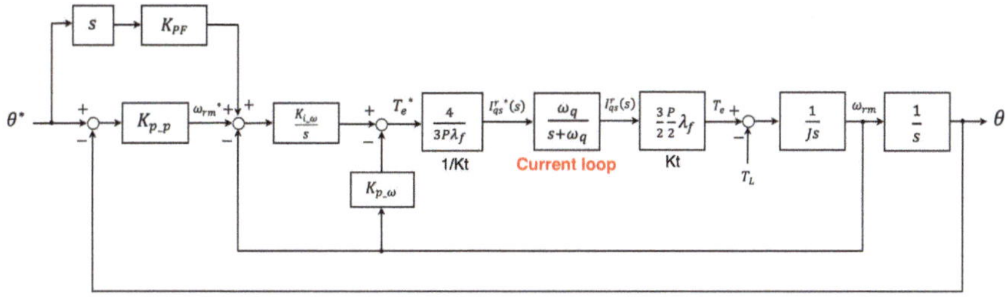

Fig. 2.78 Block diagram of the position control loop with feedforward compensation

- **Feedforward Compensation in PMSM Position Control Loop**

Next, we will add a **feedforward path** to the position control loop to evaluate whether it can effectively improve the position response performance.

Figure 2.78 shows the position control loop with the feedforward path included.

In Fig. 2.78, the position feedforward path consists of a differentiator s and a feedforward gain K_{PF}. Due to the inclusion of this feedforward path, it is necessary to re-derive the closed-loop transfer function of the position control loop. Using Mason's Gain Formula, the closed-loop transfer function of the position loop can be derived as follows:

$$\frac{\theta}{\theta *} = \frac{K_{i\_\omega} K_{PF} s + K_{p\_p} K_{i\_\omega}}{J s^3 + K_{p\_\omega} s^2 + K_{i\_\omega} s + K_{p\_p} K_{i\_\omega}} \tag{2.5.4}$$

Comparing Eq. (2.5.4) with Eq. (2.5.2), we observe that after adding the position feedforward path, a differentiation term $K_{i\_\omega} K_{PF} s$ is introduced in the numerator of the closed-loop transfer function of the position loop, while the denominator remains unchanged. This additional differentiation term $K_{i\_\omega} K_{PF} s$ helps to improve the response speed to the reference command.

Next, we use SIMULINK to perform a computer simulation. Figure 2.79 shows the constructed SIMULINK block diagram. In this setup, we set the feedforward gain K_{PF} to 0.5 to avoid overshoot in the position response. After running the SIMULINK simulation, we obtain:

- Figure 2.80: the position response waveform
- Figure 2.81: the position error waveform.

As clearly shown in Fig. 2.80, adding the position feedforward path significantly enhances the speed of the position response and greatly reduces the transient position error.

In this example, setting the position feedforward gain $K_{PF} = 0.5$ is a conservative approach designed to prevent overshoot. You are encouraged to experiment with different settings—note that the maximum value of K_{PF} is 1. A larger K_{PF} results in faster transient response, but it may also introduce overshoot.

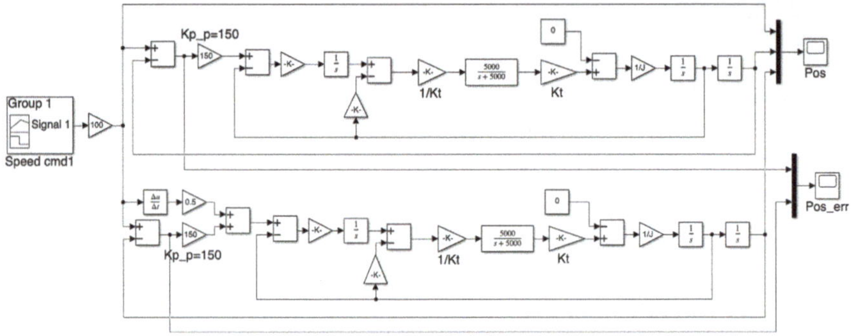

Fig. 2.79 SIMULINK simulation of position control with and without feedforward compensation, Example model: mdl_positionLoop_2.slx

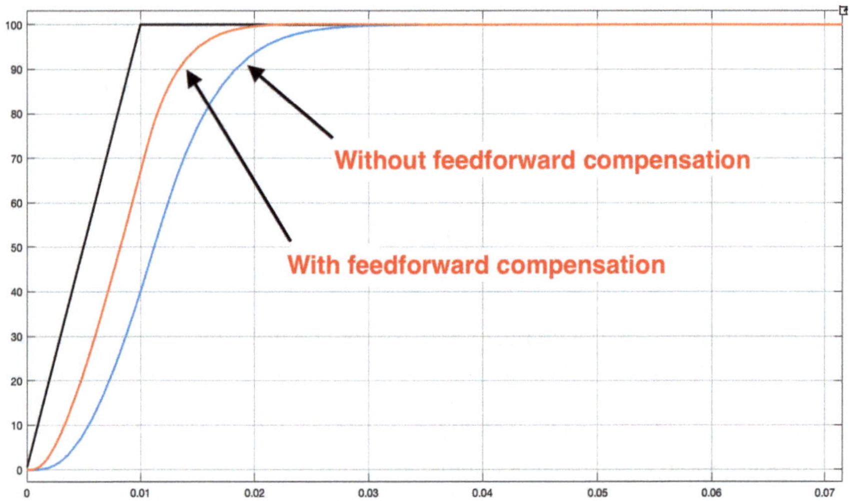

Fig. 2.80 Position response waveforms with and without feedforward compensation

You can use example program m2_5_2 to plot the Bode diagram of the position closed-loop system with the feedforward path, and compare it with that of the system without the feedforward path, as shown in Fig. 2.82.

From the Bode plot in Fig. 2.82, it can be seen that the position loop bandwidth without feedforward is 211 (rad/s), whereas with feedforward, the bandwidth increases to 341 (rad/s), which is 1.6 times greater than before. If the feedforward gain K_{PF} continues to increase, for example in this case, when $K_{PF} = 0.8$, the position loop bandwidth will approach the speed loop bandwidth of 627 rad/s. A similar effect occurs in the speed loop: when a feedforward path is included, the speed loop bandwidth can approach that of the current loop.

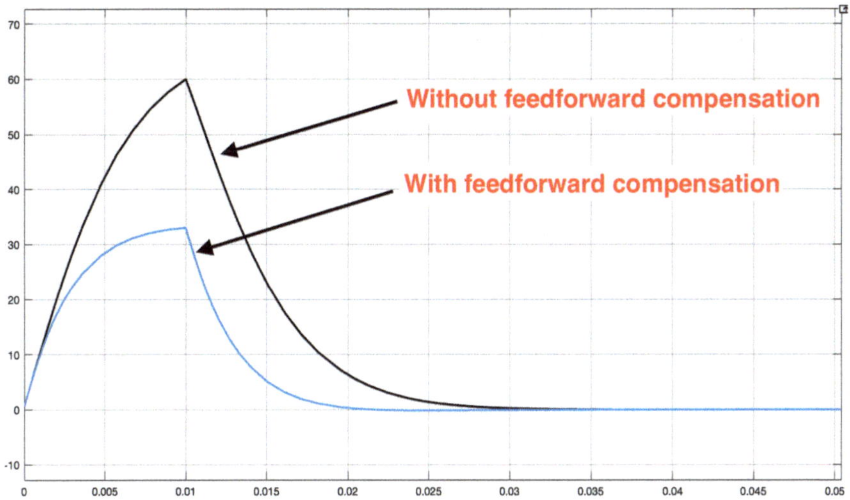

Fig. 2.81 Position error waveforms with and without feedforward compensation

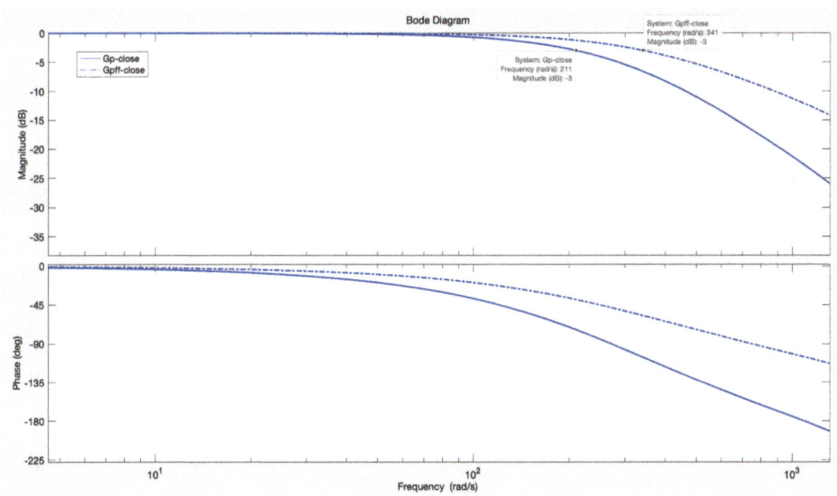

Fig. 2.82 Closed-loop Bode plots of position control with and without feedforward compensation

Although feedforward can greatly improve command response speed, it does not form a feedback loop and thus:

- Does not affect loop stability, and
- Does not respond to disturbances.

Therefore, adding a feedforward path does not improve the disturbance rejection capability of the system.

MATLAB Example Script: m2_5_2.m

```
J=0.00016;
s = tf('s');
Kp_w=0.3905; Ki_w=190.6307;
tf_cmdFilter = tf(Ki_w/Kp_w, [1 Ki_w/Kp_w]);
tf_pi = tf([Kp_w Ki_w],[1 0]);
tf_pc = tf([5000], [1 5000]);
tf_plant = tf([1], [J 0]);
Gw_open = tf_pi*tf_pc*tf_plant;
Gw_close = tf_cmdFilter*Gw_open/(1+Gw_open);
Kp_p = 150;
tf_plant_p = tf(1,[1 0]);
Kpf = 0.5;
Gp_open = Kp_p*Gw_close*tf_plant_p;
Gp_close = Gp_open/(1+Gp_open);
Gp_close_ff = tf([Ki_w*Kpf Kp_p*Ki_w], [J Kp_w Ki_w Kp_p*Ki_w]);
h=bodeoptions; h.PhaseMatching='on';
h.Title.FontSize = 14;
h.XLabel.FontSize = 14;
h.YLabel.FontSize = 14;
h.TickLabel.FontSize = 14;
bodeplot(Gp_close,'-b',Gp_close_ff,'-.b',{1,10000},h);
legend('Gp-close', 'Gpff-close');
h = findobj(gcf,'type','line');
set(h,'linewidth',2);
grid on;
```

2.6 Summary

The key points of this chapter are summarized as follows:

- The larger the delay in a control system, the smaller the allowable loop gain. This is because delay effects erode the system's original phase margin, and increasing loop gain further reduces phase. Therefore, in high-performance applications, it is essential to increase the speed of the control loop and use sensors with minimal phase delay to raise the maximum allowable gain and achieve high-bandwidth, high-performance operation.
- Although feedforward compensation greatly improves command response speed, it does not form a closed loop, and therefore does not affect loop stability nor respond to disturbances. Consequently, adding a feedforward path cannot improve the system's disturbance rejection capability.
- For a controller, plant parameters are considered unknown, while feedforward compensation usually requires reasonably accurate plant parameters. In general, when parameter variation is within $\pm$ 20%, feedforward compensation can still perform well. For applications where certain plant parameters (such as mechanical inertia) vary significantly, it is recommended to use estimators for real-time parameter estimation or predefine the maximum expected variation to design appropriate feedforward gains and prevent overshoot.
- In practice, the velocity signal required for position feedforward and the acceleration signal required for velocity feedforward are both provided by the motion controller. Therefore, these two feedforward paths can avoid direct differentiation. Internally, the motion controller first generates an acceleration command, then obtains the velocity command via integration, and the position command via double integration. This method provides nearly noise-free command signals for use in feedforward paths. When using the velocity command from the motion controller in the position loop's feedforward path, it can effectively eliminate position steady-state error.
- Since the current loop transfer function is not a perfect unity gain, setting both feedforward gains to 1 may cause overshoot. Therefore, in practice, feedforward gains are typically adjusted within the range of 0.4–0.8. Higher values result in faster command response speeds, but also carry a higher risk of overshoot.

References

1. Sul, Seung-Ki, Control of Electric Machine Drive Systems, Wiley-IEEE Press, 2011.
2. F. Blaschke, "The principle of field orientation as applied to the NEW Transvector closed-loop control system for rotating-field machines," Siemens Rev., vol. 34, pp. 217–220, 1972.
3. Chang-Hwan Liu, AC Motor Control: Principles of Vector Control and Direct Torque Control, Taipei: Dong Hua Book Co., 2001.
4. George Ellis, Control System Design Guide: Using Your Computer to Understand and Diagnose Feedback Controllers, Butterworth-Heinemann, 2016.

5. Chih-Chun Yeh, AC Motor Control and Simulation Techniques: Mastering Core Algorithms of EV and Inverter Technology, Taipei: Wu-Nan Book Inc., 2023.
6. H. Tajima and Y. Hori, “Speed sensorless field-orientation control of the induction machines,” IEEE Trans. Ind. Appl., vol. 29, no. 1, pp. 175–180, Jan./Feb. 1993.
7. R. Krishnan, Permanent Magnet Synchronous and Brushless DC Motor Drives, CRC Press, Boca Raton, Florida, 2010.
8. P. L. Jansen, R. D. Lorenz and D. W. Novotny, “Observer-based direct field orientation: analysis and comparison of alternative methods,” IEEE Trans. Ind. Appl., vol. 30, no. 4, pp.945–953, July/Aug. 1994.
9. P. L. Jansen and R. D. Lorenz, “A physically insightful approach to the design and accuracy assessment of flux observers for field oriented induction machine drives,” IEEE Trans. Ind. Appl., vol. 30, no. 1, pp. 101–110, Jan/Feb. 1994.
10. J. Bocker, S. Beineke, and A. Bahr, “On the control bandwidth of servo drives,” in Proc. Eur. Conf. Power Electron. Appl., pp. 1–10, 2009.
11. Kessler, C. (1958). Das symmetrische Optimum. Regelungstechnik, 6, pp. 395–400 and 432–436, 1958.

Chapter 3
Modulation Strategies for Three-Phase Inverters

In Chap. 1, we derived the mathematical models for three-phase induction motors and three-phase permanent magnet synchronous motors. In Chap. 2, we designed control loops based on these AC motor models. In this chapter, we will introduce the **power converter used in AC motor control**: the **three-phase inverter**. Without the three-phase inverter, the controllers designed in Chap. 2 cannot be practically implemented. This is because motor control algorithms operate on a microcontroller, and in order for the controller's output signal to drive the motor, the signal must be **amplified** using a three-phase inverter. Only then can the motor be successfully driven.

The simulation structure shown in Fig. 3.1 is identical to that of Fig. 2.11. When performing numerical simulations of control systems, we can directly use the control loop's output voltage as the input to the motor model. However, in real-world applications, the control loop is executed by a microcontroller, whose output is typically a **low-power signal** (usually in the range of **0–5 V** or **0–3.3 V**). This type of low-power signal **cannot drive an AC motor**. Therefore, in practice, we must use a **three-phase inverter** to amplify the microcontroller's output, converting it into a **three-phase AC voltage signal** capable of driving the motor [1–3].

Strictly speaking, a PWM inverter [4] is a DC-AC converter, but it typically includes a front-end AC-DC converter (rectifier module) that converts three-phase AC power into DC. Then, the back-end (DC-AC converter) transforms the DC voltage into a three-phase voltage with variable amplitude and frequency to drive the motor. Therefore, a typical "inverter" has a two-stage structure (AC-DC converter + DC-AC converter).

Figure 3.2 shows the structure of a typical three-phase Voltage Source Inverter (VSI). Its front-end is a three-phase AC-DC converter (also known as the rectifier module), which converts three-phase AC voltage into DC voltage and temporarily

---

Supplementary Information The online version contains supplementary material available at https://doi.org/10.1007/978-981-95-3261-2_3.

C-C. Yeh, *AC Motor Control Loop Design*, Power Systems,
https://doi.org/10.1007/978-981-95-3261-2_3

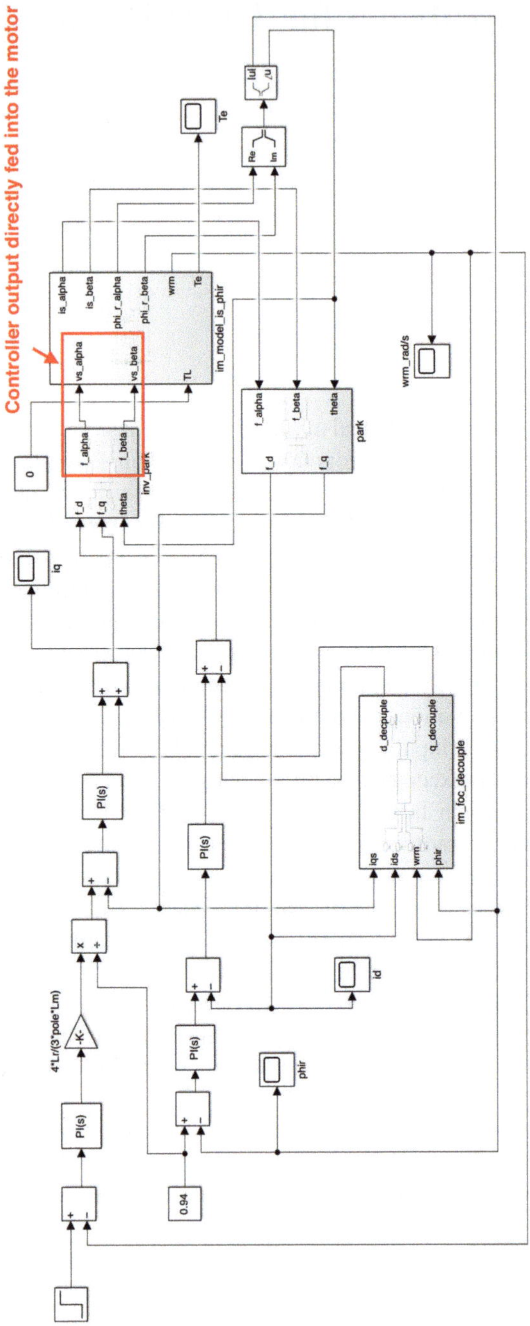

Fig. 3.1 Field-oriented control (FOC) block diagram of the induction motor

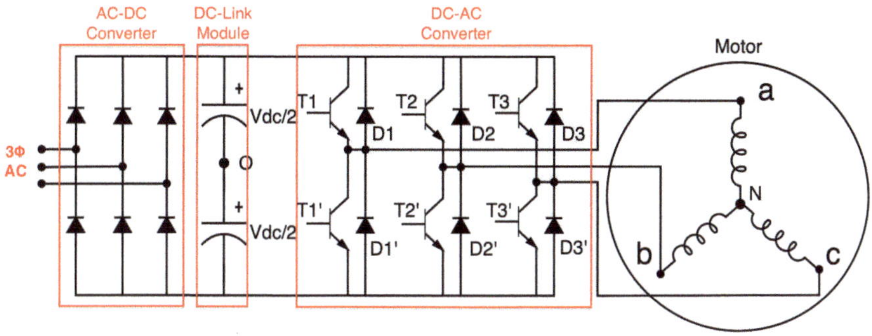

Fig. 3.2 Block diagram of a typical three-phase voltage source inverter (VSI) motor drive architecture

stores the energy in the DC link (consisting of large capacitors). The back-end DC-AC converter uses PWM techniques to switch the power transistors (T1, T1′, T2, T2′, T3, T3′) and convert the stored DC voltage into a three-phase AC voltage to supply the AC motor.

The diodes D1, D1′, D2, D2′, D3, and D3′ serve as freewheeling diodes for the inductive load current. Without them, the discontinuous inductor current could generate high voltage spikes that may damage the inverter.

Sinusoidal Pulse Width Modulation (SPWM) is one of the most classic PWM modulation techniques. It uses a high-frequency carrier and compares it with the three-phase voltage commands generated by the control loop to switch the transistors in the inverter.

Taking phase **a** as an example, as shown in Fig. 3.3: when the **a-phase voltage command** is higher than the carrier voltage, the output is set to HIGH, turning **T1 ON**, while **T1′** must be OFF simultaneously to prevent a short circuit.

During three-phase SPWM modulation, the three-phase voltage commands are compared with the high-frequency carrier **simultaneously**, as in the comparison operation shown in Fig. 3.3. In practice, SPWM modulation is typically implemented using a **microcontroller (MCU)** or an **FPGA (Field Programmable Gate Array)**.

Explanation

In practice, although the voltage command output from the vector control loop is a sinusoidal waveform, the PWM period is much shorter than the sinusoidal period. Therefore, within each PWM cycle, the voltage command can be approximated as a DC value [4].

This chapter will introduce several commonly used three-phase inverter modulation strategies and use SIMULINK for system simulation:

- Sinusoidal PWM (SPWM) modulation strategy

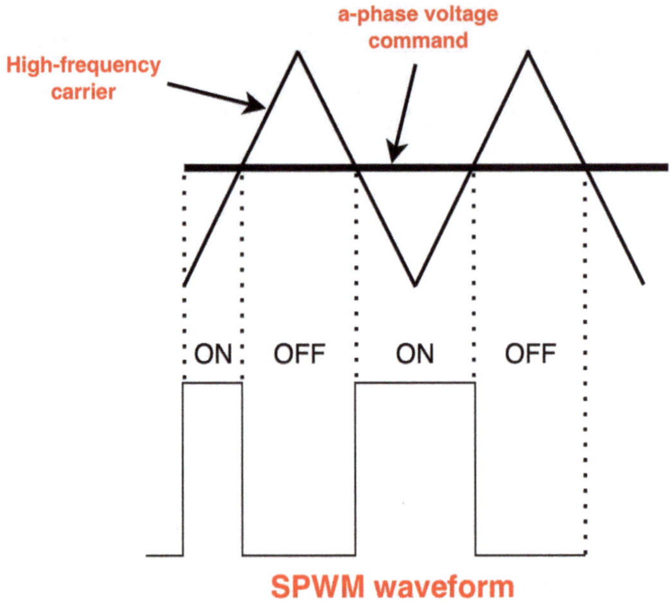

Fig. 3.3 Schematic diagram of SPWM operating principle

- Third-harmonic injection modulation strategy [2, 3]
- Modulation strategy with added offset [3]
- Space Vector PWM (SVPWM) modulation strategy [2, 3].

Finally, in Sect. 3.5, you will learn how to use the inverter and motor modules from SIMULINK's built-in Simscape library and integrate the control loops designed in Chap. 2 to build a vector control system for permanent magnet synchronous motors with more realistic physical characteristics.

Before introducing modulation techniques, it is necessary to define two indicators that are closely related to modulation. The first is called the **amplitude modulation index**, denoted as m_a, and it is defined as follows [4]:

$$m_a = \frac{|v_m|}{|v_c|} \tag{3.0.1}$$

where $|v_m|$ is the peak absolute value of the modulation wave, and $|v_c|$ is the peak absolute value of the carrier wave (Note: the phase-a voltage command in Fig. 3.3 is the modulation wave).

When the peak of the modulation wave is less than or equal to the peak of the carrier wave, i.e., $m_a \leq 1$, the system is said to be in the **linear modulation region**. If the modulation wave exceeds the peak of the carrier wave, i.e., $m_a > 1$, the system is operating in the **nonlinear modulation region**, also known as the **overmodulation region**.

This book primarily focuses on the linear modulation region. Readers interested in nonlinear or overmodulation regions may refer to relevant resources [3, 4].

Next, we define the **second indicator**, known as the **frequency modulation index**, denoted as m_f [4], also referred to as the **switching ratio**. It is defined as:

$$m_f = \frac{f_c}{f_m} \tag{3.0.2}$$

where f_c is the frequency of the **carrier wave**, and f_m is the frequency of the **modulation wave**.

When $m_f \leq 21$, the **carrier** and **modulation wave** must be synchronized—meaning m_f must be an integer—to avoid significant **subharmonic** effects. In this case, the carrier (typically a triangular wave) must adjust its frequency in real time to match changes in the modulation frequency [4].

When $m_f > 21$, the subharmonic effects caused by **asynchronous PWM** are less severe. Therefore, the carrier frequency can be set as a constant. However, for applications sensitive to subharmonics, **synchronous PWM** (i.e., synchronized carrier and modulation waves) may still be used to reduce subharmonic distortion [4].

The carrier frequency f_c determines the switching frequency of the inverter's power transistors. A **higher** f_c increases switching frequency, leading to greater switching losses. On the other hand, a **lower** f_c reduces switching losses but degrades voltage waveform resolution, thereby impairing drive performance. In motor drive applications, f_c is typically set between **4 and 10 kHz**.

3.1 Sinusoidal PWM (SPWM)

Figure 3.4 shows a typical **three-phase voltage source inverter (VSI)** structure. Points **a**, **b**, and **c** are connected to the stator windings of the motor. We need to use circuit theory concepts to derive the relationship between the inverter leg voltages v_{ao}, v_{bo}, and v_{co}, and the motor phase voltages v_{aN}, v_{bN}, and v_{cN}.

First, we know

$$v_{ao} = v_{aN} + v_{No} \tag{3.1.1}$$

$$v_{bo} = v_{bN} + v_{No} \tag{3.1.2}$$

$$v_{co} = v_{cN} + v_{No} \tag{3.1.3}$$

Assuming the motor has a **three-phase balanced winding**, and the input voltage to the motor is **three-phase balanced**, i.e.,

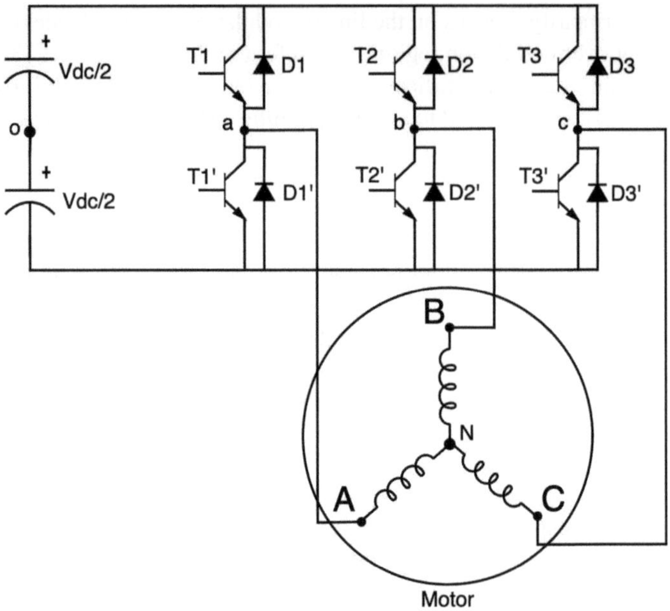

Fig. 3.4 Three-phase VSI motor drive system architecture

$$v_{aN} + v_{bN} + v_{cN} = 0 \tag{3.1.4}$$

By adding (3.1.1), (3.1.2), and (3.1.3), and applying the condition given in Eq. (3.1.4), we obtain:

$$v_{No} = \frac{1}{3}(v_{ao} + v_{bo} + v_{co}) \tag{3.1.5}$$

By substituting Eq. (3.1.5) back into Eqs. (3.1.1), (3.1.2), and (3.1.3), we can obtain:

$$v_{aN} = \frac{1}{3}(2v_{ao} - v_{bo} - v_{co}) \tag{3.1.6}$$

$$v_{bN} = \frac{1}{3}(2v_{bo} - v_{co} - v_{ao}) \tag{3.1.7}$$

$$v_{cN} = \frac{1}{3}(2v_{co} - v_{ao} - v_{bo}) \tag{3.1.8}$$

Next, we use S_a, S_b, and S_c to represent the switching states of the transistors in the a, b, and c legs, respectively. Taking phase a as an example: if the upper transistor in phase a is ON, then $S_a = 1$; if the lower transistor is ON, then $S_a = 0$. Similarly, S_b and S_c represent the switching states of the transistors in phases b and c, respectively.

Therefore, we can express v_{aN}, v_{bN}, and v_{cN} as functions of S_a, S_b, S_c, and V_{dc}:

$$v_{aN} = \frac{V_{dc}}{3}(2S_a - S_b - S_c) \quad (3.1.9)$$

$$v_{bN} = \frac{V_{dc}}{3}(2S_b - S_c - S_a) \quad (3.1.10)$$

$$v_{cN} = \frac{V_{dc}}{3}(2S_c - S_a - S_b) \quad (3.1.11)$$

When the switching sequence in Table 3.1 is applied to the three-phase VSI switches, the output voltage waveforms v_{ao}, v_{bo}, and v_{co} will be as shown in Fig. 3.5. If the connected motor is an induction motor, it will rotate smoothly, and the resulting motor phase voltages v_{aN}, v_{bN}, and v_{cN} are also listed in Table 3.1.

Next, let us review the space vector formulas from Chap. 2.

$$V_{abc} = \frac{2}{3}\left[v_a(t) + e^{j\frac{2\pi}{3}} \times v_b(t) + e^{j\frac{4\pi}{3}} \times v_c(t)\right] \quad (3.1.12)$$

We can substitute the motor phase voltages (v_{aN}, v_{bN}, and v_{cN}) from the six switching states in Table 3.1 into Eq. (3.1.12) to obtain six voltage vectors. These vectors can then be plotted on a two-dimensional space vector plane, as shown in Fig. 3.6.

The switching sequence in Table 3.1 is also known as the "six-step square wave control" for induction motors [2, 3]. Although this method can provide a relatively high motor phase voltage ($2V_{dc}/3$), it also introduces significant harmonic content. High harmonic distortion means greater energy loss. Moreover, six-step square wave control does not support variable voltage and variable frequency operation.

Therefore, this switching method is not commonly used in practical applications. Instead, in real-world implementations, each switching interval T/6 is further subdivided into much smaller PWM periods to perform Sinusoidal PWM (SPWM) modulation. This approach meets the requirements of Variable Voltage Variable

Table 3.1 Motor phase and inverter leg voltages under different switching states [2, 3]

Switching state	v_{ao}	v_{bo}	v_{co}	v_{aN}	v_{bN}	v_{cN}
101	$\frac{V_{dc}}{2}$	$-\frac{V_{dc}}{2}$	$\frac{V_{dc}}{2}$	$\frac{V_{dc}}{3}$	$-\frac{2V_{dc}}{3}$	$\frac{V_{dc}}{3}$
100	$\frac{V_{dc}}{2}$	$-\frac{V_{dc}}{2}$	$-\frac{V_{dc}}{2}$	$\frac{2V_{dc}}{3}$	$-\frac{V_{dc}}{3}$	$-\frac{V_{dc}}{3}$
110	$\frac{V_{dc}}{2}$	$\frac{V_{dc}}{2}$	$-\frac{V_{dc}}{2}$	$\frac{V_{dc}}{3}$	$\frac{V_{dc}}{3}$	$-\frac{2V_{dc}}{3}$
010	$-\frac{V_{dc}}{2}$	$\frac{V_{dc}}{2}$	$-\frac{V_{dc}}{2}$	$-\frac{V_{dc}}{3}$	$\frac{2V_{dc}}{3}$	$-\frac{V_{dc}}{3}$
011	$-\frac{V_{dc}}{2}$	$\frac{V_{dc}}{2}$	$\frac{V_{dc}}{2}$	$-\frac{2V_{dc}}{3}$	$\frac{V_{dc}}{3}$	$\frac{V_{dc}}{3}$
001	$-\frac{V_{dc}}{2}$	$-\frac{V_{dc}}{2}$	$\frac{V_{dc}}{2}$	$-\frac{V_{dc}}{3}$	$-\frac{V_{dc}}{3}$	$\frac{2V_{dc}}{3}$

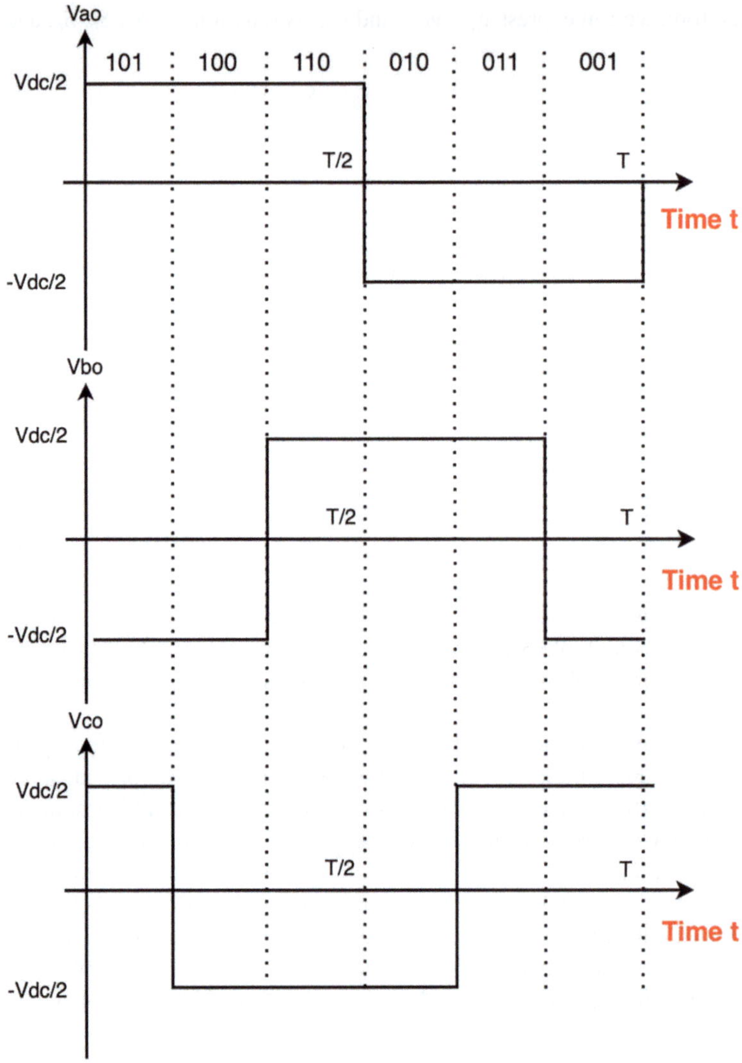

Fig. 3.5 Waveforms of v_{ao}, v_{bo}, and v_{co} under different switching states

Frequency (VVVF) control, significantly reduces harmonic distortion, and minimizes unnecessary energy losses.

- **SIMULINK Simulation**

We have now provided a detailed explanation of the operating principles of a three-phase inverter. Next, we will use MATLAB/SIMULINK to perform a system simulation of three-phase inverter SPWM (Sinusoidal Pulse Width Modulation) control.

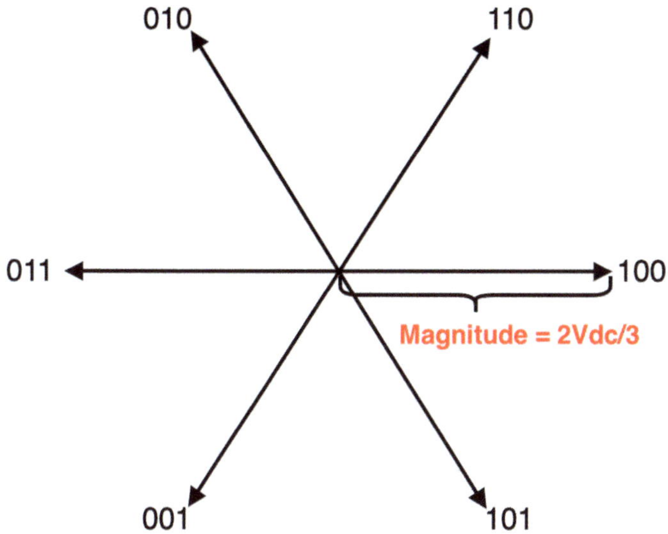

Fig. 3.6 Two-dimensional representation of the six active voltage vectors

Step 1

To implement SPWM modulation for a three-phase inverter, you need to use three sine modulation waves with a 120° phase shift to perform the SPWM modulation. Please use SIMULINK blocks to construct the three-phase SPWM modulator as shown in Fig. 3.7. The carrier wave should be set as a triangular wave with a frequency of 2 kHz and an amplitude of 1. After the construction is complete, select all blocks (you can use CTRL + A), right-click, and choose "Create Subsystem from Selection" to create a single subsystem block, as shown in Fig. 3.8. Name it "SPWM_modulator_VSI" and save it.

Step 2

After creating the "SPWM_modulator_VSI" subsystem, create a new blank SIMULINK file and build the block diagram as shown in Fig. 3.9. The "SPWM_modulator_VSI" block should be placed in the lower-left corner.

Please configure the blocks as follows:

- For the two "DC Voltage Source" blocks in the diagram, set the voltage of each to **100 V** (which represents $V_{dc}/2 = 100$ V).
- Double-click the **Series RLC Branch** block:
  - Set **Branch type** to **RL**
  - Set the **Resistance** to **5 Ω**
  - Set the **Inductance** to **150e − 3 H (i.e., 150 mH)**
- Double-click the **powergui** block:

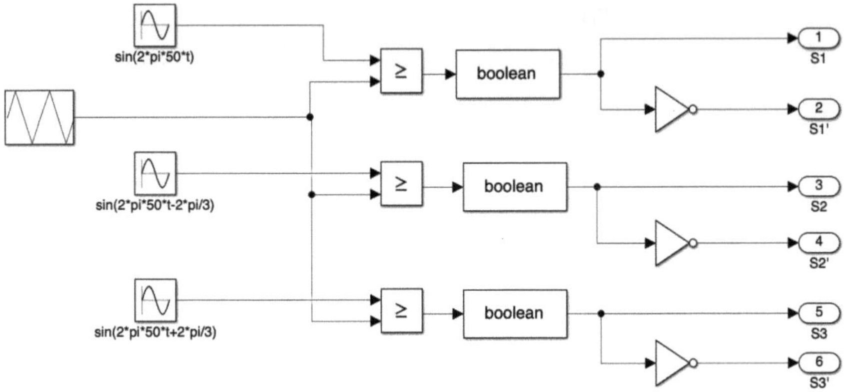

Fig. 3.7 SIMULINK modeling of a three-phase SPWM modulator, Example model: SPWM_modulator_VSI.slx

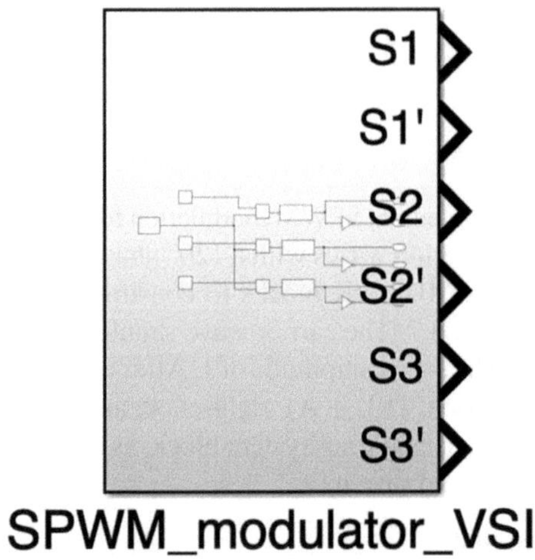

Fig. 3.8 Subsystem diagram of the three-phase SPWM modulator, Example model: SPWM_modulator_VSI.slx

- Set **Simulation Type** to **Discrete**
- Set the **Sample time** to **5e − 5**.

This configuration will allow you to simulate the SPWM-controlled three-phase inverter in a discretized power system environment.

Step 3

Set the SIMULINK simulation solver as follows:

- **Solver type**: Fixed-step
- **Fixed-step size**: auto
- **Stop time**: 0.5 s.

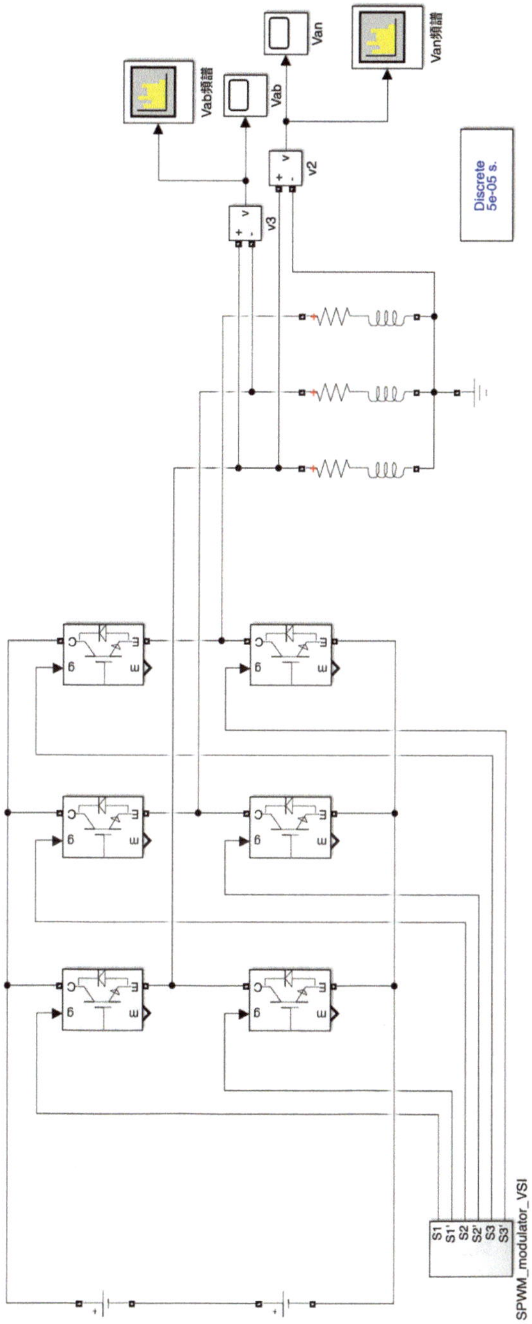

Fig. 3.9 SIMULINK simulation of the three-phase SPWM modulator, Example model: SPWM_3ph_VSI.slx

After completing the settings, click the "**Run**" button to execute the system simulation.

Step 4

If the simulation is successfully completed, first double-click the $\mathbf{V_{ab}}$ and $\mathbf{V_{an}}$ oscilloscope blocks to observe the waveforms of the line-to-line voltage $\mathbf{V_{ab}}$ and the motor phase voltage $\mathbf{V_{an}}$, respectively, as shown in Figs. 3.10 and 3.11. From the waveforms, it can be seen that both $\mathbf{V_{ab}}$ and $\mathbf{V_{an}}$ are PWM voltage waveforms:

- $\mathbf{V_{ab}}$ switches between $\mathbf{V_{dc}}$ **(200 V)** and – $\mathbf{V_{dc}}$ **(– 200 V)**,
- $\mathbf{V_{an}}$ is the motor phase voltage generated based on Eqs. (3.8) to (3.10), as shown in Fig. 3.11.

Step 5

Next, double-click the $\mathbf{V_{an}}$ **Spectrum Analyzer** block to observe the frequency spectrum of the $\mathbf{V_{an}}$ voltage (Note: set the **Window** to **Rectangular** to reduce spectral leakage), as shown in Fig. 3.12.

The fundamental frequency component at **50 Hz** in the $\mathbf{V_{an}}$ spectrum has an amplitude of approximately **70.5** $\mathbf{V_{rms}}$. Since the unit of the spectrum is RMS, we need to convert it to a peak value, which gives approximately **99.69 V**.

In the three-phase SPWM modulator, the peak of the sine modulation wave is set equal to the peak of the carrier wave, i.e., $m_a = 1$. Therefore, the theoretical peak value of the fundamental component of the output phase voltage is: $\hat{V}_{aN1} = 1 \times \frac{V_{dc}}{2} = 100$ (V). The measured result from the Spectrum Analyzer is **99.69 V**, which is very close to the theoretical value.

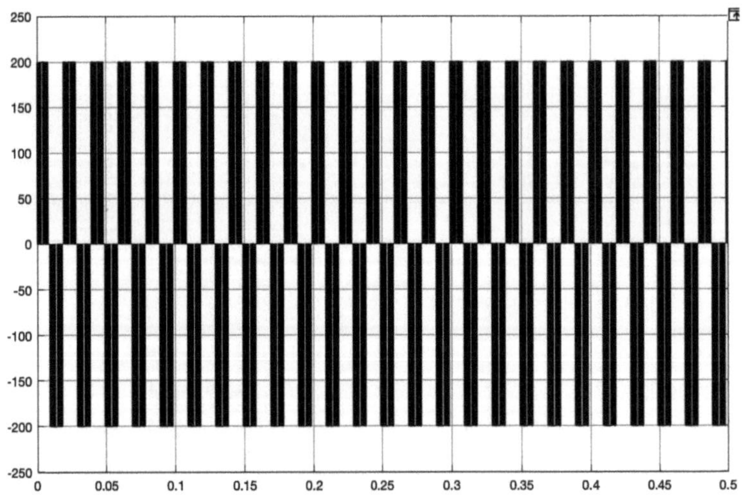

Fig. 3.10 Waveform of the line-to-line voltage V_{ab}

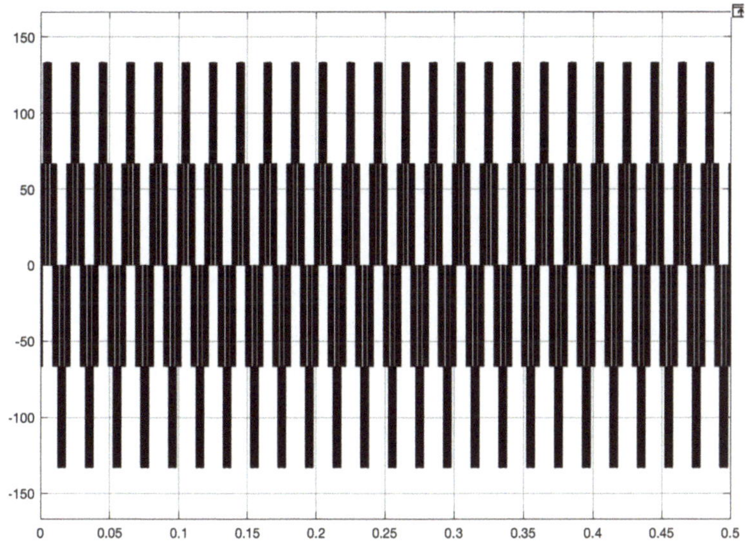

Fig. 3.11 Waveform of the motor phase voltage V_{an}

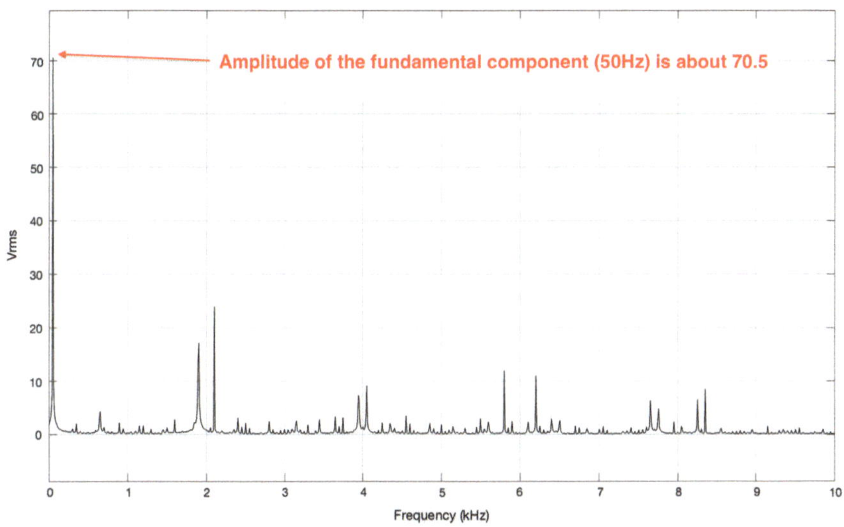

Fig. 3.12 Spectrum of the motor phase voltage V_{an}

Step 6

Next, double-click the V_{ab} spectrum block to observe the frequency spectrum of the line voltage V_{ab} (note: set the window to Rectangular to reduce spectral leakage), as shown in Fig. 3.13. The RMS value of the fundamental component of V_{ab} at 50 Hz

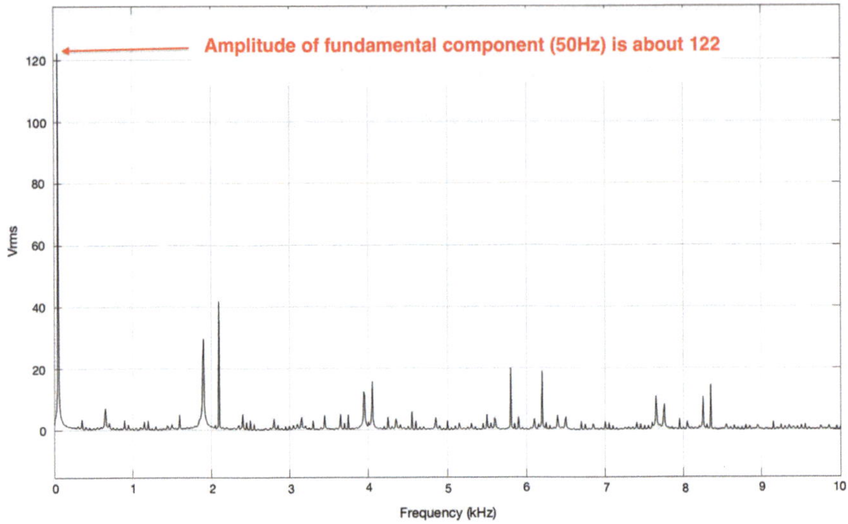

Fig. 3.13 Spectrum of the line-to-line voltage V_{ab}

is approximately 122 $\mathbf{V_{rms}}$. Since the spectrum analyzer displays values in RMS, it must be converted to the peak value, which is 172.5 V.

In the SPWM modulator, the amplitude of the sinusoidal modulation wave is set equal to that of the carrier, i.e., $m_a = 1$. Therefore, the theoretical peak value of the fundamental component of the line voltage is:

$$\hat{V}_{ab1} = \sqrt{3} \times \hat{V}_{aN1} = 172.6 \text{ (V)}$$

The result observed using the Spectrum Analyzer is 172.5 (V), which is very close to the theoretical value. (Note: In a Y-connected system, the line voltage is $\sqrt{3}$ times the phase voltage.)

Explanation

The discrepancy between the spectrum magnitude observed using the Spectrum Analyzer block and the theoretical value is caused by **spectral leakage**, which is an unavoidable phenomenon in practice. Since the number of sampling points in a computer is finite—assuming the number of points is N and the sampling frequency is f_s—the frequency resolution is $\Delta f = f_s/N$. If the signal frequency of interest is not an integer multiple of the frequency resolution Δf, spectral leakage will occur.

3.2 Third-Harmonic Injection PWM

In Sect. 3.1, we simulated the functionality of three-phase SPWM modulation. Based on spectral analysis, we verified that the fundamental peak of the phase voltage output by the three-phase SPWM is $\mathbf{V_{dc}/2}$. However, in practical applications, many motors have relatively high rated voltages. If modulation techniques can be used to increase the inverter's output voltage, the motor will be able to deliver higher torque.

Therefore, in this section, we introduce the **third-harmonic injection modulation method**, which can be considered an enhanced version of SPWM. By injecting a third-harmonic component into the three-phase SPWM modulation signals, the output voltage of the three-phase VSI can be effectively increased by **15.47%**, giving it a higher voltage output capability than the conventional SPWM inverter. The injected third-harmonic components cancel out at the output terminals and thus **do not appear** in the motor's terminal voltages.

First, we list the three-phase SPWM modulation waveforms used in Sect. 3.1 as follows:

$$v_{am}(t) = V_{am}\sin(\omega t) \tag{3.2.1}$$

$$v_{bm}(t) = V_{bm}\sin\left(\omega t - \frac{2\pi}{3}\right) \tag{3.2.2}$$

$$v_{cm}(t) = V_{cm}\sin\left(\omega t + \frac{2\pi}{3}\right) \tag{3.2.3}$$

where $V_{am} = V_{bm} = V_{cm} = V_{tri}$ and V_{tri} is the carrier amplitude.

Next, we will add a third harmonic component to the three-phase SPWM modulation waves, as shown below:

$$v'_{am}(t) = V_{am}\sin(\omega t) + V_{m3}\sin(3\omega t) \tag{3.2.4}$$

$$v'_{bm}(t) = V_{bm}\sin\left(\omega t - \frac{2\pi}{3}\right) + V_{m3}\sin(3\omega t) \tag{3.2.5}$$

$$v'_{cm}(t) = V_{cm}\sin\left(\omega t + \frac{2\pi}{3}\right) + V_{m3}\sin(3\omega t) \tag{3.2.6}$$

Since the amplitude of the sine wave decreases after injecting the third harmonic, we want the maximum value of the three-phase modulation wave with the third harmonic to remain consistent with the carrier amplitude. Therefore, we first differentiate Eq. (3.2.4) with respect to ωt to find its extreme value and determine the relationship between V_{m3} and V_{am} [3].

$$\frac{d}{d\omega t}v'_{am}(t) = V_{am}\cos(\omega t) + 3V_{m3}\cos(3\omega t) = 0 \tag{3.2.7}$$

Therefore, when $\omega t = \frac{\pi}{3}$, Eq. (3.2.7) holds. At this point, V_{m3} is:

$$V_{m3} = \frac{1}{3} V_{am} \cos\left(\frac{\pi}{3}\right) \tag{3.2.8}$$

We substitute Eq. (3.2.8) into Eq. (3.2.4), and set the absolute value of $v'_{am}(t)$ equal to the carrier amplitude V_{tri}.

$$\left|v'_{am}\right| = \left|V_{am}\sin(\omega t) + \frac{1}{3}V_{am}\cos\left(\frac{\pi}{3}\right)\sin(3\omega t)\right| = V_{tri} \tag{3.2.9}$$

By substituting $\omega t = \frac{\pi}{3}$ into Eq. (3.2.9), we obtain:

$$V_{am} = \frac{V_{tri}}{\sin\left(\frac{\pi}{3}\right)} \tag{3.2.10}$$

Therefore, when using the third harmonic injection modulation method, if the carrier amplitude is V_{tri}, the amplitudes of V_{am}, V_{bm}, and V_{cm} should be set to $\frac{V_{tri}}{\sin\left(\frac{\pi}{3}\right)}$, and the amplitude of the third harmonic component V_{m3} should be set to $\frac{1}{3}V_{am}\cos\left(\frac{\pi}{3}\right)$.

If the carrier amplitude V_{tri} is set to 1, then

$$V_{am} = V_{bm} = V_{cm} = \frac{1}{\sin\left(\frac{\pi}{3}\right)} \text{ and } V_{m3} = \frac{1}{3} \times \frac{\cos\left(\frac{\pi}{3}\right)}{\sin\left(\frac{\pi}{3}\right)}$$

Figure 3.14 shows the difference between a conventional SPWM sine modulation waveform and a modulation waveform with third harmonic injection for phase a.

- **SIMULINK Simulation**

We have now thoroughly explained the operating principle of the third-harmonic injection modulation method. Next, we will use MATLAB/SIMULINK to perform a system simulation of a VSI using this modulation technique.

Step 1

To implement the third-harmonic injection modulation method, we need to add a third-harmonic component to the modulation waves of the three-phase SPWM. Therefore, please modify the three-phase SPWM modulator from Sect. 3.1 to match the structure shown in Fig. 3.15. After completing the modifications, select all blocks (you can use CTRL + A), right-click, and choose "Create Subsystem from Selection" to create a single Subsystem block, as shown in Fig. 3.16. Name it "**third_harmonic_SPWM**" and save it.

Step 2

After completing the creation of the "**third_harmonic_SPWM**" block, open a new blank SIMULINK model and construct the system as shown in Fig. 3.17. Except for

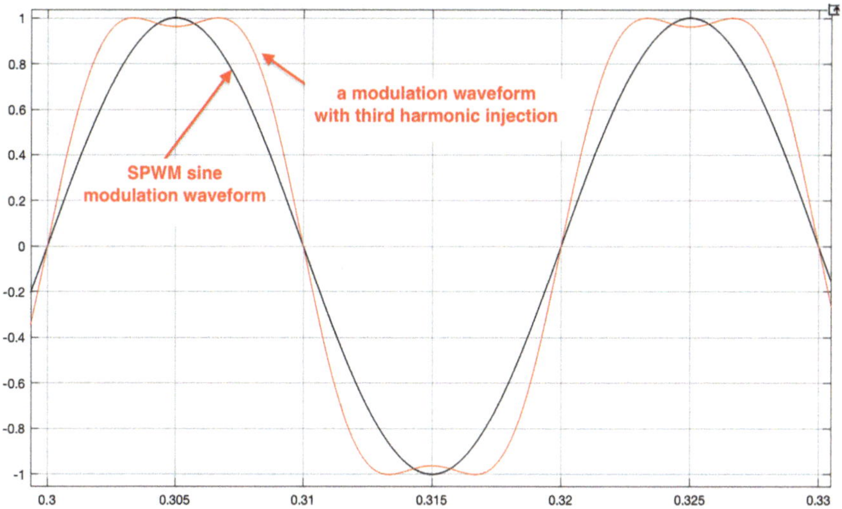

Fig. 3.14 Comparison of the SPWM modulation wave and the third harmonic injection modulation wave

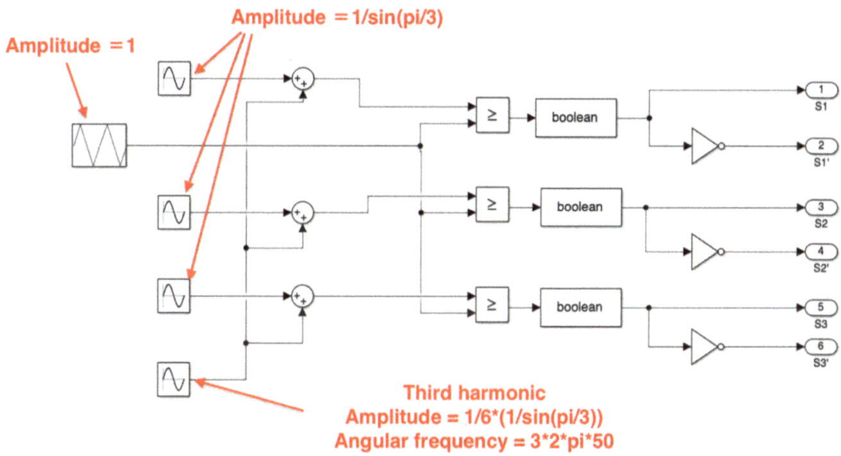

Fig. 3.15 SIMULINK modeling of a third harmonic injection modulator, Example model: 3rd_harmonic_SPWM.slx

the "**third_harmonic_SPWM**" block, all system block settings in Fig. 3.17 should be identical to those used in the **SPWM_3ph_VSI** model from Sect. 3.1.

Step 3

Set the SIMULINK simulation solver to "**Fixed-step**", and set the "**Fixed-step size**" to **auto**. Then, set the **total simulation time** to **0.5** s.

After completing these settings, click "**Run**" to start the system simulation.

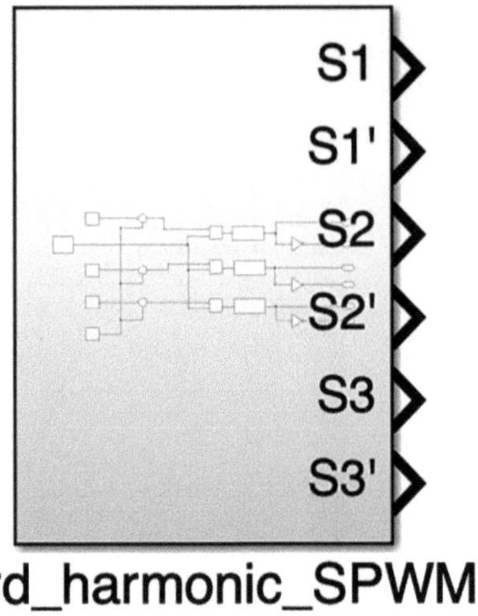

Fig. 3.16 Subsystem diagram of the third harmonic injection modulator, Example model: third_harmonic_SPWM.slx

Step 4

Double-click the $\mathbf{V_{an}}$ **spectrum** block to observe the frequency spectrum of the V_{an} voltage (Note: set the window to **Rectangular** to reduce spectral leakage). As shown in Fig. 3.18, the fundamental frequency component at **50 Hz** has a spectrum magnitude of approximately **81.5** $\mathbf{V_{rms}}$.

Since the spectrum unit is in **RMS**, it must be converted to **peak value**, which is **115.24 V**. Compared to the maximum phase voltage output of **100 V** from conventional three-phase SPWM, the **third harmonic injection modulation method** increases the phase voltage by approximately **15.24%**, which is close to the theoretical value of **15.47%**.

Step 5

Next, double-click the V_{ab} spectrum block to observe the frequency spectrum of the line voltage V_{ab} (Note: set the window to Rectangular to reduce spectral leakage). As shown in Fig. 3.19, the fundamental frequency component at 50 Hz has a spectrum magnitude of approximately 141.5 $\mathbf{V_{rms}}$.

Since the spectrum unit is in RMS, it must be converted to peak value, which is 200 V. You can observe that the third harmonic injection modulation method fully utilizes the entire DC-link voltage, i.e., V_{dc}.

In comparison, the maximum line voltage output from conventional three-phase SPWM is 172.6 V. Therefore, the third harmonic injection method increases the line voltage by approximately 15.87%, which is very close to the theoretical value of 15.47%.

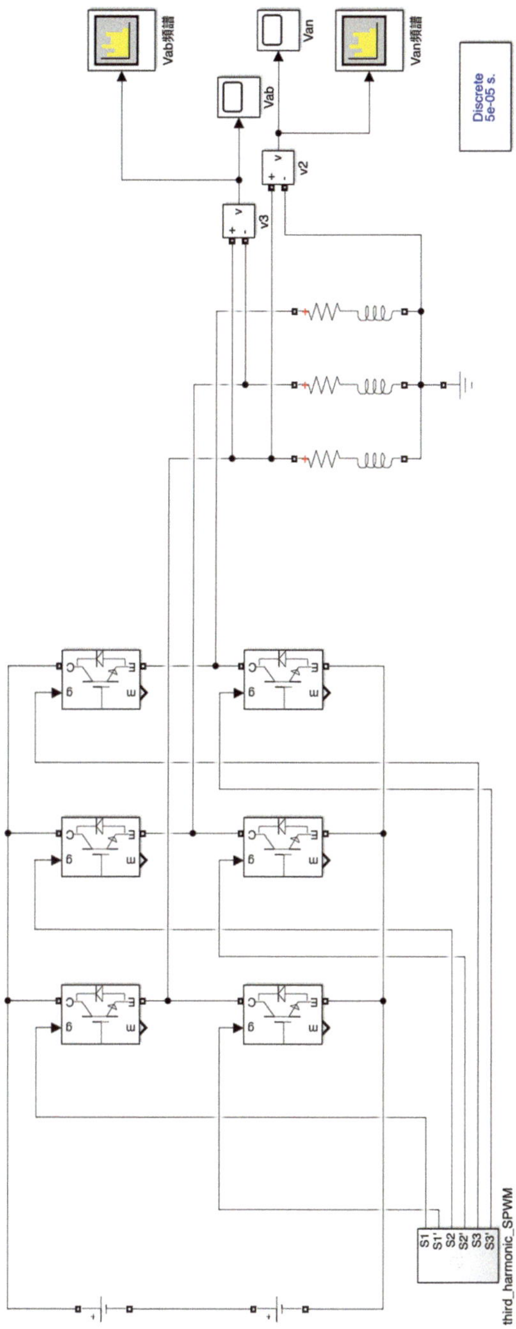

Fig. 3.17 SIMULINK Simulation of the third harmonic injection method, Example model: third_harmonic_SPWM_VSI.slx

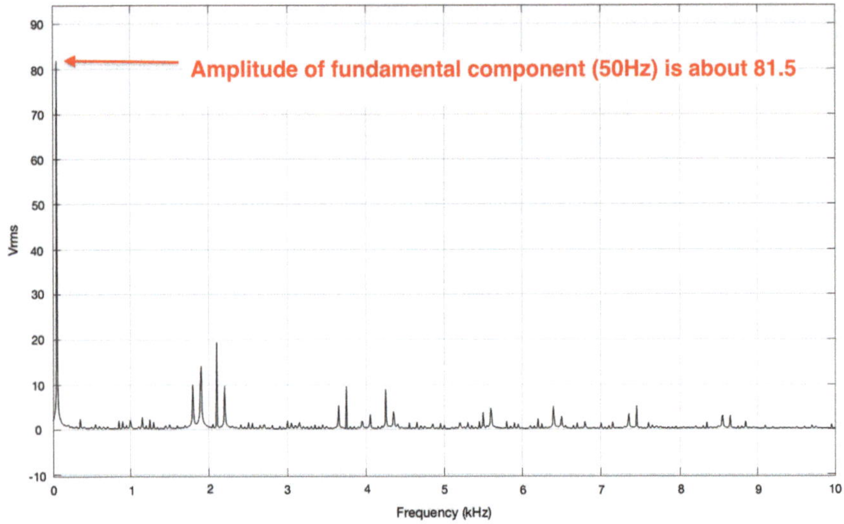

Fig. 3.18 Spectrum of the motor phase voltage V_{an}

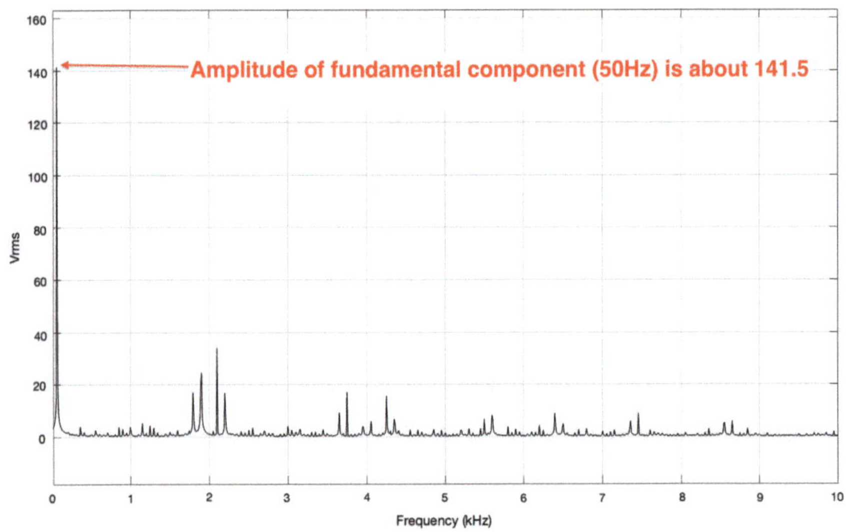

Fig. 3.19 Spectrum of the line-to-line voltage V_{ab}

3.3 Offset Addition PWM Method

Now that we understand how injecting third harmonics into the three-phase SPWM modulation waves can effectively increase the inverter's output voltage by 15.47%, giving it superior voltage output capability compared to conventional SPWM

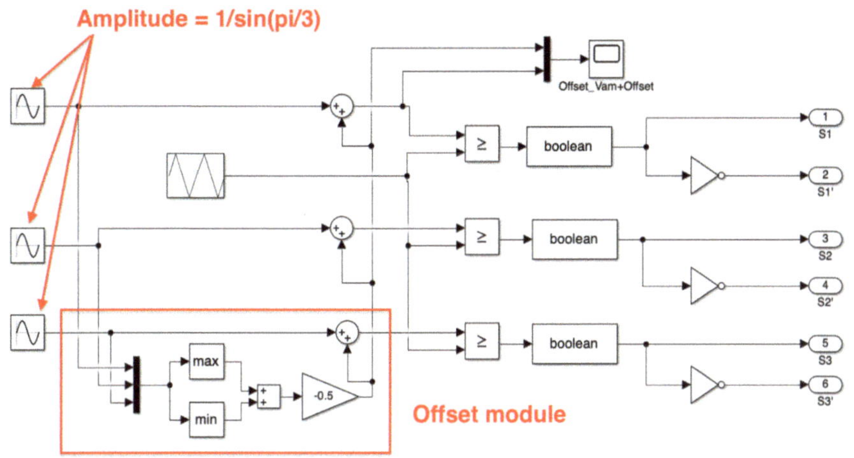

Fig. 3.20 SIMULINK modeling of the offset-addition modulator, Example model: offset_addition_SPWM.slx

inverters, we turn to practical motor control applications. In such applications, directly injecting third harmonics into the modulation waves is not easy to implement in software.

Therefore, in this section, I will introduce the "Offset addition Modulation Method", which achieves the same effect as third harmonic injection but is much easier to implement via software.

- **SIMULINK Simulation**

Step 1

To implement the "**Offset addition Modulation Method**", we need to modify the **third harmonic injection modulator** (third_harmonic_SPWM) from Sect. 3.2. Please update the modulator to match the configuration shown in Fig. 3.20.

In this diagram, the "**Offset Module**" is introduced. It produces an effect **equivalent to third harmonic injection**.

Note: The output of the Offset Module actually includes not only the third harmonic, but also **harmonics that are integer multiples of three** (i.e., 3rd, 6th, 9th, etc.).

The output of the "Offset Module" can be expressed as:

$$Offset = -\frac{\max\{v_{am}, v_{bm}, v_{cm}\} + \min\{v_{am}, v_{bm}, v_{cm}\}}{2} \tag{3.3.1}$$

After completing the modifications, select all the blocks (you can use **CTRL + A**), right-click and choose "**Create Subsystem from Selection**" to create a single subsystem block, as shown in Fig. 3.21. Name it "**offset_addition_SPWM**" and save the file.

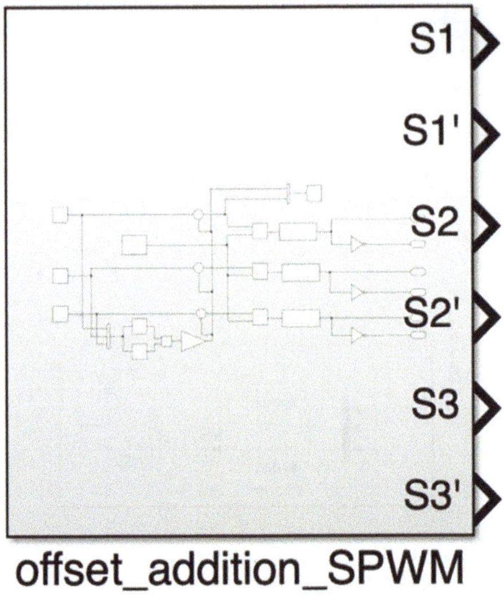

Fig. 3.21 Subsystem diagram of the offset-addition modulator, Example model: offset_addition_SPWM.slx

Step 2

After completing the creation of "**offset_addition_SPWM**", create a new blank **SIMULINK** file and build the block diagram as shown in Fig. 3.22.

Except for the "**offset_addition_SPWM**" block, the configuration settings of the other blocks in Fig. 3.22 are the same as those used in Sect. 3.2's **third_harmonic_SPWM_VSI**.

Step 3

Set the **SIMULINK solver** to "**Fixed-step**", set the "**Fixed-step size**" to **auto**, and set the **total simulation time** to **0.5** s.

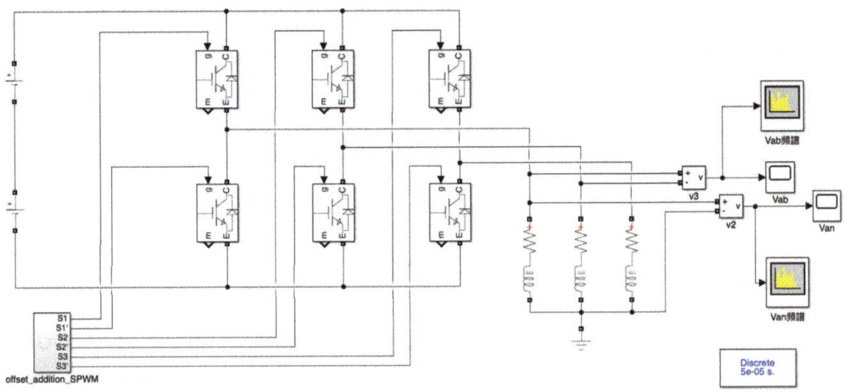

Fig. 3.22 SIMULINK simulation of the offset-addition modulation, Example model: offset_addition_SPWM_VSI.slx

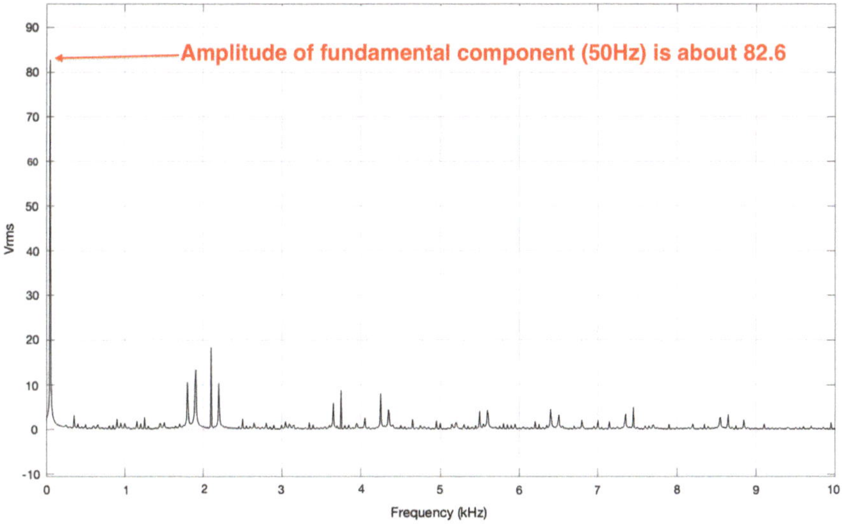

Fig. 3.23 Spectrum of the motor phase voltage V_{an}

After completing the settings, click "**Run**" to execute the simulation.

Step 4

After the simulation is complete, double-click the **V_{an}spectrum** block to observe the frequency spectrum of the **V_{an}voltage** (note: be sure to set the **window** to **Rectangular** to reduce spectral leakage). As shown in Fig. 3.23, the magnitude of the **fundamental frequency component at 50 Hz** is approximately **82.6 V_{rms}**. Since the unit of the spectrum is in **RMS (root mean square)**, it should be converted to **peak value**, yielding **116.8 V**.

This result demonstrates that the **offset addition modulation method** achieves the **same voltage boost effect** as the **third harmonic injection method**.

Step 5

Next, double-click the **V_{ab}spectrum** block to observe the **line voltage V_{ab}frequency spectrum** (note: be sure to set the **window** to **Rectangular** to reduce spectral leakage). As shown in Fig. 3.24, the **fundamental component at 50 Hz** has a spectral magnitude of approximately **142 V_{rms}**. Since the spectral unit is in **RMS**, converting it to **peak value** gives **200.8 V**.

This confirms that the **offset addition modulation method** provides the **same voltage boost effect** as the **third harmonic injection method**.

Step 6

Next, double-click the Offset_Vam + Offset oscilloscope block inside the offset_addition_SPWM subsystem to observe the output of the Offset module and the a-phase modulating wave after the offset has been added. As shown in Fig. 3.25, we can

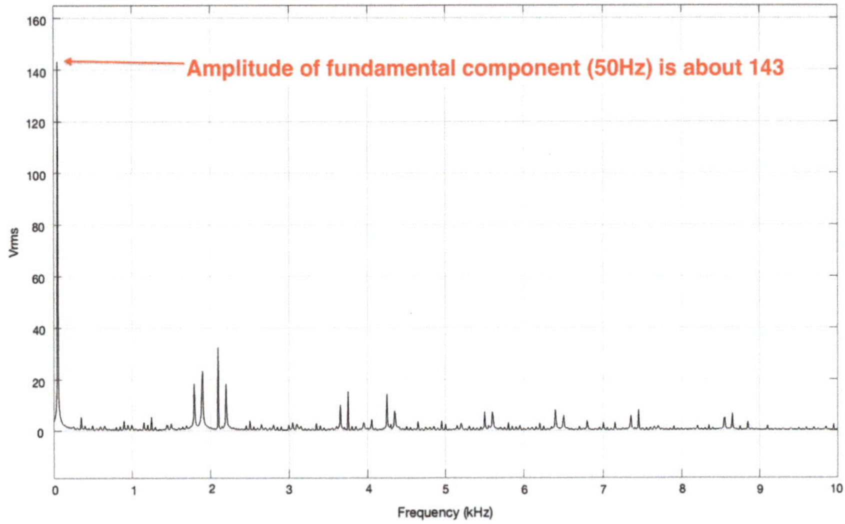

Fig. 3.24 Spectrum of the line-to-line voltage V_{ab}

see that the a-phase modulating waveform with offset is very similar to the a-phase modulating waveform of the third harmonic injection method. The Offset waveform itself is a symmetric triangular waveform. Unlike the third harmonic injection method, the offset addition modulation method injects not only the third harmonic but also all triplen harmonics (multiples of three). These injected triplen harmonics will still cancel each other out at the output and will not appear in the motor's terminal voltage.

3.4 Space Vector PWM (SVPWM)

Whether it is **SPWM modulation**, **third harmonic injection modulation**, or **offset addition modulation**, the essence of these modulation techniques is to use three-phase modulating waves compared with a carrier wave to generate PWM signals for switching the transistors of the three inverter legs. In practical motor control applications, the three-phase sinusoidal modulation signals are obtained from the voltage commands generated by the d-axis and q-axis current controllers through coordinate transformations (inverse Park and inverse Clarke transformations), as shown in Fig. 3.26.

In this section, I will introduce another widely adopted PWM modulation method known as **Space Vector Modulation (SVM or SVPWM)**. The operating principle of SVM is fundamentally different from the three previously discussed modulation techniques (SPWM, third harmonic injection, and offset addition modulation).

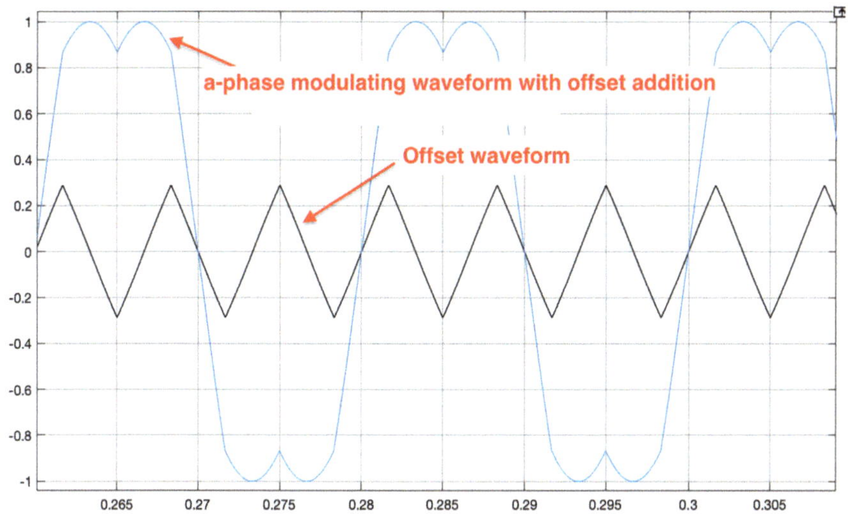

Fig. 3.25 Offset waveform and phase-a modulation waveform

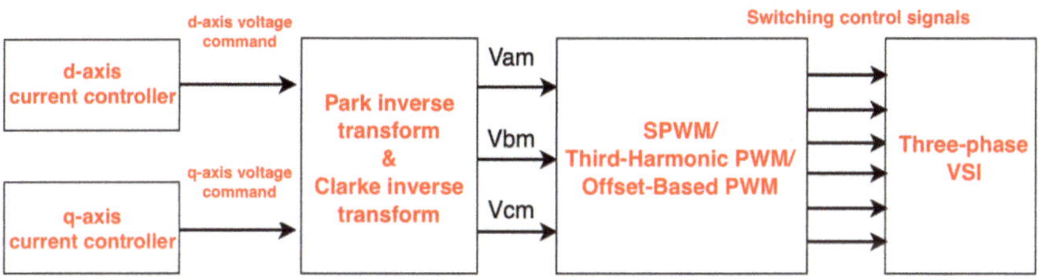

Fig. 3.26 Modulation system block diagram of the three-phase VSI motor drive system

SVM is based on **vector synthesis**, and its operation in motor control applications is illustrated in Fig. 3.27.

Space Vector Modulation (SVM) synthesizes the desired voltage vector based on its magnitude and angle using *active voltage vectors*. What are active voltage vectors? In Sect. 3.1, we introduced the six-step waveform technique, where six

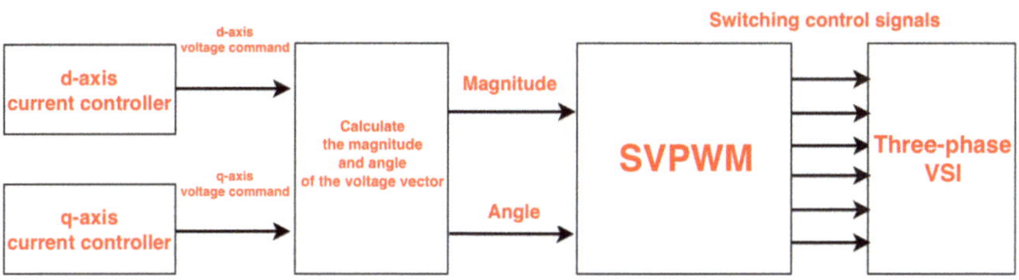

Fig. 3.27 SVPWM modulation system block diagram of the three-phase VSI motor drive system

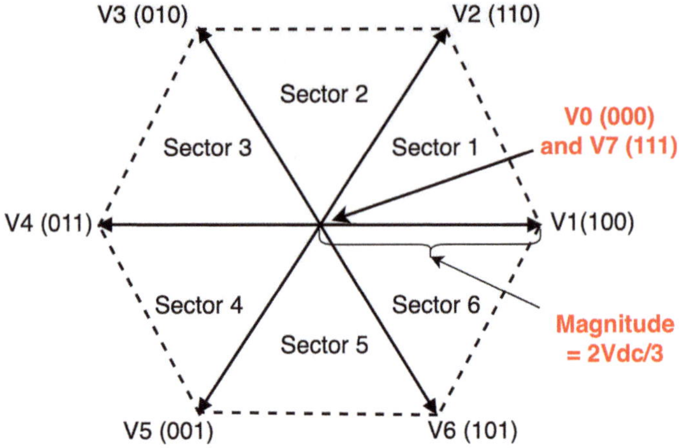

Fig. 3.28 SVPWM voltage vector diagram

non-zero voltage vectors can be generated from six switching combinations—these are called **active voltage vectors**, as shown in Fig. 3.6.

In addition to the active voltage vectors, there are also switching states (0, 0, 0) and (1, 1, 1), which produce voltage vectors $\mathbf{V_0}$ and $\mathbf{V_7}$, respectively. These vectors have zero magnitude and are referred to as **zero voltage vectors**. All eight voltage vectors (six active and two zero vectors) can be plotted in the space vector plane, as shown in Fig. 3.28.

Figure 3.28 shows a total of eight voltage vectors in the space vector plane, including six **active voltage vectors** (V_1 to V_6) and two **zero voltage vectors** (V_0 and V_7). The plane is divided into six sectors (Sector 1 to Sector 6) by the active voltage vectors, with each sector spanning 60°. Each active voltage vector has a magnitude of $\mathbf{2V_{dc}/3}$.

If the three-phase VSI switches sequentially through V_1 to V_6, this corresponds to the **six-step waveform control method**. In contrast, **space vector modulation (SVM)** determines which sector the input voltage vector lies in based on its position (i.e., its angle).

Assuming the input voltage vector is V_s^*, with an angle α, it lies within **Sector 1**, as shown in Fig. 3.29.

From Fig. 3.29, we can deduce—using basic trigonometric relationships—that the radius of the inscribed circle within the regular hexagon formed by the six active voltage vectors is $\frac{V_{dc}}{\sqrt{3}}$.

(Note: The angle between the diameter of the inscribed circle and vector V_1 is 30°, and the magnitude of V_1 is $\mathbf{2V_{dc}/3}$.)

The value $\frac{V_{dc}}{\sqrt{3}}$ also represents the **maximum phase voltage** that space vector modulation (SVM) can provide within its **linear modulation region**. If the magnitude of the reference voltage vector exceeds $\frac{V_{dc}}{\sqrt{3}}$, it enters the **nonlinear modulation**

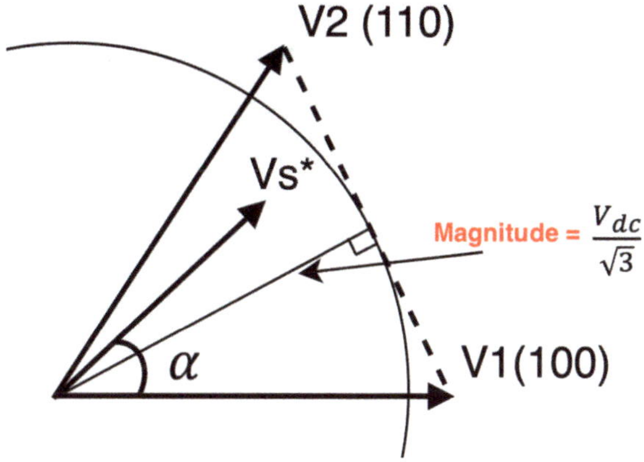

Fig. 3.29 Geometric relationship of the inscribed circle in SVPWM

region, also known as the **overmodulation region**, which is beyond the scope of this book.

Assuming a Y-connected motor winding, the corresponding maximum **line-to-line voltage** is $\mathbf{V_{dc}}$, meaning that **SVM provides the same voltage boosting capability** as the **third harmonic injection method**. As shown in Fig. 3.29, suppose the input voltage vector to the space vector modulator is V_s^*, which lies in **Sector 1**.

The **space vector modulation (SVM)** technique synthesizes this voltage vector by combining the **two adjacent active voltage vectors**. For **Sector 1**, V_1 and V_2 are used to synthesize V_s^*. Assuming the sampling period is T_s, the application durations of the voltage vectors V_1, V_2, and the **zero voltage vector** are denoted as T_a, T_b, and T_0, respectively.

Using the **volt-second balance principle**, we can write the following equation:

$$V_s^* \times T_s = V_1 \times T_a + V_2 \times T_b + V_0 \times T_0 \tag{3.4.1}$$

where $T_a + T_b + T_0 = T_s$.

Additionally, we know that

$$V_s^* = \left|V_s^*\right| e^{j\alpha} \tag{3.4.2}$$

$$V_1 = \frac{2V_{dc}}{3} \tag{3.4.3}$$

$$V_2 = \frac{2V_{dc}}{3} e^{j\frac{\pi}{3}} \tag{3.4.4}$$

By substituting Eqs. (3.4.2) through (3.4.4) into Eq. (3.4.1), the real and imaginary components can be organized as follows:

$$\left|V_s^*\right| \cos(\alpha) T_s = \frac{2V_{dc}}{3} \times T_a + \frac{2V_{dc}}{3} \times \cos\left(\frac{\pi}{3}\right) \times T_b \tag{3.4.5}$$

$$\left|V_s^*\right| \sin(\alpha)T_s = \frac{2V_{dc}}{3} \times \sin\left(\frac{\pi}{3}\right) \times T_b \tag{3.4.6}$$

The expressions for T_a, T_b, and T_0 can be summarized as follows:

$$T_a = \frac{\sqrt{3}\left|V_s^*\right|}{V_{dc}} \sin\left(\frac{\pi}{3} - \alpha\right)T_s \tag{3.4.7}$$

$$T_b = \frac{\sqrt{3}\left|V_s^*\right|}{V_{dc}} \sin(\alpha)T_s \tag{3.4.8}$$

$$T_0 = T_s - T_a - T_b \tag{3.4.9}$$

Equations (3.4.7)–(3.4.9) can be transformed into general formulas applicable to all sectors (Sector 1 to Sector 6).

$$T_a = \frac{\sqrt{3}\left|V_s^*\right|}{V_{dc}} \sin\left(S\frac{\pi}{3} - \alpha\right)T_s \tag{3.4.10}$$

$$T_b = \frac{\sqrt{3}\left|V_s^*\right|}{V_{dc}} \sin\left(\alpha - (S-1)\frac{\pi}{3}\right)T_s \tag{3.4.11}$$

$$T_0 = T_s - T_a - T_b \tag{3.4.12}$$

Here, $S = 1, 2, 3, \ldots, 6$ represents the sector in which the input voltage vector V_s^* is located.

After obtaining T_a, T_b, and T_0, the desired voltage vector V_s^* can be successfully synthesized by switching the three-phase inverter's transistor switches according to the durations of the adjacent active voltage vectors and the zero voltage vector. This concludes the operating principle of Space Vector Pulse Width Modulation (SVPWM). As for the SIMULINK simulation of the SVPWM method, you are encouraged to try it out yourself.

Compared to the "Offset Addition Modulation" method, the "Space Vector Modulation" (SVM) technique provides the same voltage boosting capability. However, since SVM requires the computation of Eqs. (3.4.10)–(3.4.12) in every sampling cycle, it consumes more computational resources. Therefore, in practical applications—unless a Direct Torque Control (DTC) architecture is needed or PWM modulation techniques are being studied—the author tends to prefer the "Offset Addition Modulation" method due to its simpler computational requirements.

3.5 Integrated PMSM Vector Control System Simulation with Inverter

By now, you should have a clear understanding of several important modulation strategies for three-phase inverters. Next, we will combine the vector control loop of a permanent magnet synchronous motor (PMSM) with the three-phase inverter to perform a system simulation that closely reflects actual physical characteristics. This simulation will also include delay effects to make the results more representative of a real physical system. You can open the example model file mdl_pm_foc_plus_inv. Once opened, the SIMULINK block diagram will appear as shown in Fig. 3.30.

In Fig. 3.30, we use the *Universal Bridge* and *Permanent Magnet Synchronous Machine* components from the Simscape group in SIMULINK to simulate the three-phase inverter and the three-phase Interior Permanent Magnet Synchronous Motor (IPMSM). The parameter settings of the *Permanent Magnet Synchronous Machine* component are consistent with those listed in Table 2.3, except that the friction coefficient B is set to zero. (In this simulation, the effect of friction is accounted for in the load torque, so B is set to zero.)

The PI controller parameters of the current control loop are configured according to Table 2.4, which allows the current loop to achieve a bandwidth of 1000 (rad/s). With the inclusion of a current decoupling module, the current loop can be considered as an equivalent first-order low-pass filter with a cutoff frequency of 1000 (rad/s).

For the speed control loop, we will use the parameter values designed using the classical PI controller design method for the speed loop introduced in Sect. 2.3.1, as shown below:

$$K_{p_\omega} = 0.016,\ K_{i_\omega} = 0.32 \tag{3.5.1}$$

The output of the speed PI controller will pass through a gain of $1/K_T = 11.11$, calculated as follows:

$$\frac{1}{K_T} = \frac{2}{3} \times \frac{2}{P} \times \frac{1}{\lambda_f} = \frac{2}{3} \times \frac{2}{4} \times \frac{1}{0.03} = 11.11 \tag{3.5.2}$$

where K_T is the motor torque constant, P is the number of pole pairs, and λ_f is the rotor flux linkage.

In Fig. 3.30, the computation delay, sampling delay, and feedback low-pass filter are all configured identically to those in Sect. 2.3.1. The configuration values are as follows:

- **Computation delay (Compu-Delay)**: $T_c = T_s = 2.5\text{e} - 4$ (s)
- **Sampling delay (SH-Delay)**: $T_{SH} = \frac{T_s}{2} = 1.25e - 04$ (s)
- **Feedback low-pass filter (LPF)**: First-order low-pass filter with a cutoff frequency of $\omega_{fc} = 500$ (Hz).

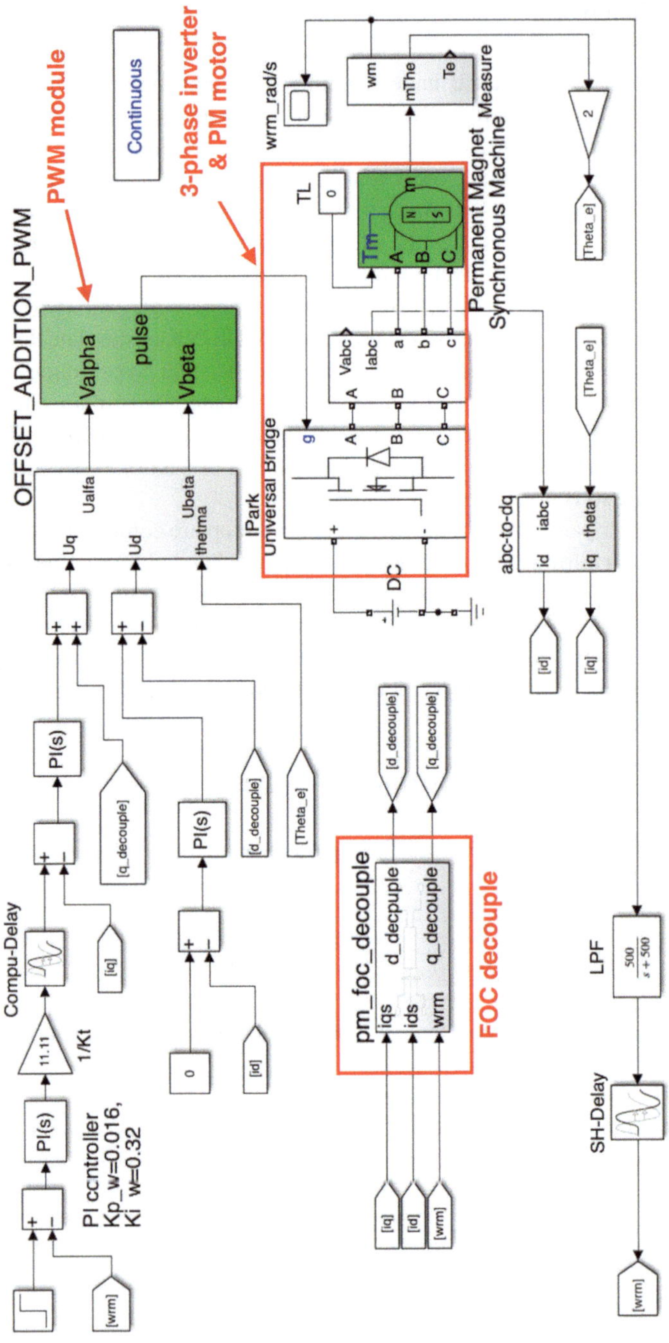

Fig. 3.30 SIMULINK simulation system integrating offset-addition modulation and Simscape components, Example model: mdl_pm_foc_plus_inv.slx

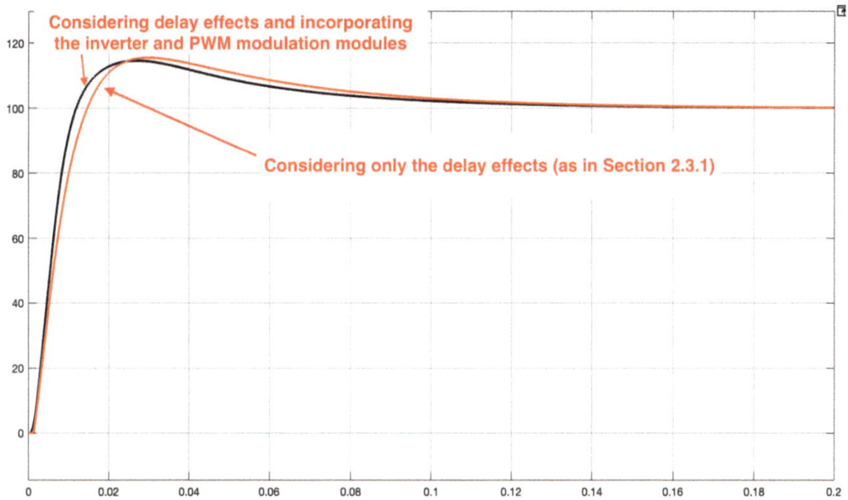

Fig. 3.31 Speed response waveforms of the simulation systems

In addition, the three-phase modulation strategy used in this simulation is the "**Offset Addition Modulation Method**", as shown in the "OFFSET_ADDITION_PWM" module in Fig. 3.30.

Run the SIMULINK simulation of this example model and compare the speed response waveform with the simulation results from Sect. 2.3.1, as shown in Fig. 3.31.

As shown in Fig. 3.31, this simulation result closely matches the simulation result that includes delay effects in Sect. 2.3.1. Compared to the simplified model in Fig. 2.34, this simulation also incorporates the three-phase inverter module and the modulation module typically used in actual motor drive systems, making it more representative of the real system's dynamic behavior. Nevertheless, the two simulation results are quite similar overall, indicating that the control loop model used in Sect. 2.3.1 is sufficient for simulation and verification during the design phase. However, if more types of signals need to be observed, a simulation system such as that shown in Fig. 3.30 must be constructed.

3.6 Summary

The key points of this chapter are summarized as follows:

- The **Spectrum Analyzer** component in **SIMULINK** can be used to observe signal spectra in real time. However, **spectral leakage** may cause measurement errors. To mitigate this, an appropriate **window function** must be selected. Based on the author's testing, for PWM signals, the **Rectangular window** provides the best reduction in spectral leakage–induced errors.

- This chapter does not delve into the **harmonic issues** caused by different PWM techniques. Interested readers may refer to related literature or use the example models provided in this chapter to simulate and analyze PWM harmonics.
- The simulation system in Fig. 3.30 does **not account for current loop delay effects**. Readers may incorporate such effects as needed and apply the current loop design methodology discussed in Chap. 2 that considers delays. Doing so should yield simulation results that more closely reflect the behavior of a real physical system.

References

1. Sul, Seung-Ki, Control of Electric Machine Drive Systems, Wiley-IEEE Press, 2011.
2. Chang-Hwan Liu, AC Motor Control: Principles of Vector Control and Direct Torque Control, Taipei: Dong Hua Book Co., 2001.
3. Haitham Abu-Rub, Atif Iqbal and Jaroslaw Guzinski, High Performance Control of AC Drives with MATLAB/SIMULINK, John Wiley & Sons, Ltd, UK, 2021.
4. N. Mohan, T. M. Undeland, and W. P. Robbins, Power Electronics: Converters, Applications and Design, Second ed. New York:Wiley, 1995.

Chapter 4
Hardware-Based Design Validation

In this chapter, we will use a hardware platform to verify the motor control loop designed in Chap. 2. Typically, the verification process for AC motor control algorithms follows the standard procedure shown in Fig. 4.1.

In the early stages, motor control algorithms are first evaluated and verified through numerical simulations on a computer [1]. Commonly used software platforms at this stage include MATLAB/SIMULINK or LabVIEW, which are commercial paid software. For cost-saving purposes, one can also use Python with the python-control and Matplotlib libraries for control system simulations [2]. If the control algorithm passes the computer-based simulation, it proceeds to the second stage: hardware-in-the-loop (HIL) verification. Since HIL uses actual hardware to emulate the behavior of a physical system, it provides more realistic verification results compared to software simulations. There are also integrated platforms available on the market that support both the second and third stages of verification, such as dSPACE or NI's RCP (Rapid Control Prototyping) platforms.

If the functionality and feasibility of the control algorithm are confirmed through HIL verification, the process advances to the final stage. At this point, the control algorithm can be implemented on a microcontroller or FPGA and tested with an actual inverter and motor. Based on the author's experience, companies may sometimes simplify the verification process in Fig. 4.1 to save time and costs. Many engineers even skip the first and second stages and move directly to the third stage of real hardware testing. However, starting with physical hardware for algorithm validation may risk hardware damage or even pose a threat to personal safety.

The purpose of the first three chapters of this book is to help readers establish the first-stage computer-based simulation tools for AC motor control. Although major companies such as dSPACE and NI offer integrated platforms for the second and

Supplementary Information The online version contains supplementary material available at https://doi.org/10.1007/978-981-95-3261-2_4.

C-C. Yeh, *AC Motor Control Loop Design*, Power Systems,
https://doi.org/10.1007/978-981-95-3261-2_4

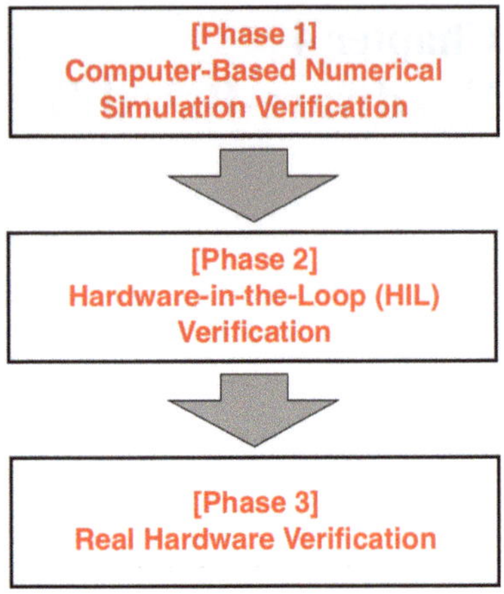

Fig. 4.1 Typical motor control algorithm verification flow

third stages of verification, their products are very expensive and not conducive to widespread learning. Therefore, this chapter will use the ODrive control platform [3] to connect to a real motor for verifying the control algorithm. Based on the author's actual testing, the ODrive platform offers good accessibility and usability. Within a reasonable scope, it should help readers learn how to validate the performance and feasibility of their designed motor control algorithms.

4.1 The Evolution of ODrive

ODrive was founded in 2017 by its current CEO, Oskar Weigl. In 2014, he completed his master's thesis on "Supercapacitor Energy Storage in Brushless Motor Applications," during which he built a prototype brushless motor driver that later became the first version of ODrive.

After completing his thesis, he developed a more advanced version of ODrive and announced it as an open-source project in 2016, aiming to make building robots easier and more affordable. As ODrive continued to develop and improve, version v3.6 gained widespread popularity.

Oskar Weigl has over 10 years of industry experience, having worked at ARM, ABB Robotics, and Rapyuta Robotics. He saw the opportunity to develop motor drivers for industrial and commercial applications and believed in the potential of widely available consumer-grade brushless motors and sensors. As a result, he began developing ODrive full-time and officially founded the ODrive company in 2017. Since then, ODrive Pro has been developed and released. ODrive S1 was created as a companion project, serving as a replacement for the now-obsolete ODrive v3.6. It delivers high-performance motor control at a lower cost than the ODrive Pro, providing a more powerful and cost-effective servo motor solution for the market.

Today, the company continues to develop new products and solutions, bringing high-performance motor control to robotics and industrial applications. The company's mission is to make high-performance robotic and industrial motion solutions as accessible as possible—enabling diverse, high-performance, and low-cost robotics and automation by maximizing performance at any given price point, while building flexible products that are easy to integrate. ODrive's core goal is to make it as easy as possible for people to build new robots and solutions without having to worry about motor control—hence the motto: "Effortless Robotic Motion."

You can find open-source resources related to ODrive v3.6 at the following website.

https://github.com/odriverobotics/ODrive.

4.2 ODrive System Configuration and Setup [3]

The ODrive control platform consists of the following components:

- **ODrive Control Board (ODrive S1 used in this chapter)**:

The ODrive control board integrates the control circuit's processing unit (MCU) and the three-phase inverter module onto a single PCB. It is very compact in design, measuring only 66 × 50 mm.

- **ODrive GUI Interface (Web-based, no additional software installation required)**:

When the ODrive board is connected to a computer via USB, no additional software needs to be installed. The system will automatically prompt a connection message. If the user chooses to connect, it will open a web-based ODrive GUI interface via browser at https://gui.odriverobotics.com/. Through this interface, users can configure, set up, operate, and monitor the ODrive board online.

- **ODrive Firmware (Firmware version 0.6.7 used in this chapter)**:

The early version of ODrive (ODrive 3.6) is open-source and available on GitHub at https://github.com/odriverobotics/ODrive. However, ODrive S1 and ODrive PRO are not open-source, although their firmware is developed based on ODrive 3.6. Those interested in deeper firmware development can still work with the ODrive 3.6 platform.

The ODrive S1 control board used in this chapter is shown in Fig. 4.2. Its specifications are as follows:

- **1600W Continuous Output**: Supports 12–50 V input voltage, including 12S battery configurations, regenerative braking, or brake chopper.
- **Brake Chopper**: Uses an external braking resistor to safely handle up to 2000W of peak regenerative power.

- **Precision Control**: Offers torque, speed, position, and trajectory control with model-based feedforward.
- **Encoder Support**: Supports RS485, SPI, and onboard absolute encoders, allowing instant cold start. Dual encoder support. Filtering for incremental and Hall feedback.
- **Isolated I/O**: Electrically isolated UART, step/direction, and GPIO.
- **CAN Bus**: Designed for multi-axis CAN networks with daisy-chainable connectors.
- **User Interface**: Web GUI and Python tools for easy configuration.
- **Easy Integration**: Python, Arduino, CAN, and ROS2 libraries are available.
- **Compact Design**: Measures only 66 × 50 mm with integrated locking connectors.
- **Control Specifications**: 24 kHz PWM frequency and 8 kHz control loop frequency.
- **Motor Support**: Supports permanent magnet synchronous motors (PMSM) and induction motors.
- **Control Modes**: Supports torque, speed, position, and trajectory control modes, including sensorless control.

Table 4.1 shows the list of materials purchased by the author from the official ODrive website (https://shop.odriverobotics.com/). After assembling the ODrive S1 control board, the heat spreader plate, and the IO harness, the motor and power supply must be connected to the **Power Pads** on the ODrive S1 control board. The pin definitions are shown in Fig. 4.5.

The **DC+** and **DC−** terminals must be connected to a DC power supply or battery (ensure the power supply is turned **off** during connection). The supplied voltage must not exceed the maximum allowable voltage for the ODrive (50 V). The **A**, **B**, and **C** terminals are connected to the three-phase motor (either a Permanent Magnet

Fig. 4.2 ODrive S1 appearance photo

Table 4.1 Bill of Materials (BOM) for purchasing from the official ODrive website

Item	Price
ODrive S1 control board	149 USD
Heat spreader plate for ODrive S1 (as shown in Fig. 4.3)	12 USD
IO harness for ODrive S1 (as shown in Fig. 4.4)	9 USD

Fig. 4.3 Heat spreader plate for ODrive S1

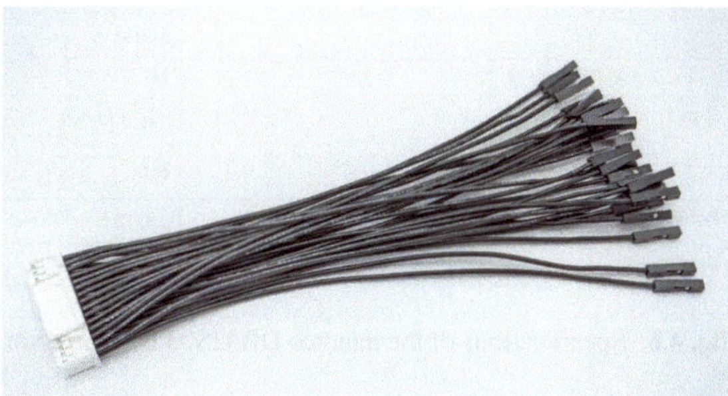

Fig. 4.4 IO harness for ODrive S1

Synchronous Motor or an Induction Motor). Finally, **R+** and **R−** should be connected to a braking resistor (in this example, a 50W 2Ω braking resistor is used).

In this chapter, we will use a three-phase SPM Permanent Magnet Synchronous Motor with a rated power of **26W** and a rated voltage of **24V** for testing. The motor model is **Nanotec DB42S03**, and its specifications are shown in Fig. 4.6

This chapter will also use a hysteresis brake as the load device (Model: Mobac HB-50 M-2). Although no load is applied during the test, it is used to alter the motor's moment of inertia. Its specifications are shown in Fig. 4.7.

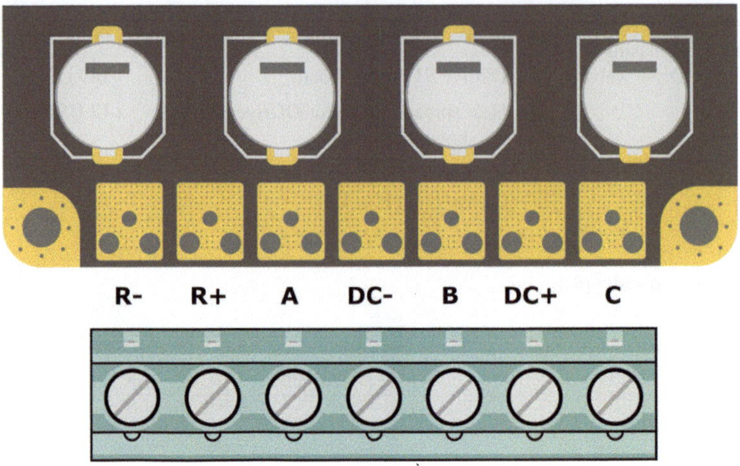

Fig. 4.5 Power pads pin definitions of ODrive S1

No. of Pol./Phases	8/3		
Voltage Rated (VDC)	24		
Current (AMP)	No load [A]	Rated [A]	Peak [A]
	0.2	1.79	5.4
Resistance / phase to phase [Ohms]@ 25°C	1.5 ± 15%		
Inductance / phase to phase [mH] @1kHz	2.1 ± 20%		
Torque Rated / Peak	Constant [Nm/A]	Rated [Nm]	Peak [Nm]
	0.035	0.0625	0.19
Power Rated [W]	26		
Speed	Rated [RPM]	No Load [RPM]	
	4000	6200	
Rotor Inertia [Kg-m^2]	2.4x10^{-6}		
Weight [Kg]	0.3		

Fig. 4.6 Specifications of the nanotec DB42S03 BLDC motor

The small dynamometer platform used in this chapter is the MOTIX Motor Bench, which includes both the motor and the hysteresis brake, as shown in Fig. 4.8. (Note: The MOTIX Motor Bench integrates the motor, brake, and base into a single unit. If you are interested in this platform, it is available for purchase online.)

If you have already connected the ODrive S1 to the power supply and motor (note: the power should remain **off** at this stage), the next step is to use a USB-A to USB Type-C cable to connect the ODrive S1 controller board to your computer. However, it is **especially important** to ensure that a **USB isolator** is used as a bridge between the computer and the ODrive S1. Without isolation, a ground loop may form due to

Torque at working current [Nm]	0.38	
Working current [mA]	270	
Resistance at 25°C ± 10% [Ohm]	95	
DC Voltage [V]	24	
Rpm max. 25°C ± 10% [min^{-1}]	15000	
Power dissipation [Watt]	Peak	Continuous
	90	23
Residual torque without current [Nm]	1.55x10^{-3}	
Rotor Inertia [Kgcm^2]	0.1670	
Weight [Kg]	0.755	

Fig. 4.7 Specfication of Mobac HB-50 M-2

Fig. 4.8 MOTIX motor bench photo

the potential difference between the computer and the ODrive S1 when the motor is running, which could damage your computer.

- **Use the ODrive GUI to perform system configuration and parameter autotuning**

Step 1:

Once the ODrive S1 is connected to the computer via a USB isolator, and since the three-phase permanent magnet synchronous motor used in this setup has a rated voltage of 24 V, turn on the power supply and adjust the voltage to 24 V. At this point, a connection message like the one shown in Fig. 4.9 should appear on the computer screen. Click the message window with your mouse to open a web browser and connect to https://gui.odriverobotics.com/configuration, as shown in Fig. 4.10.

Step 2:

After entering the configuration interface shown in Fig. 4.10, you need to configure the parameters in sequence: **power supply, motor, encoder, control mode**, etc. First, set up the power supply. The author has configured the parameters as follows:

Fig. 4.9 Notification from chrome browser detecting ODrive S1

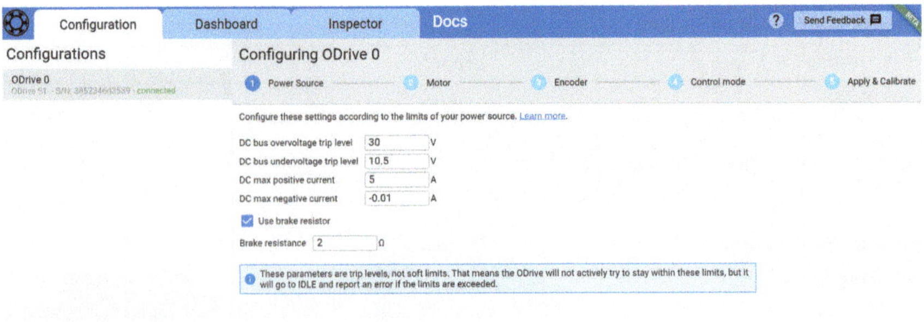

Fig. 4.10 ODrive web configuration interface

- **DC bus overvoltage trip level**: 30 V
- **DC bus undervoltage trip level**: 20 V
- **DC max positive current**: 5 A *(refer to your power supply specifications)*
- **DC max negative current**: – 0.01 A *(refer to your power supply specifications)*
- **Brake resistor enabled**, set brake resistance to **2 Ω** *(refer to the specifications of the brake resistor you are using)*

Step 3:

Next, configure the motor parameters as shown in Fig. 4.11. Since the motor used is not an official ODrive model, please select **"Other Motor"**, and set the motor parameters according to the specifications in Fig. 4.6 as follows:

- **Type**: High current
- **Pole pairs**: 4
- **KV**: 272.8 (rpm/V)

Note: The motor's torque constant KT *is 0.035, so the speed constant* KV=(1/0.035)*(30/π)=272.8 (rpm/V). *For more information about PMSM motor specifications, refer to* Sect. 5.5 *of this book.*

- **Current limit**: 5.4 A

Note: Set to the motor's peak current.

- **Motor calib. current**: 0.5 A

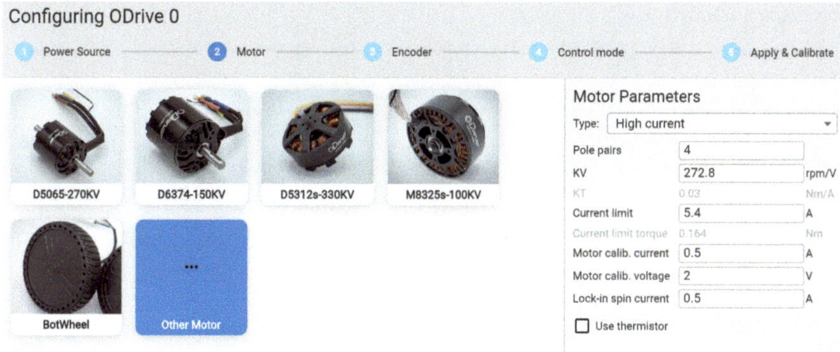

Fig. 4.11 ODrive motor parameter configuration interface

- **Motor calib. voltage**: 2 V
- **Lock-in spin current**: 0.5 A

Step 4:

Next, configure the encoder parameters. The encoder used is the **WEDS5541-B14**, which is an optical encoder with **1000 CPR** (counts per revolution). Since ODrive uses both the rising and falling edges of the encoder pulses on both channels to achieve **4 × counting**, the **Resolution** should be set to **4 times the encoder's CPR**, i.e., **4000 CPR**, as shown in Fig. 4.12.

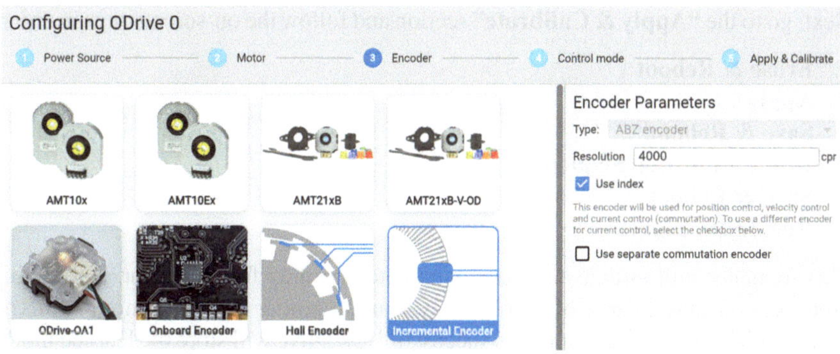

Fig. 4.12 ODrive encoder parameter configuration interface

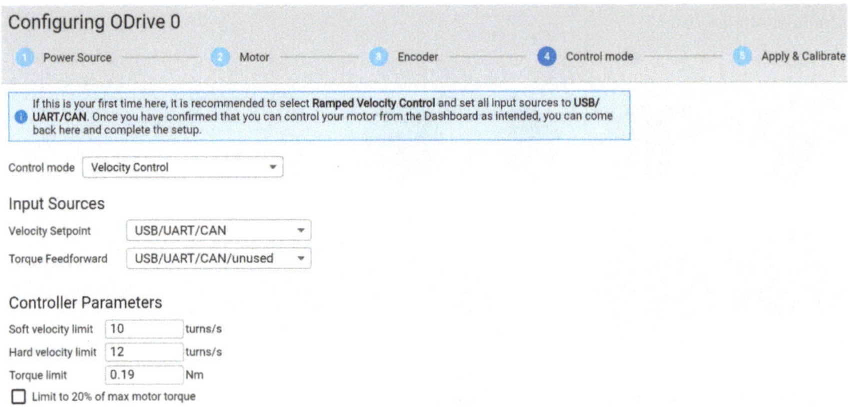

Fig. 4.13 ODrive control mode parameter configuration interface

Step 5:

Next, configure the control mode by setting the **Control Mode** to **Velocity Control**, and set the controller limit parameters as follows:

- Soft velocity limit: 10 (turns/s)
- Hard velocity limit: 12 (turns/s)
- Torque limit: 0.19 (Nm) *(Note: Set to the motor's maximum torque)* (Fig. 4.13)

Step 6:

Next, go to the **"Apply & Calibrate"** section and follow the on-screen steps in order:

1. **Erase & Reboot**
2. **Apply**
3. **Save & Reboot**
4. **Run Calibration Sequence**
5. **Save & Reboot**
6. **Run Index Search**

This sequence will initiate automatic parameter learning for the motor. During this process, you may hear a high-frequency sound coming from the motor—this is normal. If the procedure completes successfully, ODrive will store the learned motor parameters along with the configured settings into the MCU's non-volatile memory.

Step 7:

After completing the system setup, go to the **"Inspector"** tab. On the left side of the "Inspector" page, you'll see a list of all available ODrive parameters. Use the search bar on the left to locate specific parameters. Drag the following parameters into the center **"Controls"** area. Once added, the interface should resemble Fig. 4.14:

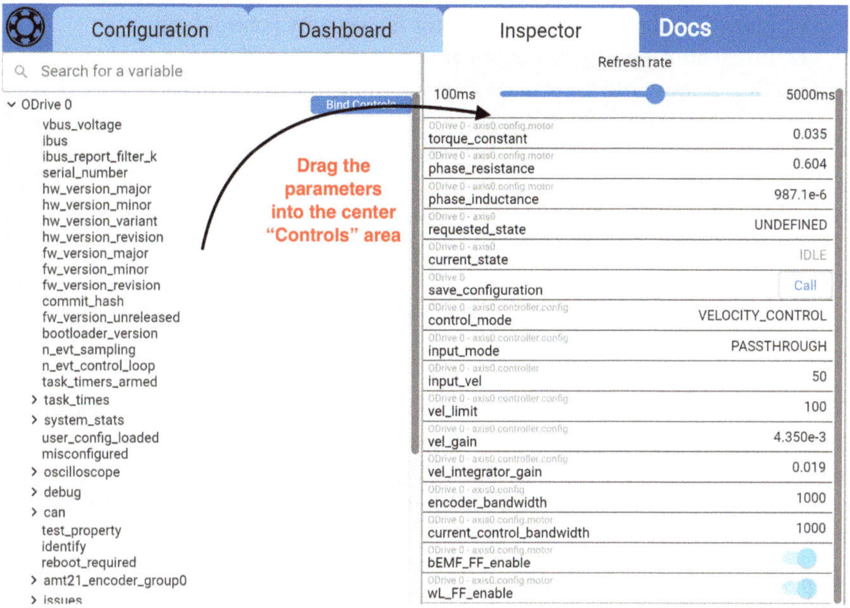

Fig. 4.14 ODrive control parameter configuration interface

- **torque_constant**: The motor torque constant K_T, automatically calculated by the system as **0.035 (Nm/A)**, as shown in Fig. 4.14.
- **phase_resistance**: The motor phase resistance measured during motor parameter calibration. The value is **0.604 (ohms)**, which is close to the motor's datasheet value of 0.75 ohms. *(Note: The datasheet only lists the phase-to-phase resistance, which is twice the phase resistance. Thus,* 1.5/2 = 0.75 ohms.*)*
- **phase_inductance**: The motor phase inductance measured by ODrive during calibration. Since the motor is an SPM type, we assume $L_d = L_q = L_s$. The value is **987.1 (μH)**, which is close to the datasheet value of 1.05 (mH). *(Note: The datasheet specifies phase-to-phase inductance, so phase inductance is* 2.1 / 2 = 1.05 (mH).*)*

The following parameters are used to control and monitor ODrive's operation:

- **requested_state**: Sets the desired operating state of ODrive.
- **current_state**: Displays ODrive's current operating state.
- **save_configuration**: Saves the current configuration to ODrive's non-volatile memory and reboots the device.
- **control_mode**: Sets the control mode; set this to **VELOCITY_CONTROL**.
- **input_mode**: Sets the input mode for the control loop; set this to **PASSTHROUGH**, which enables step command input.
- **input_vel**: Sets the target velocity for the motor.
- **vel_limit**: Sets the maximum allowable motor velocity; set this to **100**.

- **vel_gain**: Sets the proportional gain for the velocity control loop.
- **vel_integrator_gain**: Sets the integrator gain for the velocity control loop.
- **encoder_bandwidth**: Sets the bandwidth of the Luenberger observer used by the encoder; default is **1000 rad/s.**
- **current_control_bandwidth**: Sets the bandwidth of the current control loop; default is **1000 rad/s.**
- **bEMF_FF_enable**: Enables feedforward compensation for the d-axis current loop decoupling; default is **off**, please **enable** this.
- **wL_FF_enable**: Enables feedforward compensation for the q-axis current loop decoupling; default is **off**, please **enable** this.

Step 8:

Since we need to monitor the motor speed command and the speed response in real-time, please drag the **vel** parameter (under ODrive0.axis0.pos_vel_mapper, representing the motor's feedback speed) and **input_vel** (representing the motor's speed command) into the **plotting area on the right**. Once completed, the setup should look like Fig. 4.15.

Step 9:

After completing the above steps, click the **"Call"** button to the right of **save_configuration** to save the parameters and reboot the ODrive.

Step 10:

Before performing closed-loop speed control, ODrive requires an encoder calibration each time it is rebooted. Therefore, make sure the motor shaft is not connected to any load. Then, set the **requested_state** parameter to **ENCODER_OFFSET_CALIBRATION**. The motor shaft will rotate in both forward and reverse directions. If the calibration completes successfully, a success message will be displayed in the lower-left corner of the screen (Fig. 4.16).

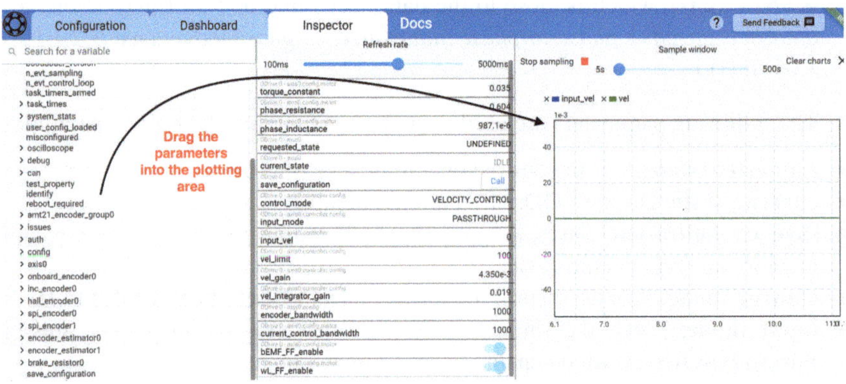

Fig. 4.15 Drag the desired control parameters to the plotting area for viewing

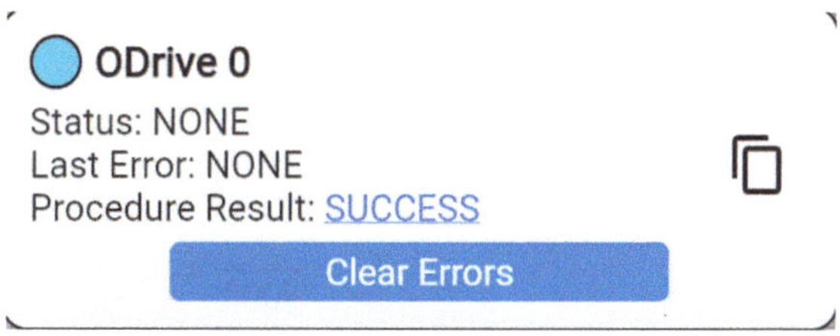

Fig. 4.16 Calibration successful message window

That completes all calibration procedures for the motor and encoder using ODrive. In the next section, we will use ODrive to test and verify the control loop design results from Chap. 2.

4.3 Validating Control Loops with ODrive

4.3.1 Verification Using Different PI Controller Parameters

Step 1:

Next, we will begin using the ODrive to test and verify the control loop design results from Chap. 2. First, let's try to illustrate the speed control loop architecture of the ODrive, as shown in Fig. 4.17.

Figure 4.17 is quite similar to Fig. 2.36. The main difference lies in the unit used for speed in ODrive, which is r/s (rounds per second). Therefore, in Fig. 4.17, we add a feedback gain of $1/(2\pi)$ to convert the motor speed unit from rad/s to r/s.

Regarding the current loop configuration, ODrive internally compensates for the nonlinear coupling terms of the current, and it only allows users to set the current loop bandwidth. Based on the set bandwidth, ODrive automatically calculates the corresponding PI controller parameters for the current loop. In our model, we use a first-order low-pass filter to approximate ODrive's current loop behavior. Since the default current loop bandwidth in ODrive is 1000 (rad/s), this implies that $\omega_q = 1000$ (rad/s) in Fig. 4.17. As a result, the delay caused by the current loop is $T_{curr} = \frac{1}{\omega_q} = 1e-03$.

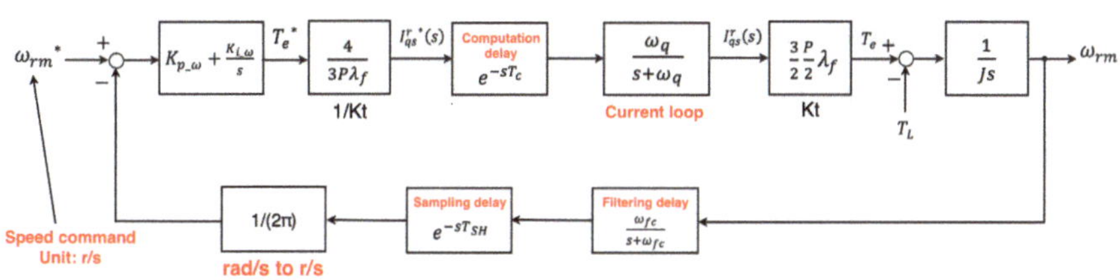

Fig. 4.17 ODrive speed loop control block diagram

Step 2:

Since all control loops in ODrive—position, speed, and current—operate under the same 8 kHz interrupt frequency, we have:

$$T_s = \frac{1}{8000} = 1.25e - 04$$

Therefore, the computation delay can be equivalently modeled as:

$$T_c = T_s = 1.25e - 04.$$

Step 3:

For calculating the feedback speed, ODrive does **not** directly compute the speed from encoder pulses. Instead, it uses a **Luenberger observer** that estimates the speed in real time by processing the position signal. This method effectively **eliminates the delay** introduced by traditional feedback low-pass filters. However, the Luenberger observer still introduces **phase delay**.

ODrive allows configuration of the observer's bandwidth through the encoder_bandwidth parameter, with a default value of **1000 rad/s**. For analysis purposes, the Luenberger observer can be **modeled as a first-order low-pass filter**, representing the **filter delay** block in Fig. 4.17. Thus:

$$\omega_{fc} = 1000(\text{rad/s}).$$

Additionally, since the Luenberger observer runs in the same **8 kHz interrupt** loop as the other control tasks, it still samples the position signal, introducing a **sampling delay**. This delay is modeled as:

$$T_{SH} = T_s/2 = 6.25e - 05\ (\text{s})$$

Step 4:

Next, we apply the **"Symmetrical Optimum Method"** (as described in Sect. 2.3.2) [4] to design the PI controller parameters for the speed control loop. The parameters for this experiment are as follows:

- **Motor moment of inertia**:

  $J = 2.4 \times 10^{\ 6}$ (kg-m^2) (as shown in Fig. 4.6)

- **Total delay time**:

$$T_{d\_total} = T_{curr} + T_c + T_{SH} = 0.0012\ (\text{s})$$

- **Feedback gain difference**:

Comparing the control loop in Fig. 4.7 with Fig. 2.29, ODrive includes an **additional feedback gain** of $1/(2\pi)$.

To compensate for this, the calculated PI controller parameters must be **multiplied by** 2π.

- **Tuning factor**:

Let's choose $\alpha = 4$. Using Eq. (2.3.19) from Sect. 2.3.2 and multiplying by 2π, we get:

Proportional gain:

$$K_{p_\omega} = \frac{J}{\alpha T_{d_{total}}} = \frac{2.4 \times 10^{-6}}{4 \times 0.0012} \times 2\pi = 0.0031$$

Integral gain:

$$K_{i\_\omega} = \frac{K_{p\_\omega}}{\alpha^2 T_{d\_total}} \times 2\pi = 0.1634$$

- **Implementation**:

These parameters are then entered into **SIMULINK**, and the control block is constructed as shown in Fig. 4.18. Upon running the simulation, the **speed response waveform** is shown in Fig. 4.19.

(**Note**: Typically, the Symmetrical Optimum method includes the **filter delay** when calculating total delay time. In this case, if filter delay were included, $T_{d\_total} = 0.0022$ (s). However, this would lead to a more oscillatory transient response, which is less ideal for comparison purposes.

Therefore, **filtering delay is omitted** in this particular PI design to produce a clearer and smoother response.)

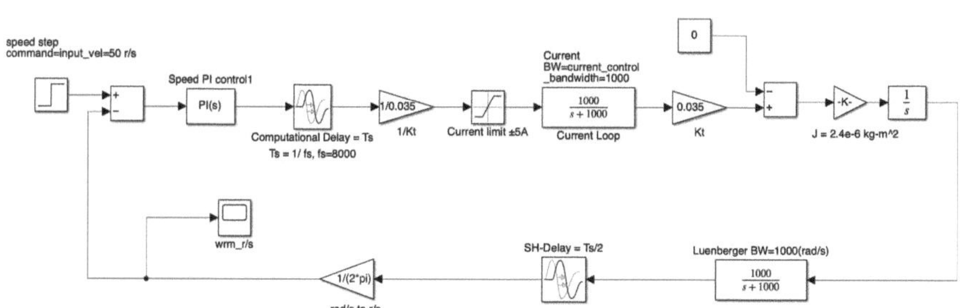

Fig. 4.18 ODrive speed loop SIMULINK simulation, example model: mdl_odrive_speed_control_1.slx

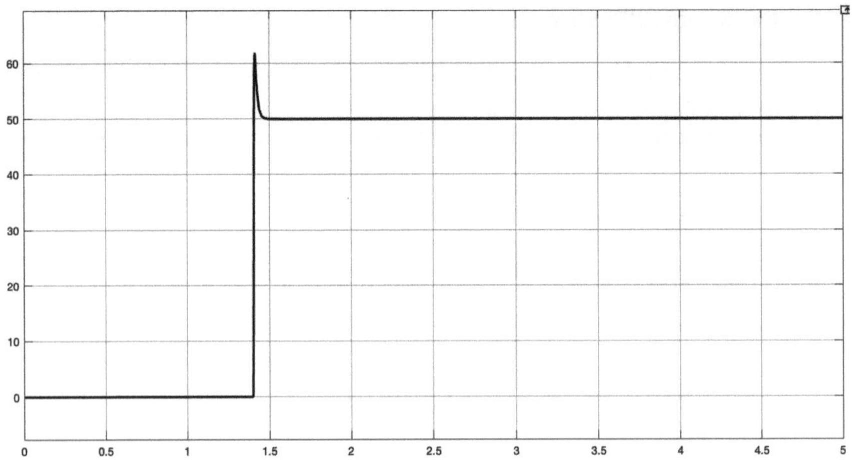

Fig. 4.19 Simulated speed step response

Step 5:

Next, we will input the PI controller parameters designed in **Step 4** into ODrive. Please go to the **ODrive GUI** and configure the following parameters:

- **vel_gain**: 0.0031
- **vel_integrator_gain**: 0.1634

After setting the parameters, set requested_state to CLOSED_LOOP_CONTROL. At this point, ODrive will enter **closed-loop control mode**. Then, set the speed command parameter input_vel to **50 (r/s)**. This will apply a speed step command to the motor, and you should observe the motor's step response output as shown in Fig. 4.20.

Step 6:

As you can see, the waveform in Fig. 4.20 closely matches the computer simulation result shown in Fig. 4.19, indicating that the control loop constructed in our computer simulation closely reflects the behavior of the actual motor control system.

Step 7:

We can further verify the accuracy of the computer-simulated control loop by testing with another set of PI controller parameters. The PI controller parameters are as follows:

$$K_{p_\omega} = 4.9338\text{e} - 04$$

$$K_{i\_\omega} = 0.0260$$

The computer simulation result using this set of parameters is shown in Fig. 4.21.

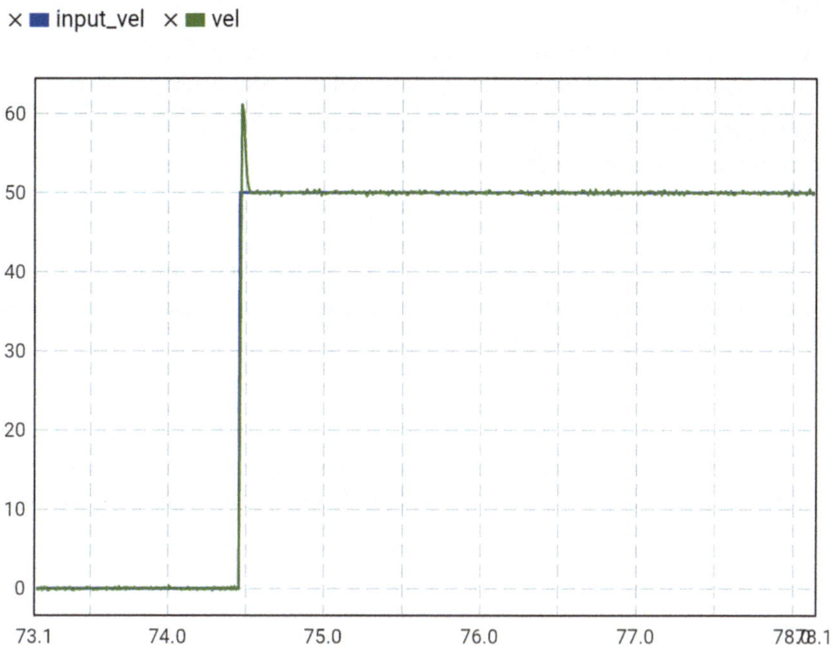

Fig. 4.20 Actual speed step response of the ODrive system

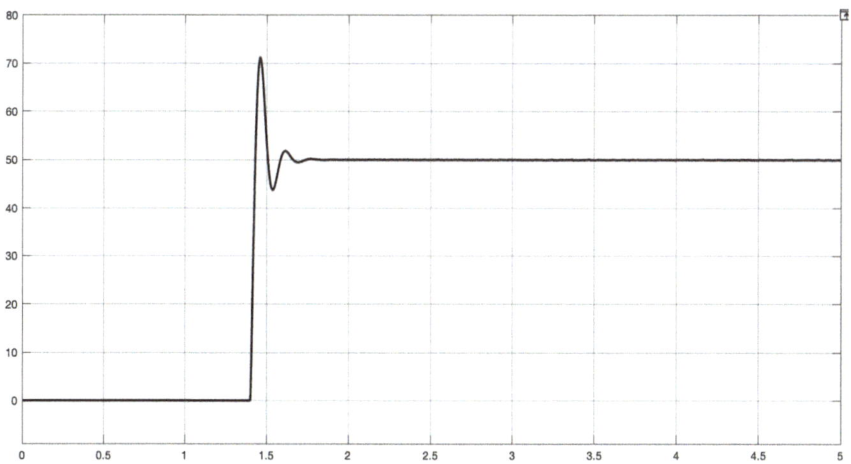

Fig. 4.21 Simulated speed step response

Step 8:

Next, we will input the PI controller parameters designed in STEP 7 into ODrive. Please navigate to the ODrive GUI and set the ODrive parameters as follows:

- **vel_gain**: 4.9338e-4
- **vel_integrator_gain**: 0.026

After completing the settings, set the parameter requested_state to CLOSED_LOOP_CONTROL. At this point, ODrive will enter closed-loop control mode. Then, set the speed command parameter input_vel to **50 (r/s)**. This will cause ODrive to apply a step speed command to the motor, and you will obtain the motor speed step response waveform as shown in Fig. 4.22.

Step 9:

As you can see, the waveform in Fig. 4.22 closely matches the computer simulation result shown in Fig. 4.21. Through the above verification process, we can confirm that the computer-simulated control system we built closely resembles a real motor control system.

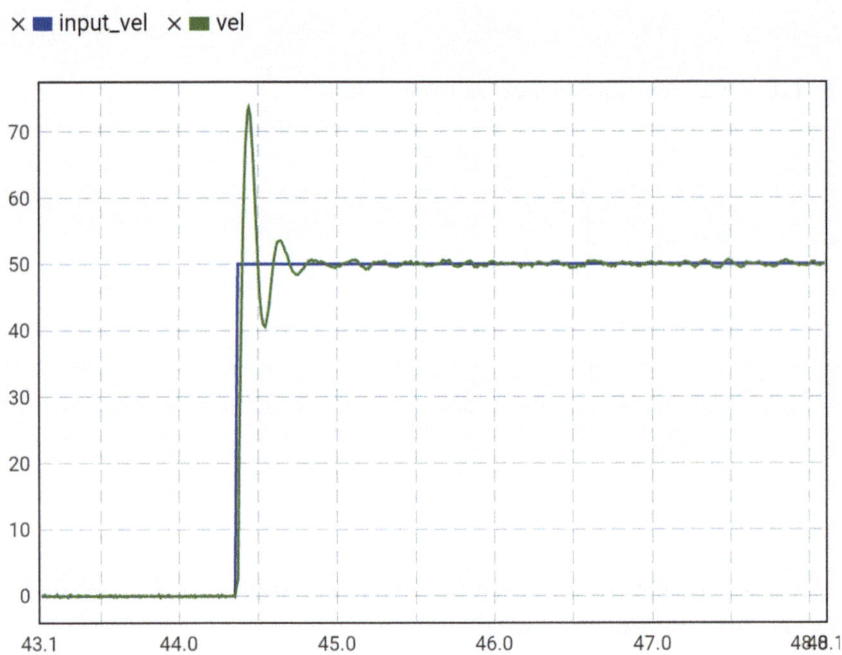

Fig. 4.22 Actual speed step response of the ODrive system

Additionally, from the ODrive operation, it is evident that the **PI controller parameters in ODrive's velocity loop**—namely vel_gain and vel_integrator_gain—correspond directly to the **S-domain PI controller parameters**, K_{p_ω} and K_{i_ω}, respectively.

The ODrive firmware internally converts these **Laplace domain PI controller parameters** into their **Z-domain equivalents** based on the system's sampling time and the discretization method used. For more information on how to convert PI controller parameters from the S-domain to the Z-domain, please refer to **Appendix A.4** of this book.

4.3.2 Influence of Mechanical Inertia on System Response

Step 1:

Next, we will try **modifying the motor's inertia** and use the **first set of PI controller parameters in** 4.3.1 for validation.

First, connect the **motor shaft** to the **brake shaft** using a mechanical coupling (make sure the coupling is securely tightened), as shown in Fig. 4.23.

Before connecting the brake, the **motor inertia** is 2.4×10^{-6} (kg-m^2), while the **brake inertia** is 1.67×10^{-5} (kg-m^2). Clearly, the brake's inertia is about **7 times greater** than that of the motor.

For this experiment, we will assume that the **total mechanical inertia** after coupling is approximately equal to the brake's inertia, that is:

$$\mathrm{J} = 1.67 \times 10^{-5}\left(\mathrm{kg} - \mathrm{m}^2\right).$$

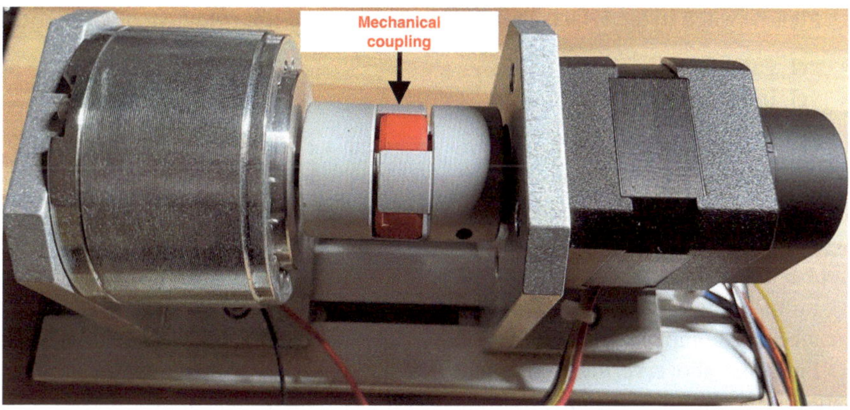

Fig. 4.23 Connect the motor and the brake using a mechanical coupling

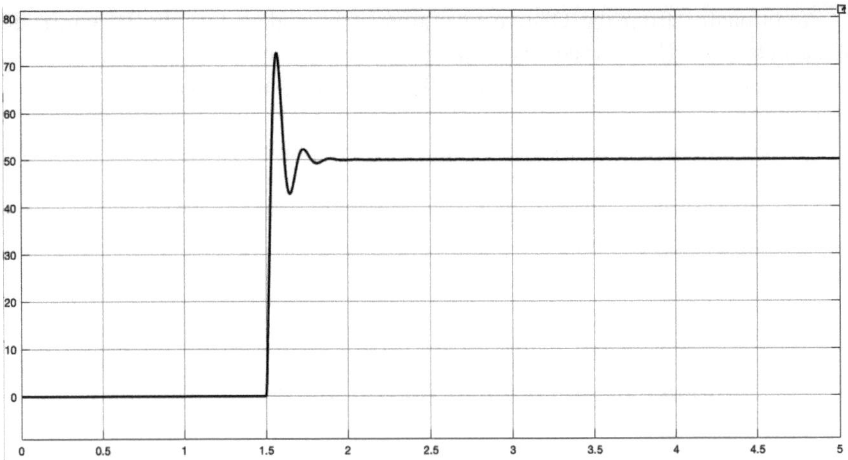

Fig. 4.24 Simulated speed step response

Step 2:

Input the new mechanical inertia $J = 1.67 \times 10^{-5}$ (kg-m^2) into the SIMULINK model shown in Fig. 4.18. (example file: mdl_odrive_speed_control_2.slx).

Then, apply the first set of PI controller parameters from Sect. 4.3.1:

- $K_{p_\omega} = 0.0031$
- $K_{i\_\omega} = 0.1634$

Keep all other parameters unchanged.

Run the simulation, and the resulting waveform will be as shown in Fig. 4.24.

Step 3:

Next, we will input the PI controller parameters from **STEP 2** into ODrive.

Please switch to the **ODrive GUI** and configure the following parameters:

- **vel_gain**: 0.0031
- **vel_integrator_gain**: 0.1634

Once configured, set the parameter requested_state to CLOSED_LOOP_CONTROL. ODrive will now enter closed-loop control mode.

Then, set the speed command parameter input_vel to 50 (r/s). At this point, ODrive will apply a step command to the motor. The motor's step response output should appear as shown in Fig. 4.25.

From the waveform, we can see that the actual motor speed response closely matches the computer simulation result.

From the above verification results, we can conclude that the computer simulation model we developed demonstrates strong predictive capability. This reflects the essence of *simulation*: simulation results should exhibit a reasonable degree of agreement with real-world behavior.

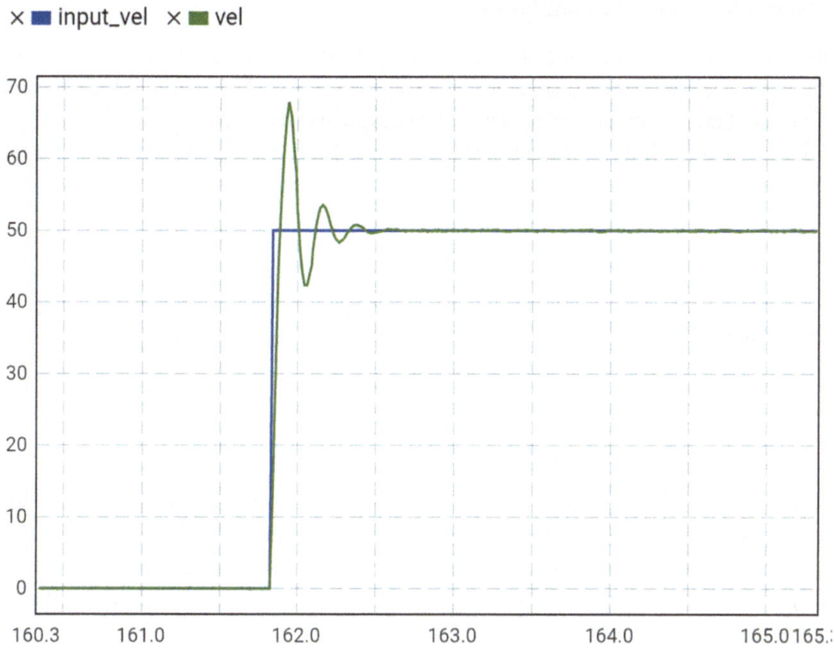

Fig. 4.25 Actual speed step response of the ODrive system

Although the simulated waveforms closely match the actual measurements, there are still minor discrepancies. The possible reasons for these discrepancies are summarized as follows:

- **Luenberger Observer Approximation**:

ODrive uses a Luenberger observer to estimate speed. Theoretically, this observer could be a second-order system, which would introduce greater phase lag. Since the official documentation does not specify its exact implementation, the author approximated it as a first-order low-pass filter in the simulation. This simplification may have introduced error in the simulation results.

- **Current Loop Approximation**:

In the simulation, the ODrive current loop is modeled as a first-order low-pass filter. However, the actual current loop implementation in ODrive may be of a higher order. If it is in fact a second-order filter, this could affect the fidelity of the simulation model.

- **Moment of Inertia Assumption**:

The discrepancy between Fig. 4.10 and the simulation may stem from the assumption about total mechanical inertia. In the simulation, only the brake's inertia was used to approximate the combined system inertia after coupling. In reality, the motor's shaft and coupling add some inertia, introducing a source of mismatch.

- **Neglecting Noise in Simulation**:

The simulation model does not include noise, whereas a real motor control system inevitably experiences electrical and measurement noise. This unmodeled effect can also contribute to deviations between simulation and reality.

These factors highlight the importance of careful modeling and approximation in simulation work and also explain the small but noticeable differences between the simulated and experimental responses.

4.4 Summary

The key points of this chapter are summarized as follows:

- **Validation of Simulation Tools**:

While the first three chapters of this book aim to help readers build a simulation tool for AC motor control, this chapter demonstrates the *effectiveness and validity* of the simulation tool through experimental verification. Readers can thus use the methodologies introduced to build their own customized simulation tools, thereby increasing R&D productivity.

- **Purpose of Modeling**:

There will always be some degree of error between simulation and physical systems. The goal is not necessarily to eliminate all discrepancies, but rather to **understand and identify their sources**. This allows for the development of **accurate yet simple simulation models** that improve design efficiency. Overly complex models are more costly to develop and harder to use.

- **Limitations Due to Unknown Details in ODrive**:

Since the author does not have access to key design details of the ODrive current control loop—such as filter parameters for current feedback, loop controller gains, and PWM output delay—it's currently not possible to simulate the full ODrive motor driver system using the third chapter's Simulink model (mdl_pm_foc_plus_inv.slx).

However, if more design parameters were known, the simulation could be improved to more closely match real-world responses. For developers using MCUs to build their own motor control systems, these internal design details are fully accessible. Therefore, you can modify the Chap. 3 simulation model and combine it with the delay-aware loop design methods introduced in Chap. 2 to achieve highly realistic simulation results.

- **Powerful Software Ecosystem from ODrive**:

ODrive provides robust software tools (e.g., odrivetool) and extensive libraries for Python and Arduino. These tools support online tuning, secondary development, and integration with other robotic controllers via ROS and CAN. All tools are well-documented online. Interested readers are encouraged to explore ODrive's official resources for more information.

References

1. George Ellis, Control System Design Guide: Using Your Computer to Understand and Diagnose Feedback Controllers, Butterworth-Heinemann, 2016.
2. Chih-Chun Yeh, The Self-Training of IoT Experts, Taiwan: Drmaster. Press Co. LTD., 2023.
3. https://docs.odriverobotics.com/.
4. J. Bocker, S. Beineke, and A. Bahr, "On the control bandwidth of servo drives," in Proc. Eur. Conf. Power Electron. Appl., pp. 1–10, 2009.

Chapter 5
Practical Issues in Control Implementation

In real-world motor control systems, achieving robust and high-performance operation requires more than theoretical design—it demands careful attention to practical implementation challenges. This chapter addresses these issues by bridging control theory with engineering practice, providing readers with actionable techniques and insights.

The chapter begins with the per-unit system, a normalization method that enables fair comparison and analysis of motors with different ratings. It then introduces anti-integral windup strategies for PI controllers, ensuring stable operation under saturation conditions. The self-tuning of motor parameters section presents both electrical and mechanical parameter identification methods, including automated tuning for PMSMs and inertia estimation.

Subsequent sections examine PWM sampling techniques and the impact of delays on control performance, followed by an overview of PMSM specifications critical for system design. The chapter also covers the design of essential filters—Butterworth, notch, and digital filters—as well as a structured digital filter design workflow.

In the latter part, readers will learn practical methods for speed estimation using encoder pulse counts and the MT method, along with the theory and application of Luenberger observers for speed and disturbance estimation. The chapter concludes with sensorless PMSM control using sliding mode observers, demonstrating advanced observer-based techniques for modern drive systems.

By integrating these topics, This Chapter equips engineers and researchers with the tools needed to translate control concepts into reliable, real-world motor drive implementations.

Supplementary Information The online version contains supplementary material available at https://doi.org/10.1007/978-981-95-3261-2_5.

C-C. Yeh, *AC Motor Control Loop Design*, Power Systems,
https://doi.org/10.1007/978-981-95-3261-2_5

5.1 The Per-unit System

In practice, motor parameters are often given in MKS units (meter–kilogram–second), as these directly reflect measurable physical quantities. However, MKS-based values alone are not ideal for comparing the characteristics and performance of different motors, because the magnitudes of these parameters scale with motor size, rating, and base values.

For example, the stator resistance of a 2.2 kW, 440 V induction motor might be 3 Ω, whereas a 110 kW, 220 V motor might have a stator resistance of only 0.1 Ω. This large numerical difference does not imply that the smaller motor suffers thirty times higher copper losses [1]. Instead, it reflects the fact that each motor's parameters are tied to its own rated voltage, current, and power base.

By converting parameters to a **per-unit (p.u.)** representation—where quantities are normalized against appropriate base values—we can eliminate the influence of rating differences, enabling a fair and dimensionless comparison of motors of varying sizes and ratings.

Similarly, in systems composed of motors with **different power ratings and voltages**, relying solely on MKS parameters can lead to misunderstandings and confusion. In practice, this issue is commonly addressed by using **per-unit (pu) values** [1–3].

By selecting a **base power**, **base voltage**, and **base frequency**, motor parameters can be normalized—i.e., **expressed as ratios** relative to these base values. These per-unit parameters **are dimensionless** and represent how a quantity compares to the base value. With a common base, comparisons between different motors become **meaningful and consistent**, enabling more insightful analysis across varying systems.

For a three-phase motor, it is traditional to define the **base values** for per-unit (pu) normalization as follows:

- **Base power**: P_b (typically the motor's rated power)
- **Base phase voltage (rms)**: $V_{b,rms}$
- **Base phase current (rms)**: $I_{b,rms}$

These base quantities are related by the fundamental power equation for a three-phase system:

$$P_b = 3V_{b,rms}I_{b,rms} \tag{5.1.1}$$

(Note: For a 5-phase motor, Eq. (5.1.1) becomes $P_b = 5V_{b,rms}I_{b,rms}$).

However, from the perspective of field-oriented control (FOC), the physical quantities used in the **d-q reference frame** are not RMS values but **peak values**. Therefore, it may be more appropriate to use the **peak values** of voltage and current as the base values. Typically, the **phase voltage peak** V_b and **phase current peak** I_b are selected as the base values.

The relationship between peak and RMS values is as follows:

Note: Here, V_b and I_b specifically refer to the **peak** values of phase voltage and current.

$$V_{b,rms} = \frac{V_b}{\sqrt{2}} \tag{5.1.2}$$

$$I_{b,rms} = \frac{I_b}{\sqrt{2}} \tag{5.1.3}$$

Therefore, Eq. (5.1.1) can be expressed as:

$$P_b = \frac{3}{2} V_b I_b \tag{5.1.4}$$

The base value of impedance is:

$$Z_b = \frac{V_b}{I_b} \tag{5.1.5}$$

The base value of torque is:

$$T_b = \frac{P_b}{(2/P)\omega_b} \tag{5.1.6}$$

The base angular frequency of the motor, denoted as ω_b,, is typically calculated from the motor's rated speed. For example, if the motor is a 4-pole machine ($P = 4$) and the rated speed is 1800 (rpm), the base angular frequency is 377 (rad/s). The calculation is as follows:

$$1800 \times \frac{P}{2} \times \frac{2\pi}{60} = 377(\text{rad/s})$$

The base value of the motor flux linkage, denoted as λ_b, can be expressed as:

$$\lambda_b = \frac{V_b}{\omega_b} \tag{5.1.7}$$

5.1.1 *Per-unit Model of Permanent Magnet Synchronous Motor (PMSM) in dq Axes*

We will now use the dq-axis mathematical model of a permanent magnet synchronous motor (PMSM) as an example to convert a mathematical model in MKS units into a per-unit (p.u.) model.

Let us first express the dq-axis mathematical model of a PMSM in MKS units:

$$v_{ds}^r = R_s i_{ds}^r + L_d p i_{ds}^r - \omega_r L_q i_{qs}^r \tag{5.1.8}$$

$$v_{qs}^r = R_s i_{qs}^r + L_q p i_{qs}^r + \omega_r \left(L_d i_{ds}^r + \lambda_f \right) \tag{5.1.9}$$

$$T_e = \frac{3}{2}\frac{P}{2}\left[\lambda_f i_{qs}^r + \left(L_d - L_q \right) i_{ds}^r i_{qs}^r \right] \tag{5.1.10}$$

where $p = \frac{d}{dt}$.

Define $X_d = \omega_b L_d$ and $X_q = \omega_b L_q$. We first normalize Eq. (5.1.8) into per-unit form.

$$\frac{v_{ds}^r}{V_b} = \frac{R_s}{Z_b} \cdot \frac{i_{ds}^r}{I_b} + \frac{X_d}{\omega_b} \cdot p i_{ds}^r \cdot \frac{1}{I_b} \cdot \frac{1}{Z_b} - \frac{\omega_r}{\omega_b} \cdot \frac{X_q}{\omega_b} \cdot i_{qs}^r \cdot \frac{1}{I_b} \cdot \frac{\omega_b}{Z_b} \tag{5.1.11}$$

It can be expressed as:

$$\begin{aligned} \overline{v_{ds}^r} &= \overline{R_s} \cdot \overline{i_{ds}^r} + \frac{\overline{X_d}}{\omega_b} p \overline{i_{ds}^r} - \overline{\omega_r} \cdot \overline{X_q} \cdot \overline{i_{qs}^r} \\ &= \overline{R_s} \cdot \overline{i_{ds}^r} + \overline{L_d} p \overline{i_{ds}^r} - \overline{\omega_r} \cdot \overline{X_q} \cdot \overline{i_{qs}^r} \end{aligned} \tag{5.1.12}$$

where $\overline{v_{ds}^r} = \frac{v_{ds}^r}{V_b}$, $\overline{R_s} = \frac{R_s}{Z_b}$, $\overline{i_{ds}^r} = \frac{i_{ds}^r}{I_b}$, $\overline{X_d} = \frac{X_d}{Z_b}$, $\overline{\omega_r} = \frac{\omega_r}{\omega_b}$, $\overline{i_{qs}^r} = \frac{i_{qs}^r}{I_b}$, $\overline{X_q} = \frac{X_q}{Z_b}$ and $\overline{L_d} = \frac{\overline{X_d}}{\omega_b}$, – represents the per-unit (p.u.) value.

Equation (5.1.12) is the per-unit (p.u.) d-axis voltage equation of a permanent magnet synchronous motor (PMSM).

Using the same method, the per-unit q-axis voltage equation of a PMSM can be obtained as follows:

$$\begin{aligned} \overline{v_{qs}^r} &= \overline{R_s} \cdot \overline{i_{qs}^r} + \frac{\overline{X_q}}{\omega_b} p \overline{i_{qs}^r} + \overline{\omega_r} \cdot \overline{X_d} \cdot \overline{i_{ds}^r} + \overline{\omega_r} \cdot \overline{\lambda_f} \\ &= \overline{R_s} \cdot \overline{i_{qs}^r} + \overline{L_q} p \overline{i_{qs}^r} + \overline{\omega_r} \cdot \overline{X_d} \cdot \overline{i_{ds}^r} + \overline{\omega_r} \cdot \overline{\lambda_f} \end{aligned} \tag{5.1.13}$$

where $\overline{v_{qs}^r} = \frac{v_{qs}^r}{V_b}$, $\overline{i_{qs}^r} = \frac{i_{qs}^r}{I_b}$, $\overline{X_q} = \frac{X_q}{Z_b}$, $\overline{\lambda_f} = \frac{\lambda_f}{\lambda_b}$ and $\overline{L_q} = \frac{\overline{X_q}}{\omega_b}$.

Next, we normalize the torque equation from Eq. (5.1.10) into per-unit form.

$$\overline{T_e} = \overline{\lambda_f} \cdot \overline{i_{qs}^r} + \left(\overline{X_d} - \overline{X_q} \right) \cdot \overline{i_{ds}^r} \cdot \overline{i_{qs}^r} \tag{5.1.14}$$

where $\overline{T_e} = \frac{T_e}{T_b}$.

Next, we normalize the motor mechanical equation. First, write the motor mechanical equation in MKS units as follows:

$$T_e = Jp\omega_{rm} + B\omega_{rm} + T_L \tag{5.1.15}$$

We normalize Eq. (5.1.15) as follows:

$$\frac{T_e}{T_b} = J\frac{\omega_b}{T_b} \cdot p\omega_{rm} \cdot \frac{1}{\omega_b} + B\frac{\omega_b}{T_b} \cdot \omega_{rm} \cdot \frac{1}{\omega_b} + \frac{T_L}{T_b} \tag{5.1.16}$$

The normalized motor mechanical equation can be expressed

$$\overline{T_e} = \overline{J}p\overline{\omega_{rm}} + \overline{B} \cdot \overline{\omega_{rm}} + \overline{T_L} \tag{5.1.17}$$

where $\overline{T_e} = \frac{T_e}{T_b}$, $\overline{J} = J\frac{\omega_b}{T_b}$, $\overline{B} = B\frac{\omega_b}{T_b}$ and $\overline{T_L} = \frac{T_L}{T_b}$.

Equations (5.1.12), (5.1.13), (5.1.14), and (5.1.17) represent the normalized mathematical model of the permanent magnet synchronous motor (PMSM).

5.1.2 *Normalized System Simulation of Permanent Magnet Synchronous Motor (PMSM)*

Step 1:

Next, we will use MATLAB/SIMULINK to verify the derived normalized model of the Permanent Magnet Synchronous Motor (PMSM). In this section, the simulation will be based on the PMSM parameters listed in Table 2.3, as shown in Table 5.1.

Table 5.1 PMSM parameters

Parameter	Symbol	Value	Unit
Stator resistance	R_s	1.2	Ω
Stator d-axis inductance	L_d	5.7	mH
Stator q-axis inductance	L_q	12.5	mH
Rotor flux linkage	λ_f	0.03	Wb
Number of poles	P	4	–
Moment of inertia	J	0.00016	kg m^2
Friction coefficient	B	0.0000028	N m s/rad
Bandwidth of d- and q-axis Current Control Loops	ω_d, ω_q	1000	rad/s
Speed Loop Bandwidth	ω_s	100	rad/s
Rated torque (not listed in Sect. 2.1.3)	P_b	890	W
Rated current (not listed in Sect. 2.1.3)	I_b	4	A (RMS)
Rated speed (not listed in Sect. 2.1.3)	$\omega_{rm,b}$	1800	rpm

MATLAB m-file Example Script: pm_params_pu.m

```
Rs = 1.2;
Ld = 0.0057;
Lq = 0.0125;
Lamda_f = 0.123;
pole = 4;
J = 0.00016;
B = 0.0000028;
Pb = 890;
Ibrms = 4;
wb = (2*pi*1800/60)*pole/2;
Ib = Ibrms*sqrt(2);
Vb = 2*Pb/(3*Ib);
Zb = Vb/Ib;
Tb = Pb/(wb*2/pole);
Lamdab = Vb/wb;
Rspu = Rs/Zb;
Xdpu = Ld*wb/Zb;
Xqpu = Lq*wb/Zb;
Lamda_fpu = Lamda_f/Lamdab;
Jpu = J*wb/Tb;
Bpu = B*wb/Tb;
```

Step 2:

Please open the example model pm_foc_models_allpu.slx. The Permanent Magnet Synchronous Motor (PMSM) Subsystem model in this example has been modified to the per-unit (normalized) model, as shown in Fig. 5.1.

Step 3:

Next, we need to redesign the PI controller parameters for both the current and speed control loops based on the per-unit (normalized) model. Starting with the **d-axis**

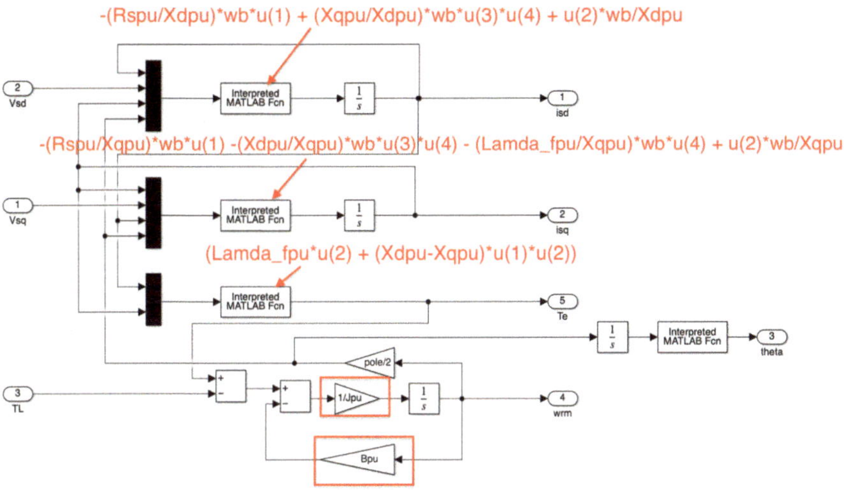

Fig. 5.1 Per-unit modeling of PMSM, example model: mdl_pm_foc_allpu.slx

current controller, and ignoring the nonlinear coupling terms, the per-unit normalized differential equation for the d-axis current of the PMSM, based on Eq. (5.1.12), can be expressed as:

$$\overline{v^r_{ds}} = \overline{R_s} \cdot \overline{i^r_{ds}} + \frac{\overline{X_d}}{\omega_b} p\overline{i^r_{ds}} \tag{5.1.18}$$

It can be expressed as

$$p\overline{i^r_{ds}} = -\frac{\omega_b}{\overline{X_d}}\overline{R_s} \cdot \overline{i^r_{ds}} + \frac{\omega_b}{\overline{X_d}}\overline{v^r_{ds}} \tag{5.1.19}$$

After loading the per-unit motor parameters ($\omega_b = 377$, $\overline{R_s} = 0.0647$, $\overline{X_d} = 0.1159$), we can obtain:

$$p\overline{i^r_{ds}} = -210.45 \cdot \overline{i^r_{ds}} + 3252.8 \cdot \overline{v^r_{ds}} \tag{5.1.20}$$

The per-unit transfer function of the d-axis current plant of the permanent magnet synchronous motor can be obtained as follows.

$$\frac{\overline{i^r_{ds}}}{\overline{v^r_{ds}}} = \frac{3252.8}{s + 210.52} \tag{5.1.21}$$

Assuming that the d-axis current loop is required to meet a bandwidth specification of $\omega_d = 1000$ (rad/s), the PI controller parameters for the d-axis current loop can be calculated using Eqs. (2.1.62) to (2.1.64), with $N = 3252.8$ and $D = 210.52$. The resulting PI controller parameters are as follows:

$$K_{p\_id} = \frac{\omega_d}{N} = 0.3074,\ K_{i\_id} = \frac{D \times \omega_d}{N} = 64.71 \tag{5.1.22}$$

Step 4:

Next, to calculate the PI controller parameters for the q-axis current loop, we start by considering the per-unit form of the q-axis current differential equation from the normalized model of the permanent magnet synchronous motor (PMSM). Ignoring the nonlinear coupling terms, the per-unit version of Eq. (5.1.13) can be written as:

$$\overline{v_{qs}^r} = \overline{R_s} \cdot \overline{i_{qs}^r} + \frac{\overline{X_q}}{\omega_b} p\overline{i_{qs}^r} \tag{5.1.23}$$

It can be expressed as

$$p\overline{i_{qs}^r} = -\frac{\omega_b}{\overline{X_q}}\overline{R_s} \cdot \overline{i_{qs}^r} + \frac{\omega_b}{\overline{X_q}}\overline{v_{qs}^r} \tag{5.1.24}$$

After loading the per-unit motor parameters ($\omega_b = 377$, $\overline{R_s} = 0.0647$, $\overline{X_q} = 0.2542$), we can obtain:

$$p\overline{i_{qs}^r} = -95.95 \cdot \overline{i_{qs}^r} + 1483 \cdot \overline{v_{qs}^r} \tag{5.1.25}$$

The per-unit transfer function of the q-axis current plant of the permanent magnet synchronous motor can be obtained as follows.

$$\frac{\overline{i_{qs}^r}}{\overline{v_{qs}^r}} = \frac{1483}{s + 95.95} \tag{5.1.26}$$

Assuming that the q-axis current loop is required to meet a bandwidth specification of $\omega_q = 1000$ (rad/s), the PI controller parameters for the q-axis current loop can be calculated using Eqs. (2.1.62) to (2.1.64), with $N = 1483$ and $D = 95.95$. The resulting PI controller parameters are as follows:

$$K_{p\_iq} = \frac{\omega_d}{N} = 0.6743,\ K_{i\_iq} = \frac{D \times \omega_d}{N} = 64.7 \tag{5.1.27}$$

Step 5:

Next, we proceed to design the PI controller for the speed loop. The per-unit (pu) form of the mechanical equation of the motor can be expressed as:

$$\overline{T_e} = \overline{J} p\overline{\omega_{rm}} + \overline{B} \cdot \overline{\omega_{rm}} + \overline{T_L} \tag{5.1.28}$$

where $\overline{T_e} = \frac{T_e}{T_b}$, $\overline{J} = J\frac{\omega_b}{T_b}$, $\overline{B} = B\frac{\omega_b}{T_b}$ and $\overline{T_L} = \frac{T_L}{T_b}$.

Assuming $\overline{T_L} = 0$, Eq. (5.1.28) can be rewritten as:

$$\overline{T_e} = \overline{J}p\overline{\omega_{rm}} + \overline{B} \cdot \overline{\omega_{rm}} \tag{5.1.29}$$

Taking the Laplace transform of Eq. (5.1.29) and substituting $\overline{J} = 0.0128$ and $\overline{B} = 2.2356e - 04$, we obtain:

$$\begin{aligned}\overline{\omega_{rm}}(s) &= \overline{T_e}'(s) \times \frac{1}{\overline{J}s + \overline{B}} = \overline{T_e}'(s) \times \frac{1}{0.0128s + 2.2356e - 04} \\ &= \overline{T_e}'(s) \times \frac{78.27}{s + 0.0175}\end{aligned} \tag{5.30}$$

Here, we design the speed loop with a target bandwidth of $\omega_s = 100$ (rad/s). Using Eqs. (2.1.84) to (2.1.86), and given $N = 78.27$ and $D = 0.0175$, the resulting PI controller parameters for the speed loop are as follows:

$$K_{p\_\omega} = \frac{\omega_s}{N} = \frac{100}{78.27} = 1.27 \tag{5.1.31}$$

$$K_{i\_\omega} = \frac{D \times \omega_s}{N} = \frac{0.0175 \times 100}{78.27} = 0.0224 \tag{5.1.32}$$

Step 6:

After completing the design of the current and speed PI controllers, we need to incorporate the nonlinear coupling terms of the d-axis and q-axis currents. According to Eq. (2.1.68), the output of the d-axis current PI controller is supplemented by a feedforward control term f_d before it is applied to the motor, as follows:

$$f_d = -L_d\left(\omega_r \frac{L_q}{L_d} i_{qs}^r\right) \tag{5.1.33}$$

After normalization, it can be expressed as:

$$f_{d,pu} = -\left(\frac{\omega_r}{\omega_b}\frac{X_q}{\omega_b}\frac{\omega_b}{Z_b}\frac{i_{qs}^r}{I_b}\right) = -\left(\overline{\omega_r}\overline{X_q}\overline{i_{qs}^r}\right) \tag{5.1.34}$$

where $X_q = \omega_b L_q$ and $\overline{X_q} = \frac{X_q}{Z_b}$.

Similarly, according to Eq. (2.1.78), the output of the q-axis current PI controller will have a feedforward control term f_q added before entering the motor, which can be expressed as:

$$f_q = L_q\left(\omega_r \frac{\left(L_d i_{ds}^r + \lambda_f\right)}{L_q}\right) \tag{5.1.35}$$

After normalization, it can be expressed as:

$$f_{q,pu} = \overline{\omega_r}\left(\overline{X_d} \cdot \overline{i_{ds}^r} + \overline{\lambda_f}\right) \tag{5.1.36}$$

where $X_d = \omega_b L_d$, $\overline{X_d} = \frac{X_d}{Z_b}$, $\overline{\lambda_f} = \frac{\lambda_f}{\lambda_b}$ and $\lambda_b = \frac{V_b}{\omega_b}$.

Step 7:

Let's revisit the entire Field-Oriented Control (FOC) loop for a Permanent Magnet Synchronous Motor (PMSM). We notice that the output gain of the speed PI controller, given by $\frac{2}{3}\frac{2}{P}\frac{1}{\lambda_f}$ has not yet been normalized (per-unitized).

Considering the normalized torque equation from Eq. (5.1.14), and ignoring the reluctance torque component, the equation can be expressed as:

$$\overline{T_e} = \overline{\lambda_f} \cdot \overline{i_{qs}^r} \tag{5.1.37}$$

From Eq. (5.1.37), we know that the normalized torque command $\overline{T_e}^*$ *multiplied by* $1/\overline{\lambda_f}$ *yields the normalized q-axis current command* $\overline{i_{qs}^r}^*$. Therefore, the output gain of the speed PI controller should be updated accordingly.

Step 8:

Finally, we need to convert the speed command into its per-unit value. The speed command input in Sect. 2.1 is 100 (rad/s), and its corresponding per-unit value is:

$$100/\omega_b = \frac{100}{377} = 0.2653(\text{pu})$$

Therefore, please set the speed command to 0.2653 (pu).

Step 9:

The parameters and per-unit decoupling blocks designed above have all been updated in the example model mdl_pm_foc_allpu.slx. As shown in Fig. 5.2, the SIMULINK system block diagram appears after opening the file. Before running the SIMULINK simulation, please first execute the example script pm_params_pu.m to load the motor per-unit parameters.

Step 10:

Figure 5.3 shows a comparison between the per-unit (pu) speed response waveform and the speed response waveform from Fig. 2.10. As can be seen from the figure, the two waveforms almost completely match. The slight discrepancy between them is due to rounding errors. This demonstrates that the per-unit system model and the system model using standard SI units are equivalent.

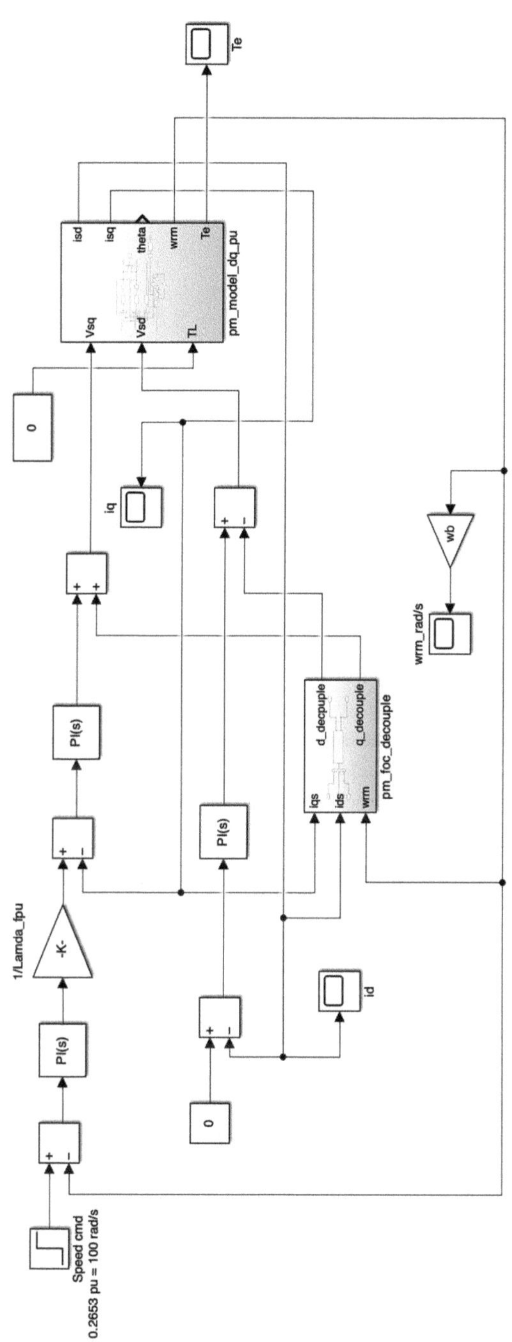

Fig. 5.2 Per-unit PMSM FOC system SIMULINK simulation, example model: mdl_pm_foc_allpu.slx

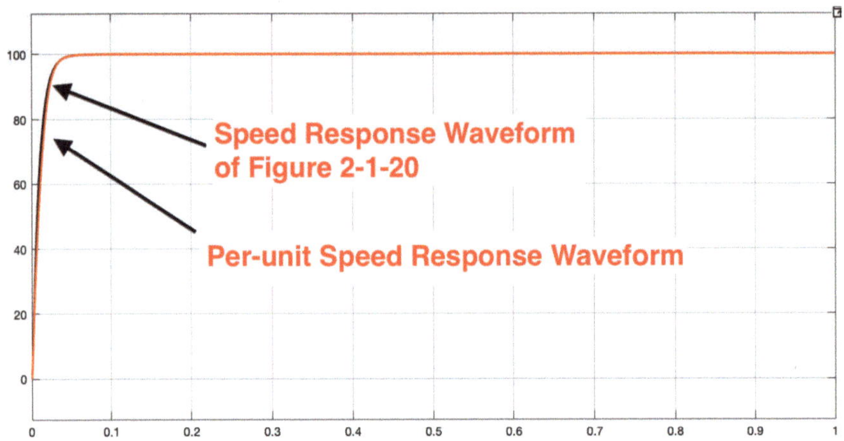

Fig. 5.3 Comparison of speed response waveforms between per-unit and non-per-unit systems

5.1.3 Summary

The conclusions of this chapter are summarized as follows:

- The normalization derivation process introduced in this section represents the standard method for motor per-unit (pu) modeling. You can apply it to normalize other types of motor systems.
- The control loop parameters designed using the per-unit motor model are not exactly the same as those derived using the MKS (SI) motor model. It is recommended to redesign the control loop based on the per-unit model to ensure control performance.
- Designing systems using per-unit models facilitates the development of universal control algorithms for motors with different rated power, voltage, and frequency, and also makes implementation on computers easier.

5.2 Anti-integral Windup Strategies for PI Control

In a control system, all physical quantities should be constrained within a reasonable range. Take a motor drive system as an example: the output voltage of the drive is limited by the inverter's DC bus voltage V_{dc}. When the inverter reaches its output voltage limit, increasing the controller's output further will not raise the actual voltage. If a steady-state error still exists at this point, the integrator in the PI controller will continue to accumulate, potentially growing to a very large value—a phenomenon known as **integrator wind-up**.

When the error eventually falls within acceptable bounds, the system requires a long time to discharge the integrator, resulting in sluggish response, oscillations, or even loss of control. To prevent this issue, an **anti-windup mechanism** should be incorporated into the PI controller design [2–4].

- **MATLAB/SIMULINK Simulation**

Step 1:

Please open the example model mdl_PI_anti_windup_1.slx. As shown in Fig. 5.4, the plant transfer function in the diagram is $\frac{1}{s+96}$. The PI controller transfer function has been designed as $\frac{1000(s+96)}{s}$ which corresponds to a proportional gain of 1000 and an integral gain of 96,000. Therefore, the entire control loop is equivalent to a first-order low-pass filter with a cutoff frequency of 1000 (rad/s).

Step 2:

After running the SIMULINK simulation, you can observe the waveform from the **Output** scope block, as shown in Fig. 5.5.

Step 3:

Next, observe the waveform of the PI controller's output signal **Vc**, as shown in Fig. 5.6. We can see that the steady-state value of the PI controller's output signal is approximately **96**.

Step 4:

To simulate the condition of **integrator windup**, we intentionally add a **Saturation** block to the output of the PI controller and set its upper and lower limits to ± **50**. After completing the setup, the system appears as shown in Fig. 5.7.

Step 5:

After completing the setup, run the system simulation again. Once the simulation is finished, double-click the Output scope block to observe the response waveform. As shown in Fig. 5.8, the steady-state value of the d-axis current response is approximately 0.52, indicating a steady-state error of around 0.48.

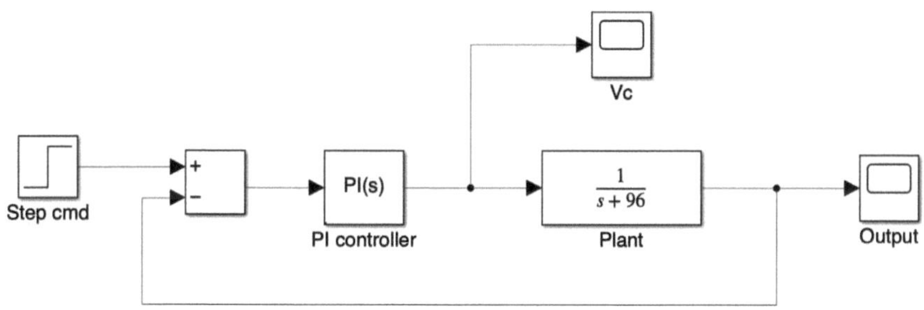

Fig. 5.4 PI control loop without anti-windup, example model: mdl_PI_anti_windup_1.slx

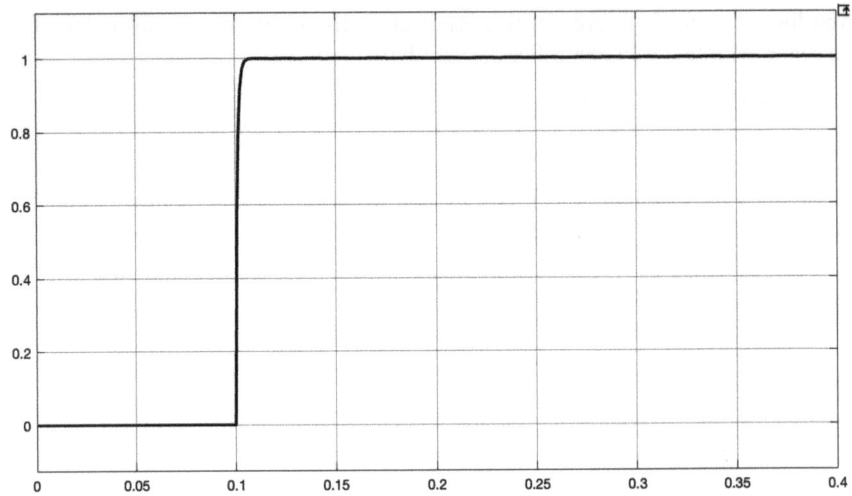

Fig. 5.5 Output response waveform

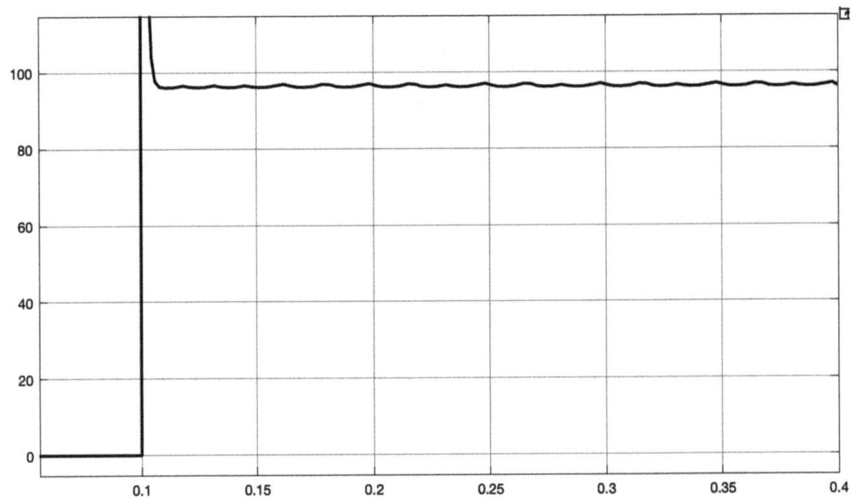

Fig. 5.6 Output waveform of the PI controller, Vc

Step 6:

Next, observe the actual **plant input signal (Vd)** and the **PI controller output signal (Vc)**, as shown in Figs. 5.9 and 5.10. From the waveforms, we can see that:

- The **actual plant input signal (Vd)** is **clamped at the saturation limit of 50**, indicating the actuator has reached its output limit.

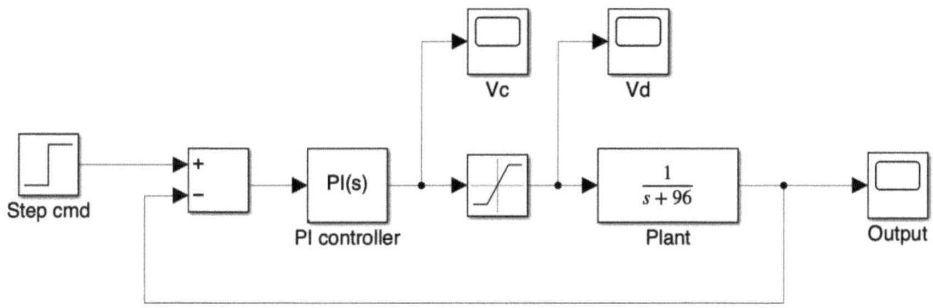

Fig. 5.7 PI control loop with a saturation block added

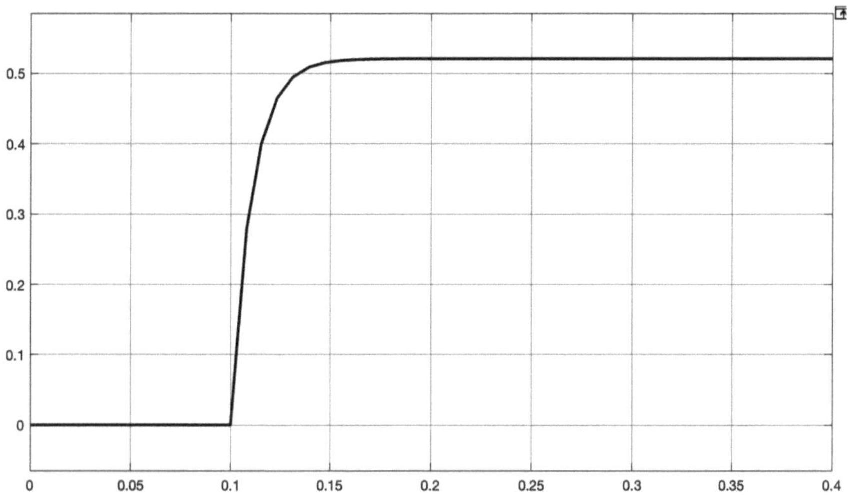

Fig. 5.8 Output response waveform

- The **PI controller output signal (Vc)** continues to accumulate due to the ongoing action of the integrator, even though the actuator output is already saturated. This results in **integrator windup**.

This phenomenon illustrates the typical behavior of **integrator saturation**, where the integrator keeps integrating the error, leading to a large accumulated value, which can cause sluggish recovery and degraded control performance once the system exits saturation.

Step 7:

Next, we will add an **anti-windup mechanism** to the PI controller. While keeping the original PI controller parameters unchanged, we incorporate an anti-windup feedback loop into the system.

Please open the example model **mdl_PI_anti_windup_2.slx**. The modified system block diagram is shown in Fig. 5.11.

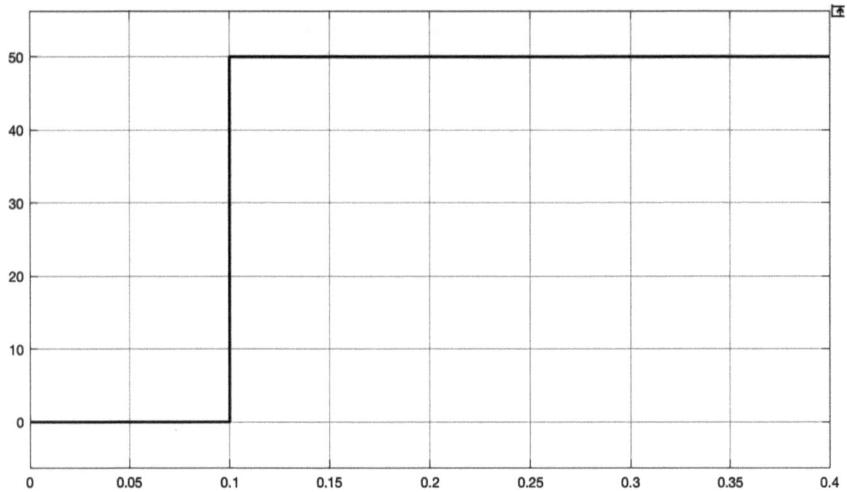

Fig. 5.9 Plant input waveform, Vd

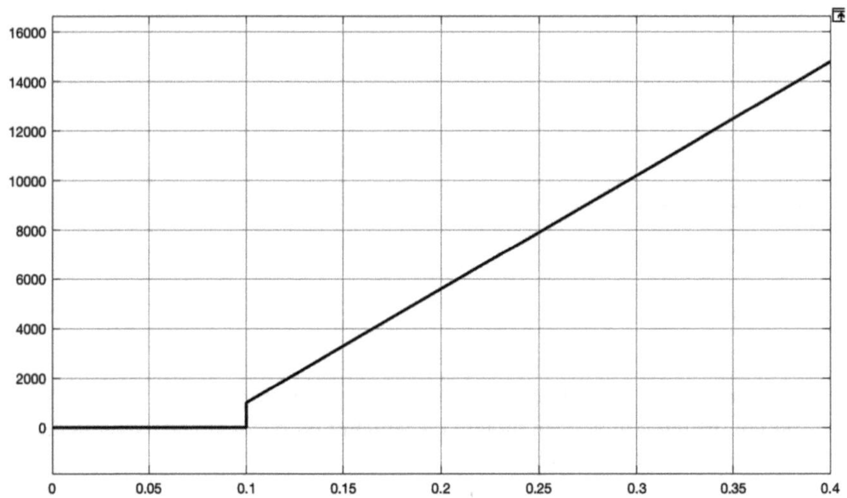

Fig. 5.10 Output waveform of the PI controller, Vc

Note: A common practice is to set the **anti-windup gain** Ka to the **inverse of the proportional gain** 1/Kp, which often yields good performance. However, feel free to experiment with different anti-windup gain values for optimization.

Step 8:

Run the SIMULINK simulation and observe the waveform on the **Output** scope. As shown in Fig. 5.12, the steady-state value of the Output waveform still remains around **0.52**.

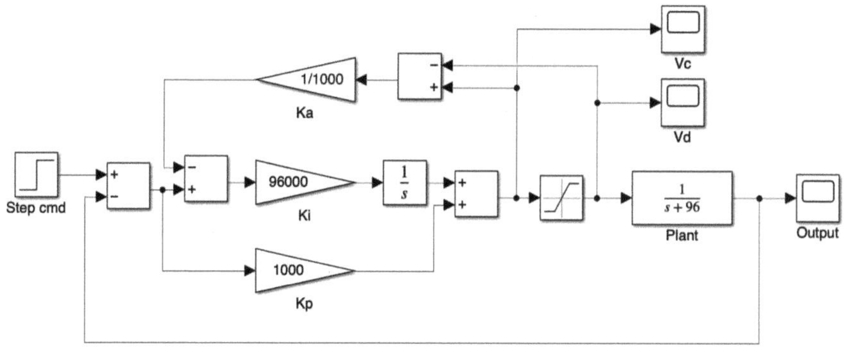

Fig. 5.11 PI control loop with anti-windup, example model: mdl_PI_anti_windup_2.slx

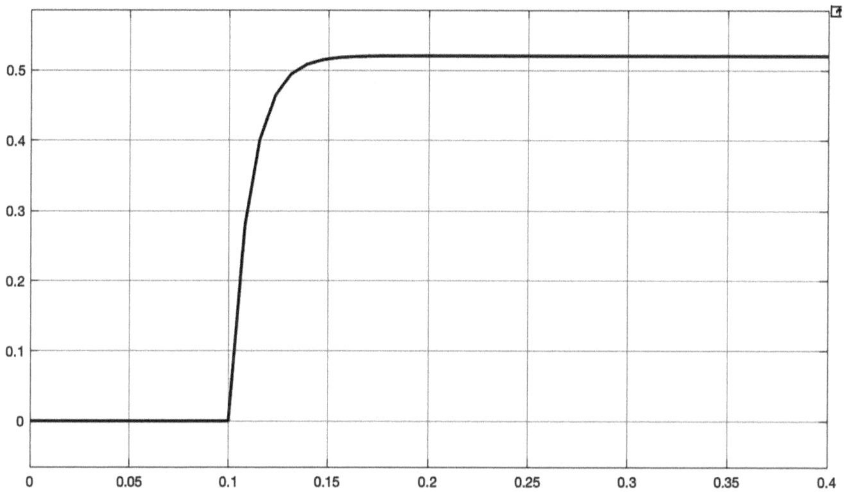

Fig. 5.12 Output response waveform

Step 9:

Next, observe the actual control plant input signal (**Vd**) and the PI controller output signal (**Vc**), as shown in Figs. 5.13 and 5.14. From the waveforms, we can see that the plant input signal remains limited at **50**. However, due to the effect of the **anti-windup mechanism**, the controller output signal is regulated to approximately **530**, effectively solving the **integrator windup** problem.

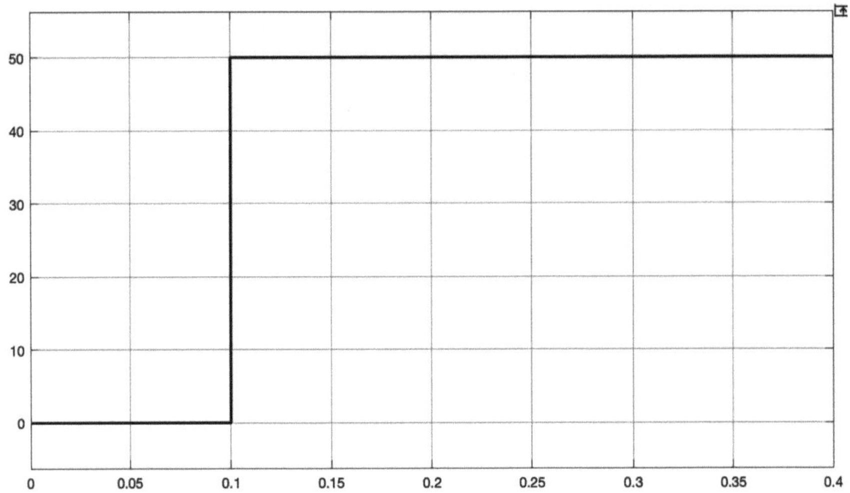

Fig. 5.13 Plant input waveform, Vd

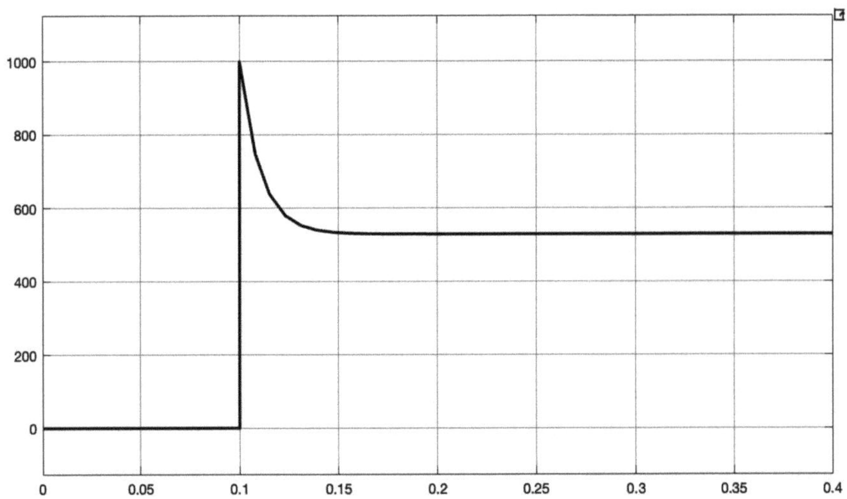

Fig. 5.14 Output waveform of the PI controller, Vc

5.3 Self-tuning of Motor Parameters

In Chap. 1, we derived and established the space vector model of AC motors. In Chap. 2, we used the AC motor model developed in that chapter to derive the field-oriented control (FOC) law and design the control loops. However, the control performance of a field-oriented system depends on the precision of the control loops, and the precision of the control loops in turn depends on the accuracy of the motor parameters [1]. Therefore, **accurate motor parameters become a necessary condition for achieving high-performance field-oriented control systems**.

In this section, we will take the permanent magnet synchronous motor (PMSM) as an example and demonstrate how to use the **d- and q-axis equivalent circuits and the mechanical equation** of the motor to develop a **motor parameter autotuning algorithm** [2, 3].

5.3.1 Auto-tuning Electrical Parameters for PMSMs

In Sect. 1.4, we previously illustrated the rotor reference frame model of a permanent magnet synchronous motor (PMSM) using an equivalent circuit, as shown in Fig. 5.15.

Assuming the rotor is stationary, i.e., $\omega_r = 0$, the circuit in Fig. 5.15 can be simplified to Fig. 5.16.

Figure 5.16 shows that when the rotor is stationary, no back electromotive force (EMF) exists. As a result, the d- and q-axis equivalent circuits become simple R–L series circuits. Therefore, we can determine the values of the stator resistance R_s, d-axis inductance L_d, and q-axis inductance L_q by analyzing the relationship between voltage and current.

- **Voltage–current relationship of an R–L series circuit**:

Figure 5.17 shows an R-L series circuit. Suppose the input voltage V is a DC voltage. In this case, the current is $i = \frac{V}{R}$. For the inductor L, its impedance is ωL. Since the frequency of the DC voltage is zero, the impedance of the inductor is also zero. Therefore, dividing the input voltage V by the current i gives the resistance R.

Now consider another scenario where the input is a sinusoidal voltage.

$$V = V_{max}\sin(\omega_1 t) \tag{5.3.1}$$

In this case, the inductor L no longer has zero impedance; its impedance becomes $\omega_1 L$. According to circuit theory, the total impedance Z of the circuit can be expressed as:

$$Z = R + j\omega_1 L = \sqrt{R^2 + (\omega_1 L)^2}\angle \tan^{-1}\frac{\omega_1 L}{R} \tag{5.3.2}$$

where Z is the total impedance of the R-L series circuit.

$|Z| = \sqrt{R^2 + (\omega_1 L)^2}$ is the magnitude of the impedance, and $\angle Z = \tan^{-1}\frac{\omega_1 L}{R}$ is the impedance phase angle in electrical degrees.

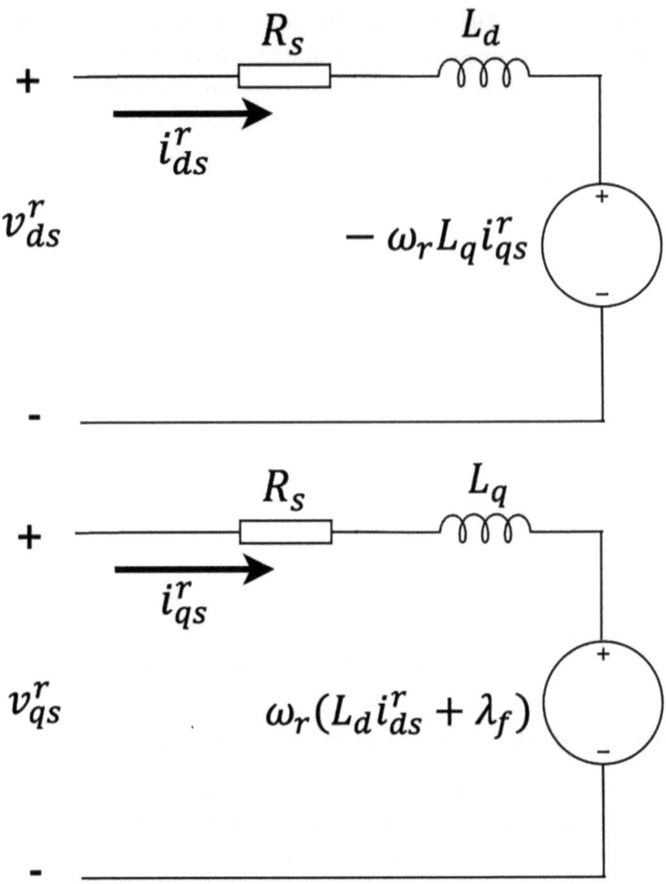

Fig. 5.15 d–q axis equivalent circuit diagram of a permanent magnet synchronous motor

The current i can be expressed as:

$$i = \frac{V}{Z} = \frac{V_{max}}{\sqrt{R^2 + (\omega_1 L)^2}} \angle - \tan^{-1}\frac{\omega_1 L}{R} \tag{5.3.3}$$

Equations (5.3.1) and (5.3.3) respectively represent the input voltage and input current shown in Fig. 5.18. If we consider only the magnitude relationship between voltage, current, and impedance, you can easily derive the following equation:

$$|Z| = \sqrt{R^2 + (\omega_1 L)^2} = \frac{Peak - to - peak\ voltage\ V_{p-p}}{Peak - to - peak\ current\ I_{p-p}} \tag{5.3.4}$$

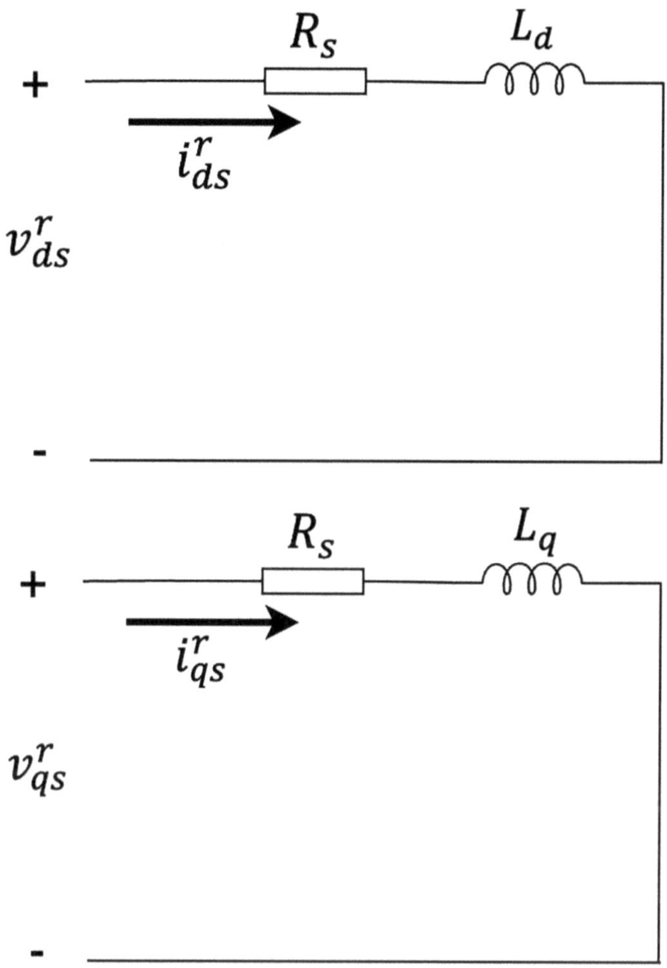

Fig. 5.16 d–q axis equivalent circuit of a permanent magnet synchronous motor at standstill

Next, we will apply Eq. (5.3.4) to identify the motor parameters of the permanent magnet synchronous motor: stator resistance R_s, d-axis inductance L_d, and q-axis inductance L_q.

Before starting the simulation, we first input the motor parameters of the permanent magnet synchronous motor model as shown in Table 5.2. Please run the example script pm_params.m provided in this section to load the motor parameters.

- **Determine the Stator Resistance** R_s

Step 1:

We can remove the speed control loop and the nonlinear current decoupling compensation blocks from the field-oriented control (FOC) system of the permanent magnet

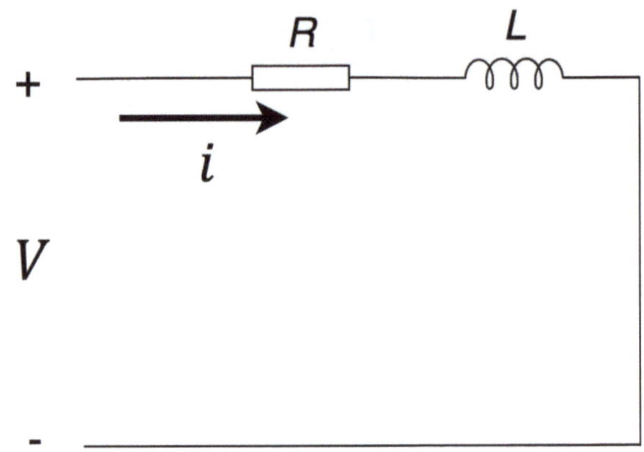

Fig. 5.17 R-L series circuit

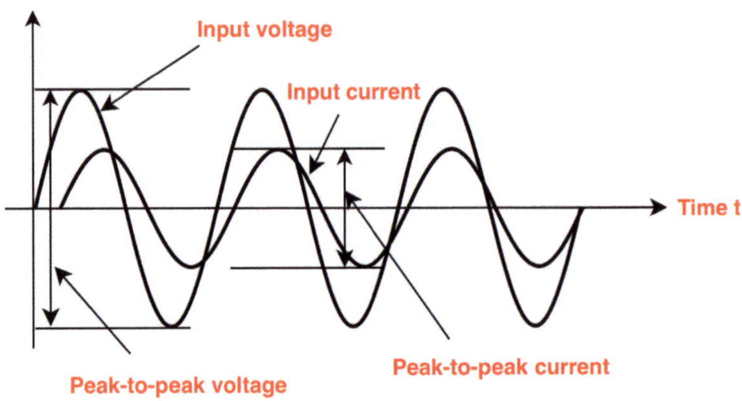

Fig. 5.18 Input voltage and current waveforms of an R-L series circuit

Table 5.2 PMSM electrical parameters

Parameter	Symbol	Value	Unit
Stator resistance	R_s	1.2	Ω
Stator d-axis inductance	L_d	5.7	mH
Stator q-axis inductance	L_q	12.5	mH

synchronous motor (PMSM) shown in Fig. 2.21. Then, set the q-axis current command to 0, the d-axis current command to 3, and the load torque T_L to 0, as shown in Fig. 5.19. All other PMSM parameters and the PI controller parameters for the d- and q-axis current loops remain the same as those in Sect. 2.1.

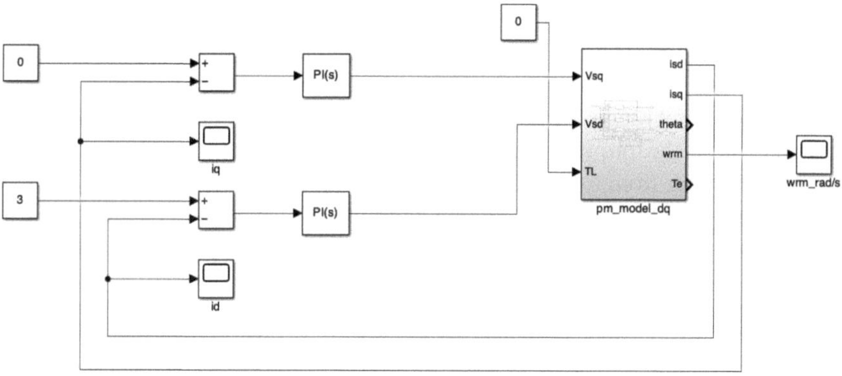

Fig. 5.19 Current loop simulation of a permanent magnet synchronous motor at standstill

Step 2:

Next, we need to keep the permanent magnet synchronous motor (PMSM) in a stationary state so that its dq-axis equivalent circuit becomes a simple R-L series circuit. How can we keep the PMSM stationary? The answer is to set the angle required for coordinate transformation to zero. Please add the Park and inverse Park transformation blocks to the system shown in Fig. 5.19, as illustrated in Fig. 5.20.

In Fig. 5.20, the Park and inverse Park transformation blocks are responsible for converting the feedback currents of the PMSM between coordinate frames. Normally, during regular motor operation, the angle required for Park transformation is obtained by integrating the rotor angular velocity of the motor. However, since we need to keep the rotor stationary, we intentionally set the angle input of the Park transformation to zero.

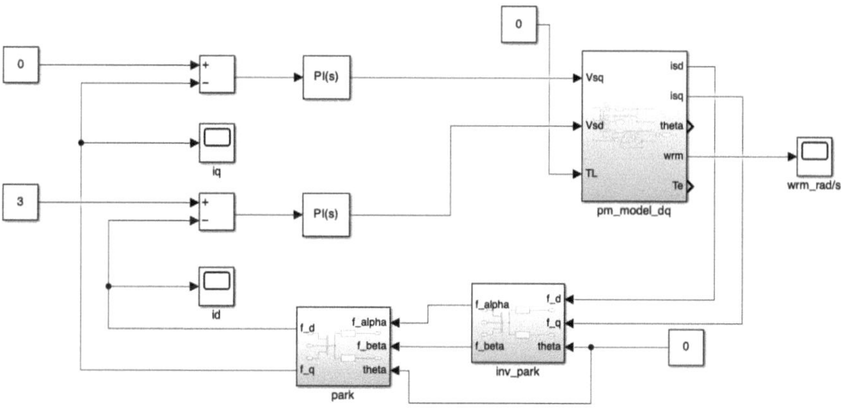

Fig. 5.20 Stator resistance autotuning simulation, example model: pm_autotune_Rs.slx

In Fig. 5.20, we set the d-axis current command to 3 (A) (Note: The current command value is not fixed—it should not exceed the motor's rated current). The purpose of this is to fix the rotor (d-axis position) to the α-axis (Explanation: since the d-axis does not rotate, it coincides with the α-axis). Additionally, since the q-axis current controls the torque magnitude, we set the q-axis current command to zero to prevent the motor from rotating.

Under this condition, the d-q axis reference coordinate system remains valid, so we can use the d-axis equivalent circuit to determine the stator resistance R_s.

Step 3:

Please set the SIMULINK solver type to "Fixed-step", and set the "Fixed-step size" to 0.00025 to simulate a 250 μs microcontroller interrupt interval. Set the simulation time to 0.05 s, and click "Run" to execute the system simulation.

(Note: Before running the simulation, please execute the example script pm_params.m in this section to load the motor parameters.)

Step 4:

After the simulation is complete, open the **wrm oscilloscope block** to observe the motor speed. As shown in Fig. 5.21, the motor is confirmed to be in a stationary state.

Step 5:

Next, observe the d-axis current and d-axis voltage waveforms, as shown in Figs. 5.22 and 5.23, respectively.

From Figs. 5.22 and 5.23, it can be observed that the steady-state values of the d-axis voltage and current are 3.6 (V) and 3 (A), respectively. By directly dividing the voltage by the current, we can obtain the stator resistance value.

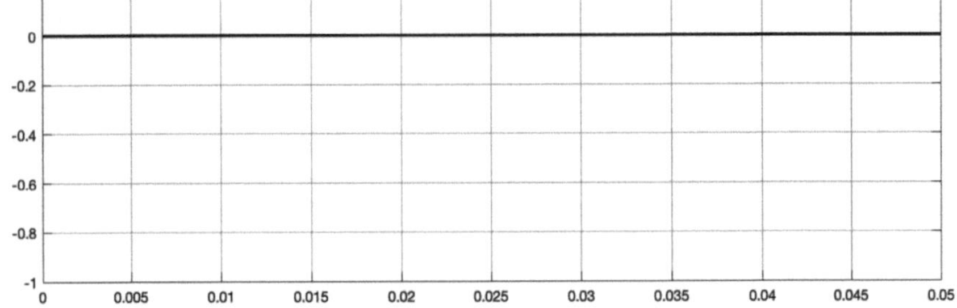

Fig. 5.21 Motor speed Waveform, ω_{rm}

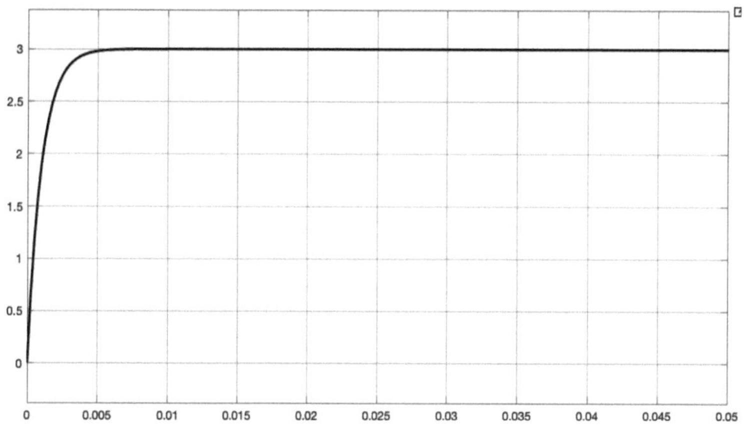

Fig. 5.22 d-axis current waveform, i_{ds}^{r}

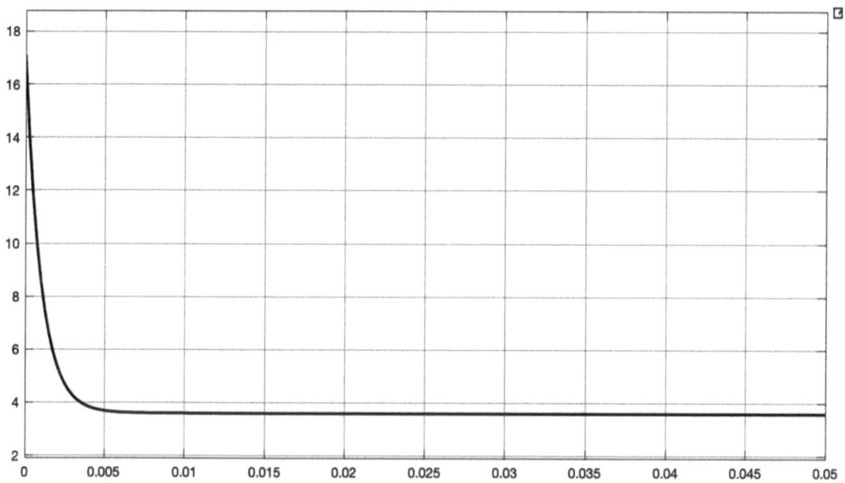

Fig. 5.23 d-axis voltage command waveform, v_{ds}^{r}

$$\text{stator resistance value } R_s = \frac{3.6}{3} = 1.2(\Omega)$$

Tips

In practice, we usually apply **two different current commands** to obtain **two corresponding voltage values**, and then calculate the **stator resistance** by dividing the voltage difference (ΔV) by the current difference (ΔI). This approach helps **eliminate any offset** that may affect the measurement accuracy.

- **Determine the d-axis Inductance L_d of the Stator**

Step 1:

After determining the stator resistance, the next step is to find the d-axis inductance. To obtain the d-axis inductance, a sinusoidal command must be applied to the d-axis current control loop. Therefore, set the d-axis current command as follows:

$$i_{d\_cmd} = 3 + 1 \times \sin(2 \times \pi \times 100) \tag{5.3.5}$$

Modify the d-axis current command in Fig. 5.20 according to Eq. (5.3.5), as shown in Fig. 5.24.

說明
The purpose of using a combination of a DC component and a sinusoidal component in the d-axis current command is that the DC component keeps the rotor aligned with the α-axis, while the sinusoidal component is used to determine the inductance value.

Step 2:

Please set the SIMULINK solver type to "Fixed-step", and set the "Fixed-step size" to 0.00025 to simulate a 250 μs microcontroller interrupt period. Set the simulation time to 0.2 s, then click "Run" to execute the system simulation.

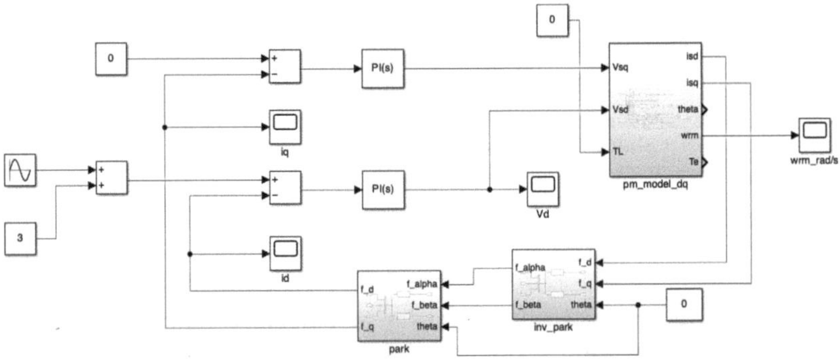

Fig. 5.24 d-axis inductance autotuning simulation, example model; pm_autotune_Ld.slx

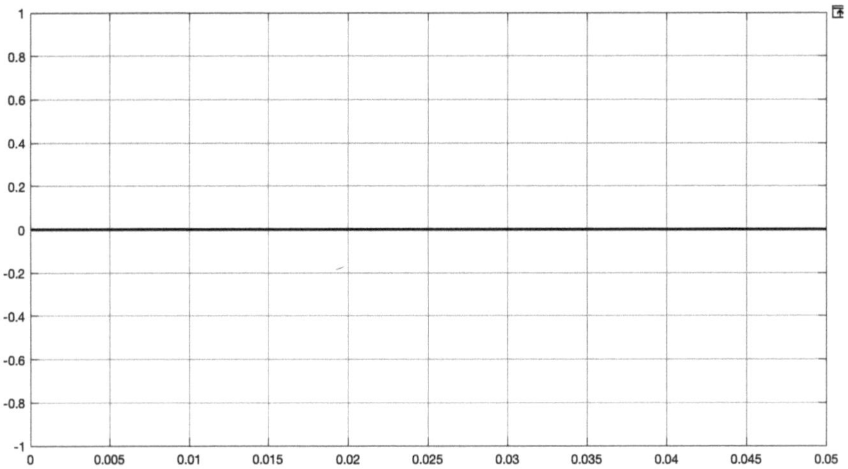

Fig. 5.25 Motor speed waveform, ω_{rm}

(Note: Before running the simulation, please execute the example script pm_params.m in this section to load the motor parameters.)

Step 3:

After completing the simulation, first open the wrm scope block to verify whether the motor remains stationary. As shown in Fig. 5.25, the motor is indeed stationary, indicating that the high-frequency sinusoidal current injected into the d-axis did not cause the rotor to rotate.

Step 4:

Next, observe the d-axis current and d-axis voltage waveforms, as shown in Figs. 5.26 and 5.27, respectively.

Step 5:

Next, we can use Eq. (5.3.4) to calculate the stator d-axis inductance L_d.

$$\text{stator d - axis inductance } L_d$$
$$= \sqrt{\left[\frac{\left(Peak - to - peakvoltage V_{p-p}\right)^2}{\left(Peak - to - peakcurrent I_{p-p}\right)^2} - R_s^2\right] \times \frac{1}{(2 \times \pi \times f)^2}}$$
$$= \sqrt{\left[\frac{(6.8 - 0.4)^2}{(3.84 - 2.16)^2} - 1.2^2\right] \times \frac{1}{(2 \times \pi \times 100)^2}} = 0.0058 \qquad (5.3.6)$$

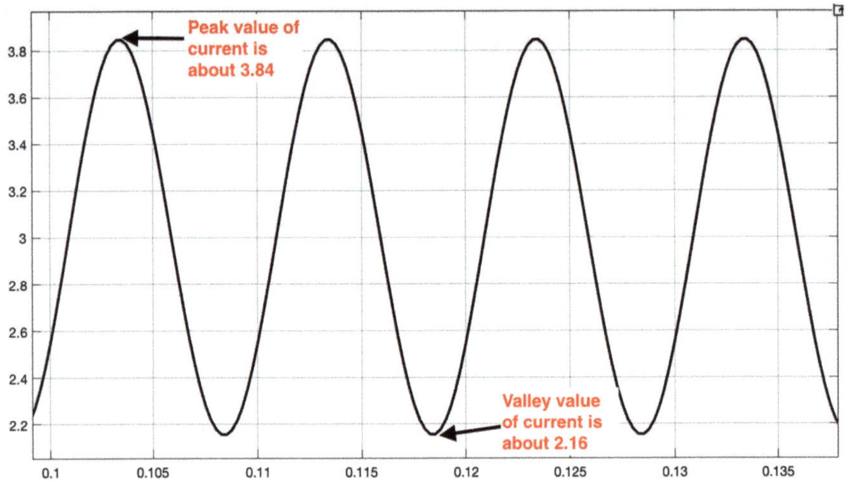

Fig. 5.26 d-axis current waveform, i_{ds}^{r}

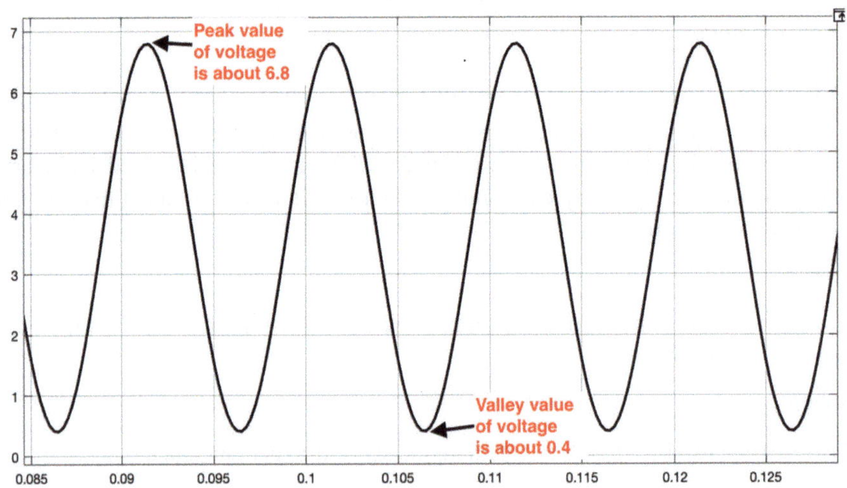

Fig. 5.27 d-axis voltage waveform, v_{ds}^{r}

We calculated the stator d-axis inductance L_d to be 0.0058 (H), which is very close to the actual value of 0.0057 (H). The estimation error is:

$$\begin{aligned} \textit{Estimation error of d-axis inductance} &= \frac{0.0057 - 0.0058}{0.0057} \times 100\% \\ &= -1.75\% \end{aligned} \tag{5.3.7}$$

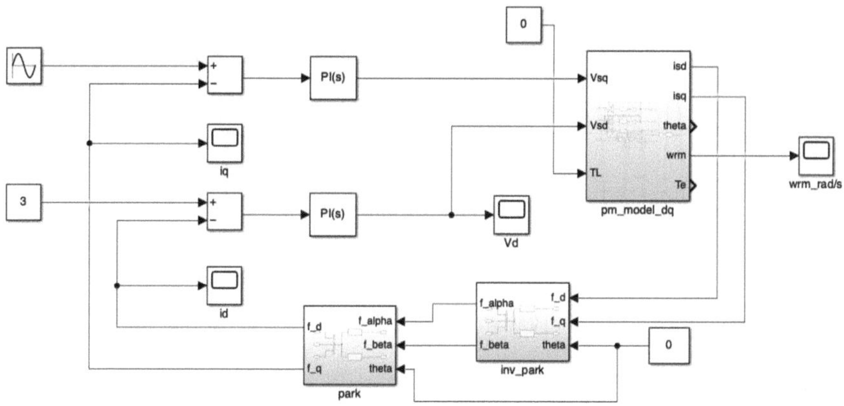

Fig. 5.28 d-axis inductance autotuning simulation, example model: pm_autotune_Lq.slx

- **Determine the q-axis Inductance L_q of the Stator**

Step 1:

Next, we will use the same method to determine the **q-axis inductance** L_q. To do this, we need to apply a sinusoidal current command to the q-axis current input.

However, since **q-axis current generates torque**, it may cause the motor rotor to rotate, resulting in **back electromotive force (EMF)** and potentially significant measurement errors. To minimize this effect, we reduce the amplitude of the sinusoidal q-axis current command to **0.2 A**, as shown below:

$$i_{q\_cmd} = 0.2 \times \sin(2 \times \pi \times 100) \tag{5.3.8}$$

At the same time, in order to maintain the decoupled state of the dq-axes, we still need to apply a 3 A DC current command to the d-axis current loop, as shown in Fig. 5.28.

Step 2:

Please set the SIMULINK solver type to **"Fixed-step"**, and set the **"Fixed-step size"** to **0.00025** to simulate a 250 μs microcontroller interrupt period. Then set the **simulation time to 0.2** s, and click **"Run"** to execute the system simulation.

(**Note**: Before running the simulation, please make sure to execute the example script pm_params.m from this section to load the motor parameters).

Step 3:

After completing the simulation, you can first open the wrm oscilloscope block to check whether the motor remains stationary. As shown in Fig. 5.29, we can observe that the input q-axis current command inevitably causes slight rotor vibration, which may generate a small amount of q-axis back electromotive force (EMF) and affect the accuracy of the measured q-axis inductance.

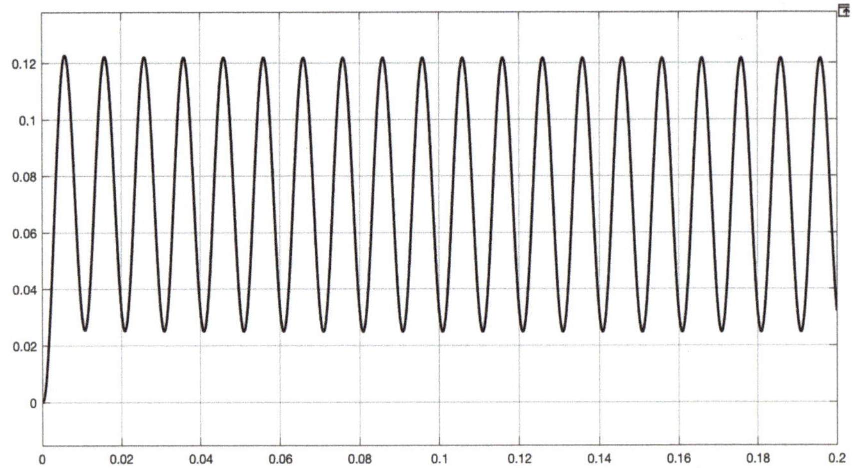

Fig. 5.29 Motor speed waveform ω_{rm}

Step 4:

Next, observe the q-axis current and q-axis voltage waveforms, as shown in Figs. 5.30 and 5.31, respectively.

Step 5:

Next, we can use Eq. (5.3.4) to calculate the q-axis inductance L_q of the stator.

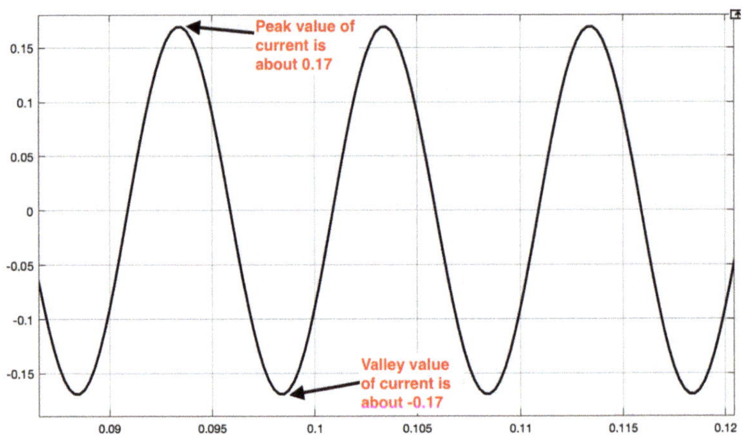

Fig. 5.30 q-axis current waveform i_{qs}^{r}

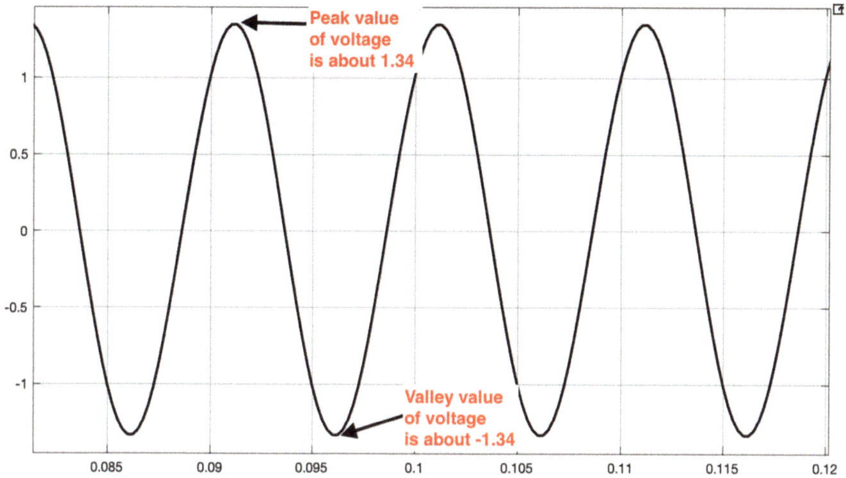

Fig. 5.31 q-axis voltage waveform v_{qs}^{r}

stator q - axis inductance L_q

$$= \sqrt{\left[\frac{(Peak - to - peakvoltageV_{p-p})^2}{(Peak - to - peakcurrentI_{p-p})^2} - R_s^2\right] \times \frac{1}{(2 \times \pi \times f)^2}}$$

$$= \sqrt{\left[\frac{(1.34 + 1.34)^2}{(0.17 + 0.17)^2} - 1.2^2\right] \times \frac{1}{(2 \times \pi \times 100)^2}} = 0.0124 \tag{5.3.9}$$

The calculated value of the stator q-axis inductance L_q is 0.0124 (H), while the actual value is 0.0125 (H). The estimation error can be computed as follows:

$$Estimation\ error\ of\ q{-}axis\ inductance = \frac{0.0125 - 0.0124}{0.0125} \times 100\% = 0.8\% \tag{5.3.10}$$

If your estimation of the q-axis inductance has a large error, you can try the following methods to improve accuracy:

- **Reduce the amplitude of the input q-axis current command** to minimize the torque ripple and rotor motion, which helps suppress the generation of back-EMF.
- **Increase the sampling frequency** to capture more accurate voltage and current waveforms, especially at higher frequencies.
- **Apply moving average filtering** to the voltage and current signals to reduce high-frequency noise and improve measurement stability.

These approaches should help enhance the accuracy of the estimated q-axis inductance value.

5.3.2 Auto-identification of Mechanical Inertia

The motor parameter identification technique introduced in the previous section enables automatic and accurate estimation of permanent magnet synchronous motor (PMSM) parameters, including the stator resistance R_s, d-axis inductance L_d, and q-axis inductance L_q. With these precise parameters, we can then design the current control loops using the methods taught in Chap. 2.

However, in addition to the current control loops, the design of the speed control loop also requires knowledge of the motor's mechanical inertia. Therefore, this section introduces a fundamental and widely used technique for estimating the motor's moment of inertia. Please note that this method may cause the motor shaft to rotate, which makes it unsuitable for applications where shaft rotation is not permitted.

Let us examine the motor's mechanical equation, as follows:

$$T_e - T_L = J\frac{d\omega_{rm}}{dt} + [B]\omega_{rm} \tag{5.3.11}$$

Assuming that the friction coefficient B can be neglected and the load torque T_L is zero, the mechanical equation of the motor can be simplified as follows:

$$T_e = J\frac{d\omega_{rm}}{dt} \tag{5.3.12}$$

Then, the motor's moment of inertia can be expressed as:

$$J = \frac{T_e}{\frac{d\omega_{rm}}{dt}} \approx \frac{T_e}{\frac{\Delta\omega_{rm}}{\Delta t}} \tag{5.3.13}$$

From Eq. (5.3.13), we can see that if the applied torque T_e is known, then the motor's moment of inertia J can be estimated by measuring the rate of change in motor speed $\frac{\Delta\omega_{rm}}{\Delta t}$.

- **SIMULINK Simulation**

Step 1:

Please open the example program **pm_autotune_J.slx**. Once opened, it should appear as shown in Fig. 5.32.

The parameters used in this simulation are the same as those listed in Table 2.3, with the motor inertia set to 0.00016 (kg m^2). In this case, the friction coefficient B is set to zero. In Fig. 5.32, the inner loop is the q-axis current control loop of the permanent magnet synchronous motor (PMSM), and the current loop parameters are the same as those used in Sect. 5.3.1.

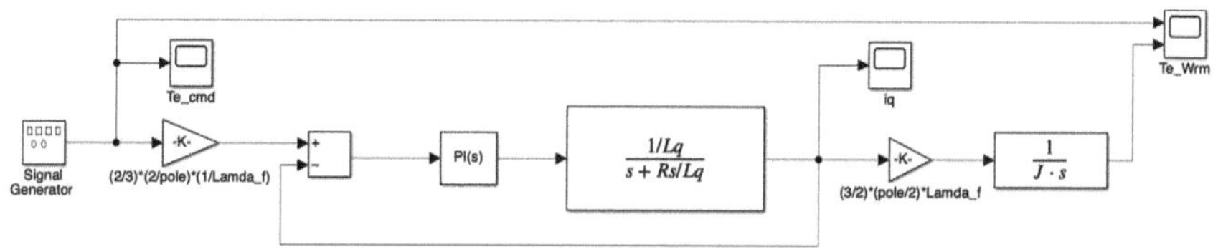

Fig. 5.32 Motor inertia autotuning simulation, example model: pm_autotune_J.slx

The motor torque command is generated by a Signal Generator, which produces a torque command with a period of 0.1 s and an amplitude of $\pm$ 0.2 (Nm), as shown in Fig. 5.33.

Step 2:

After running the simulation, open the Te_Wrm oscilloscope block to observe the waveform showing both motor torque and motor speed, as illustrated in Fig. 5.34. In the figure:

- The upper trace shows the motor torque command waveform.
- The lower trace displays the motor speed waveform.

You can see that:

- During the first half-cycle, a positive torque is applied, causing the motor to accelerate in the forward direction.
- During the second half-cycle, a negative torque is applied, causing the motor to decelerate.
- Because the positive and negative torque durations are equal, the motor decelerates exactly to zero speed at the end of the second half-cycle.

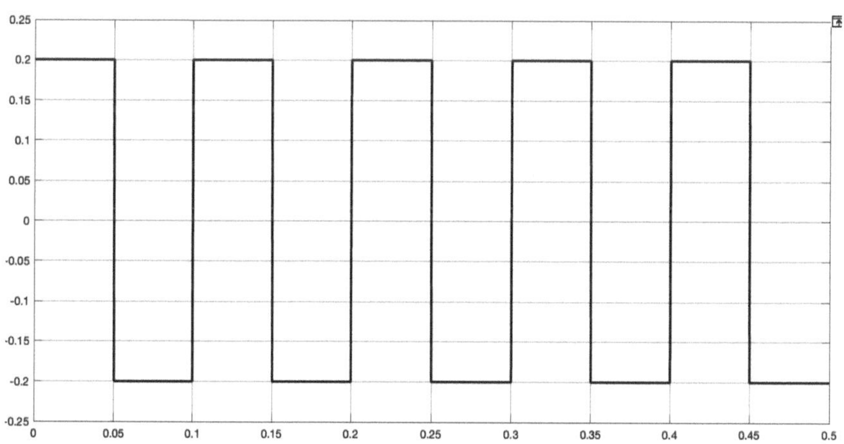

Fig. 5.33 Motor torque command waveform

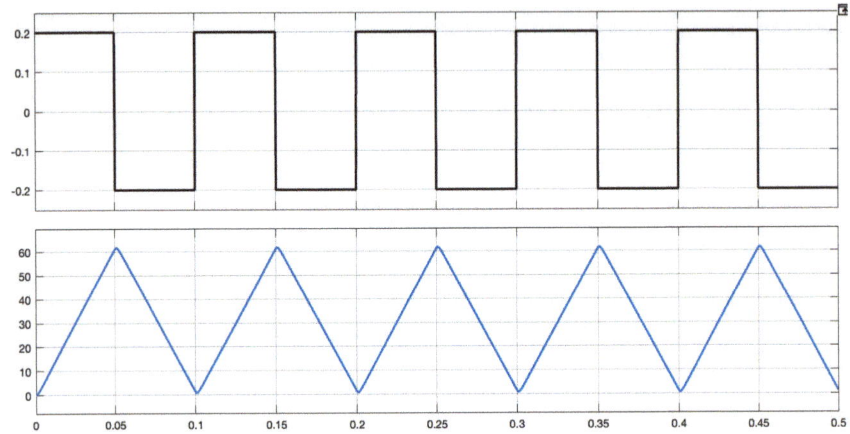

Fig. 5.34 Motor torque command and speed response waveforms

Step 3:

A specific portion of the speed ramp-up can be magnified, as shown in Fig. 5.35. In the figure, it is clearly observed that:

- At 0.021 s, the motor speed is 25 rad/s.
- At 0.025 s, the motor speed is 30 rad/s.
- During this time interval, the applied torque is 0.2 Nm.

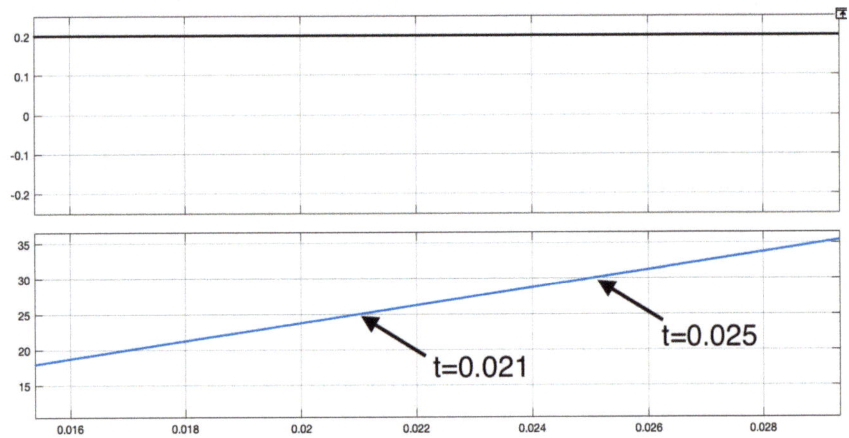

Fig. 5.35 Zoomed-in waveform during speed acceleration

Step 4:

Next, we use Eq. (5.3.13) to calculate the motor's moment of inertia:

$$J \approx \frac{T_e}{\frac{\Delta\omega_{rm}}{\Delta t}} = \frac{0.2}{\frac{30-25}{0.025-0.021}} = 1.6e - 04 \tag{5.3.14}$$

The mechanical inertia calculated using Eq. (5.3.13) is consistent with the value listed in Table 2.3. Therefore, the verification of the motor mechanical inertia self-tuning technique is successfully completed.

Key Points of This Section:

- For permanent magnet motor **d- and q-axis inductance parameters**, traditional measurement methods rely on external LCR meters and manual calculation, which are time-consuming.
- The **self-learning technique** for permanent magnet motors introduced in this section effectively identifies motor parameters. Simulation results show that the estimation error can be controlled within 5%, meeting the accuracy required for high-performance field-oriented control.
- This motor parameter self-learning method provides the **precise motor parameters** needed for designing field-oriented control loops, enabling high-performance operation of AC motors.
- As long as the motor undergoes rotational motion, the **mechanical** Eq. (5.3.11) remains applicable. Therefore, the self-learning technique for mechanical inertia described in Sect. 5.3.2 is **not limited to permanent magnet synchronous motors**, and can also be used for other motor types.
- Since motor parameters tend to **vary with operating conditions and environments** (e.g., temperature and current), to enhance the robustness of AC motor control loops against parameter variation, **a parameter estimator can be designed** if necessary to enable real-time estimation and compensation.

5.4 PWM Sampling Techniques and Delay Impacts

In Chap. 2, we learned that delay is inevitable in digital control systems, and it inevitably reduces system stability. If delay is not considered during the control loop design phase, the designed bandwidth and stability specifications will significantly deviate from the actual system behavior, which is unacceptable. Therefore, it is essential to account for delay during the control loop design phase. This section introduces four typical PWM sampling methods commonly used in motor control systems [5]. Each method introduces a different amount of delay.

Let us first revisit the q-axis current loop of a permanent magnet synchronous motor (PMSM) with delay effects considered, as shown in Fig. 5.36.

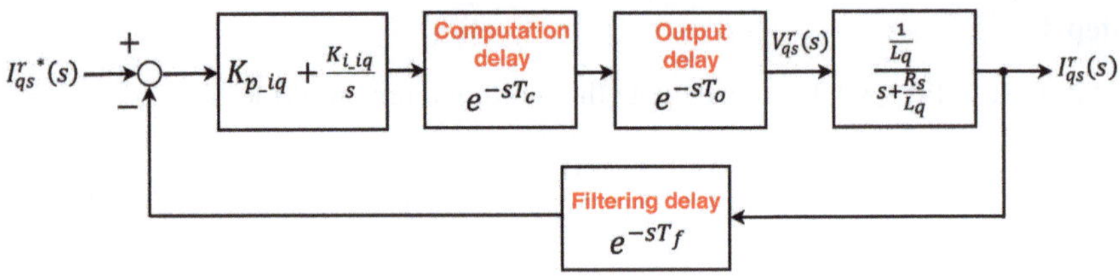

Fig. 5.36 q-axis current control loop of a permanent magnet synchronous motor considering delay effects

Among these delays, the **MCU computation delay** T_c and **MCU output delay** T_o vary depending on the PWM sampling method used. The **current sampling delay** T_{SH} is generally half of the sampling period, while the **filter delay** depends solely on the time constant of the filter.

Therefore, the following analysis will focus on how the **MCU computation delay** T_c and **MCU output delay** T_o vary with different **PWM sampling methods.**

- **Method 1: Standard Sampling Method**

Figure 5.37 shows the PWM timing diagram for the **standard sampling method**. In this method, the total delay time T_d is the sum of the **MCU computation delay** T_c and the **MCU output delay** T_o. The PWM carrier period is denoted as T_s, and the current control period is T_{con}. Assuming $T_s = T_{con}$, the delays are defined as follows:

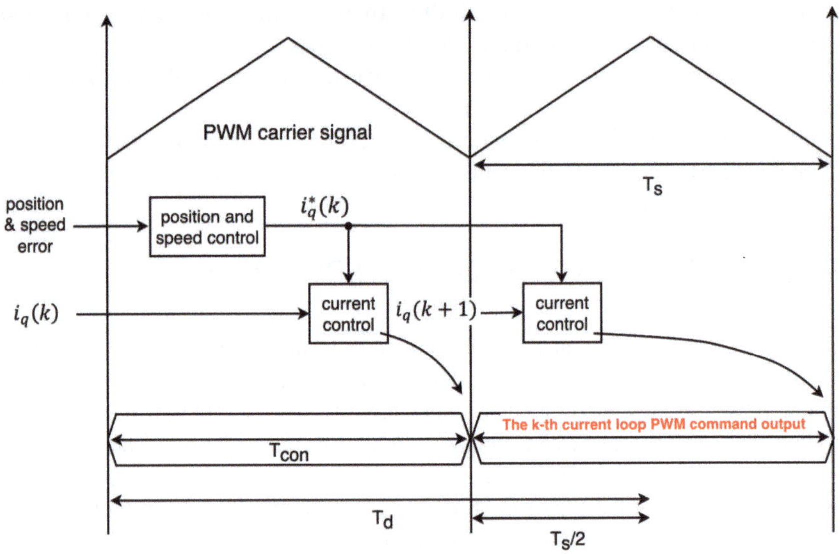

Fig. 5.37 Digital control timing diagram of the standard sampling method

- **MCU computation delay**:

$$T_c = T_s$$

because the computation is completed within the same control period.

- **MCU output delay (PWM ZOH delay)**:

$$T_o = T_s/2$$

since the output voltage command takes effect in the next PWM cycle.

Therefore, the total control delay is:

$$T_d = 1.5 \times T_s$$

(Note: Since the current sampling and computation occur within the same control cycle, the sampling delay T_{SH} is inherently included in T_c, and thus does not need to be added separately to the total delay).

- **Method 2: Modified Sampling Method**

Figure 5.38 illustrates the PWM timing diagram for the **Modified Sampling Method**, in which the total delay time T_d is the sum of the MCU computation delay T_c and the MCU output delay T_o. The PWM carrier period is T_s, and the current control period is T_{con}. Assuming $T_s = T_{con}$, in this method, the computation time required for the current control loop is only half of the control period, so the MCU computation delay is $T_c = T_s/2$.

Since the voltage command calculated in each control period is executed in the next PWM period, the MCU output delay (i.e., the ZOH output delay of the PWM) is also $T_o = T_s/2$. Therefore, the total delay is:

$$T_d = T_c + T_o = 1 \times T_s$$

(Note: Since the current sampling and computation occur within the same control cycle, the sampling delay T_{SH} is inherently included in T_c, and thus does not need to be added separately to the total delay).

- **Method 3: Double Current Sampling in a Single PWM Cycle**

Based on Method 2, if the computational capability of the MCU is enhanced to allow two current samplings and control operations within one PWM cycle, the PWM timing diagram is shown in Fig. 5.39. In this setup, the PWM carrier period is denoted as T_s, and the control period as T_{con}. Assuming that two current sampling and control operations can be completed within one PWM period, then $T_{con} = T_s/2$.

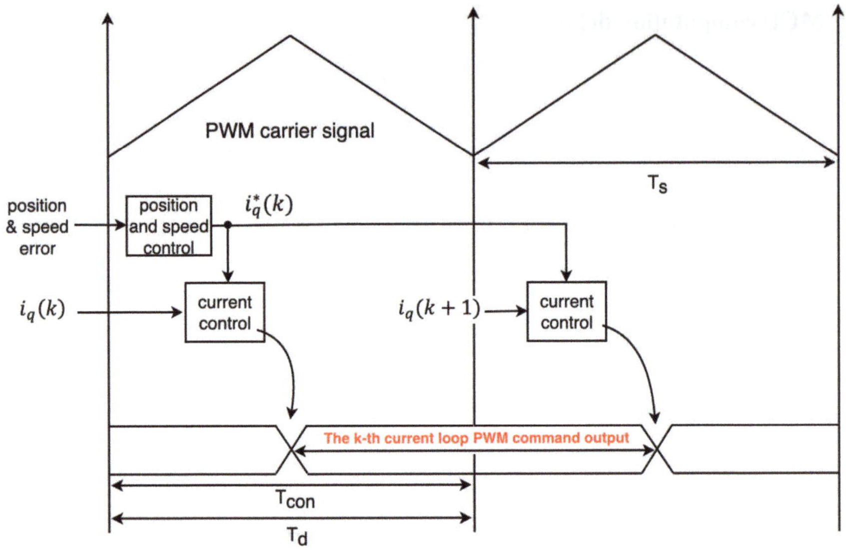

Fig. 5.38 Digital control timing diagram of the modified sampling method

For the current control loop, the MCU computation delay is $T_c = T_{con}$. Since the voltage command calculated in each control period is updated to the PWM voltage command in the next control period, the MCU output delay (i.e., the ZOH delay of the PWM) is $T_o = T_{con}/2$. Thus, the total delay time is:

$$T_d = T_c + T_o = 1.5 \times T_{con}$$

(Note: Since the current sampling and computation occur within the same control cycle, the sampling delay T_{SH} is inherently included in T_c, and thus does not need to be added separately to the total delay).

- **Method 4: Current Loop Control Using FPGA**

Based on Method 3, if the current loop is implemented using an FPGA, the current calculation delay T_c can be considered negligible, i.e., $T_c \approx 0$. Therefore, the MCU's computation delay time $T_c \approx 0$. Since the voltage command calculated by the FPGA will be updated to the PWM voltage command in the next control cycle, the MCU output delay (i.e., the ZOH output delay of the PWM) is $T_o = T_{con}/2$. Hence, the total delay time is $T_d = 0.5 \times T_{con}$, as shown in Fig. 5.40.

(*Note*: In this method, the current sampling delay time T_{SH} also approaches zero).

Key Points of This Section:

- You can use the content of this section to evaluate the delay times caused by different PWM sampling methods and then apply the method introduced in Sect. 2.2 to design the motor current control loop.

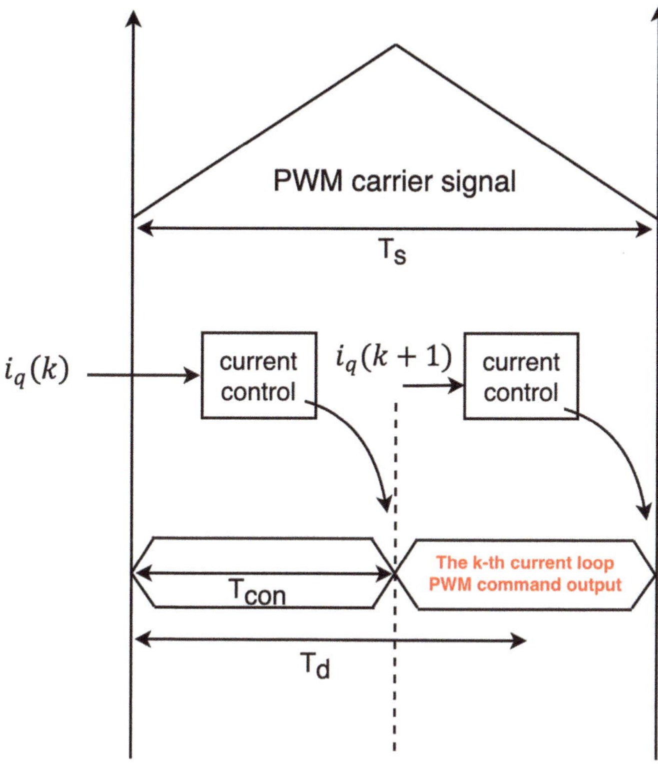

Fig. 5.39 Digital control timing diagram of the double current sampling in a single PWM

- The methods presented here are universal and not limited to AC motor current loop design—they can also be applied to other digitally controlled power electronic systems using MCUs.
- The higher the required response speed (i.e., bandwidth) of the control loop, the shorter the delay time must be. This is because higher response speed requires higher loop gain, and shorter delay increases the phase margin, which in turn allows for a higher loop gain.

5.5 Introduction to PMSM Specifications

The field-oriented control (FOC) technique for AC motors is developed based on an accurate motor model. Therefore, before applying FOC, it is essential to thoroughly understand the meaning and physical significance of the motor parameters. In this section, we will use the Permanent Magnet Synchronous Motor (PMSM) as an example to provide a comprehensive explanation of the various specification

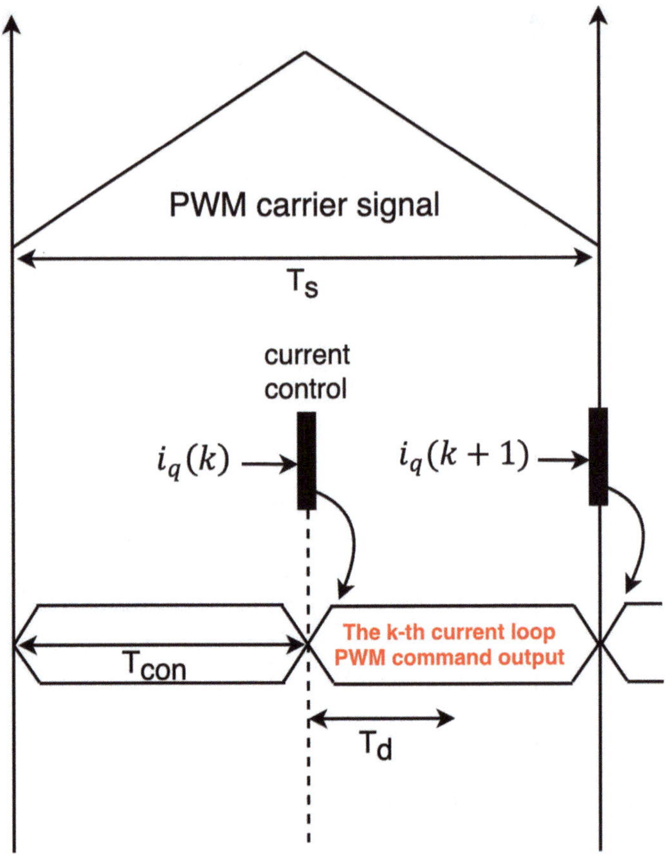

Fig. 5.40 Digital control timing diagram of the current loop control using FPGA

parameters and their physical interpretations. Let us first revisit the specifications of the PMSM used in Chap. 4, as shown in Fig. 4.6.

Below is a detailed analysis of the motor specification parameters shown in Fig. 5.41:

No. Of Poles/Phases:

This refers to the number of poles and phases of the motor. As shown in Fig. 5.41, this motor is an 8-pole, 3-phase permanent magnet synchronous motor (PMSM).

Voltage Rated (VDC):

This is the rated input voltage of the motor. In this case, the rated input voltage is 24 VDC. Typically, this voltage also limits the peak output voltage of the motor's inverter.

No. of Pol./Phases	8/3		
Voltage Rated (VDC)	24		
Current (AMP)	No load [A]	Rated [A]	Peak [A]
	0.2	1.79	5.4
Resistance / phase to phase [Ohms]@ 25°C	1.5 ± 15%		
Inductance / phase to phase [mH] @1kHz	2.1 ± 20%		
Torque Rated / Peak	Constant [Nm/A]	Rated [Nm]	Peak [Nm]
	0.035	0.0625	0.19
Power Rated [W]	26		
Speed	Rated [RPM]	No Load [RPM]	
	4000	6200	
Rotor Inertia [Kg-m^2]	2.4x10^{-6}		
Weight [Kg]	0.3		

Fig. 5.41 Nanotec DB42S03 BLDC specification

Current (AMP):

This defines the motor's input current specifications:

- **No-load Current**: 0.2 A—the input current when the motor runs at rated speed under rated voltage without any load.
- **Rated Current**: 1.79 A—the current when the motor operates at rated speed and voltage with the rated load applied.
- **Peak Current**: 5.4 A—the maximum allowable current for a short period. This corresponds to the peak torque the motor can output.

Resistance/phase-to-phase [Ohms] @ 25 °C:

This is the motor's line-to-line resistance at room temperature. In field-oriented control (FOC), we typically use the per-phase stator resistance R_s, which is half of the phase-to-phase resistance:

$$R_s = \frac{1.5}{2} = 0.75(\Omega)$$

Inductance/phase-to-phase [H] @1 kHz:

This is the motor's line-to-line inductance at room temperature. The per-phase inductance L_s is half of the phase-to-phase inductance:

$$L_s = \frac{2.1}{2} = 1.05(m\text{H})$$

The notation **@1 kHz** indicates that the inductance was measured using an AC test signal at a frequency of 1 kHz. Inductance in electrical machines is frequency-dependent because magnetic core nonlinearity, leakage flux, and eddy current effects cause the apparent inductance to vary with frequency. By specifying the measurement at 1 kHz, manufacturers provide a standardized reference value that is more representative of real inverter-driven motor operation, where switching and current dynamics typically occur in the kilohertz range.

Since this is a Surface-mounted PMSM (SPM), we have:

$$L_s = L_d = L_q = 1.05(m\mathrm{H})$$

Torque Rated/Peak:

This defines the motor's torque capabilities:

- **Rated Torque**: 0.0625 Nm
- **Peak Torque**: 0.19 Nm

The torque constant is $K_T = 0.035$, meaning each ampere of current produces 0.035 Nm of torque.

Power Rated [W]:

This is the rated mechanical output power of the motor in watts.

Speed:

The rated speeds of the motor:

- **No-load Speed**: 6200 RPM—the rotor speed when rated voltage is applied with no load.
- **Rated Speed**: 4000 RPM—the rotor speed at rated voltage under rated load.

Rotor Inertia [kg m²]:

The moment of inertia of the rotor, expressed in SI units. For this motor:

$$\mathrm{J} = 2.4 \times 10^{-6}\left(\mathrm{kg} - \mathrm{m}^2\right)$$

Weight [kg]:

The weight of the motor. For this model: weight = 0.3 (kg).

- **Torque Constant K_T of a PMSM**

It is necessary to specifically explain the **torque constant** K_T of a permanent magnet synchronous motor (PMSM). Its SI unit is **Nm/A**, which represents the amount of torque produced per ampere of current. There is another related parameter known as the **back EMF constant** K_e (also referred to as K_b), whose SI unit is **V/(rad/s)**.

This represents the DC voltage of the back electromotive force generated per unit of no-load rotational speed.

For a PMSM, in SI units:

If the motor's back EMF waveform is **trapezoidal**, then $K_T = K_e$.

However, for the PMSM in this example, the back EMF is **sinusoidal**, and it is driven using **sinusoidal control** (i.e., Field-Oriented Control or FOC).

Under such conditions, we can derive the relationship to determine the **back EMF constant** K_e of the PMSM.

In the steady state of vector control, when the **d-axis current** of the motor is zero, the **phase current** of the motor will be **in phase** with the back electromotive force (back EMF).

(*Note: This is because both the motor's q-axis current and the back EMF lead the rotor flux linkage by 90°.*)

The **input power** to the three-phase windings of the permanent magnet synchronous motor (PMSM) can be expressed as:

$$P_{sinusoidal} = 3 \times E_{p,rms} \times I_{p,rms} \tag{5.5.1}$$

where $E_{p,rms}$ is the **RMS value of the phase back EMF**, and $I_{p,rms}$ is the **RMS value of the phase current**.

Assuming that the power injected into the three-phase windings is entirely converted into mechanical power, we have:

$$3 \times E_{p,rms} \times I_{p,rms} = T \times \omega \tag{5.5.2}$$

where T is the motor output torque and ω is the motor angular velocity.

$$3 \times K_{e,rms} \times \omega \times I_{p,rms} = K_T \times I_{p,peak} \times \omega \tag{5.5.3}$$

where $K_{e,rms}$ is the RMS (root mean square) back EMF constant. Since the K_T in this example is defined using the peak phase current as the denominator, the right-hand side of Eq. (5.5.3) must be multiplied by the peak phase current $I_{p,peak}$.

Equation (5.5.3) can be rearranged as:

$$3 \times K_{e,rms} \times \omega \times I_{p,rms} = K_T \times I_{p,rms} \times \sqrt{2} \times \omega \tag{5.5.4}$$

By canceling the common terms on both sides of Eq. (5.5.4), we obtain:

$$K_{e,rms} = \frac{\sqrt{2}}{3} K_T \tag{5.5.5}$$

Substituting $K_T = 0.035$ into Eq. (5.5.5), we obtain $K_{e,rms} = 0.0165$. By multiplying $K_{e,rms}$ by $\sqrt{2}$, we get the peak phase back-EMF constant, which is:

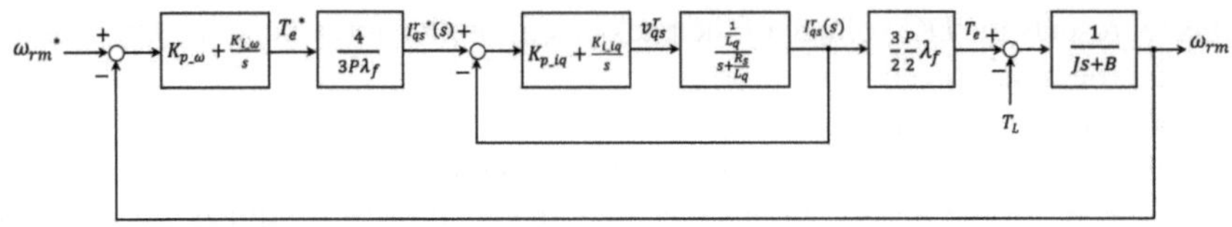

Fig. 5.42 Speed control loop of a PMSM

$$K_{e,peak} = 0.0165 \times \sqrt{2} = 0.0233 \tag{5.5.6}$$

Since the motor is Y-connected (wye connection), the peak line-to-line back-EMF constant can be expressed as:

$$K_{e,LL,peak} = 0.0233 \times \sqrt{3} = 0.0404 \tag{5.5.7}$$

Additionally, the rotor flux linkage λ_f of a permanent magnet synchronous motor (PMSM) can be calculated using the torque constant K_T.

Let us revisit the speed control loop of the PMSM shown previously in Fig. 2.18, now illustrated again in Fig. 5.42.

As shown in the figure, ω_{rm}^* represents the speed command, and T_e^* is the torque command generated by the speed loop PI controller.

The gain $\frac{4}{3P\lambda_f}$ is used to convert the torque command into a current command. This gain is the reciprocal of the motor torque constant K_T, that is:

$$\frac{4}{3P\lambda_f} = \frac{1}{K_T} \tag{5.5.8}$$

According to Eq. (5.5.8), we can substitute the motor specification parameters: $K_T = 0.035$ and the number of poles $P = 8$, to calculate the rotor flux linkage λ_f as follows:

$$\lambda_f = \frac{4K_T}{3P} = \frac{4 \times 0.035}{3 \times 8} = 0.0058(\text{Wb})$$

The rotor flux linkage λ_f is a crucial parameter required by vector control systems. Therefore, in practical applications, it is common to derive the accurate value of λ_f from the motor torque constant K_T.

Summary of This Section:

- The field-oriented control (FOC) technique for AC motors is developed based on an accurate motor model. Therefore, to successfully apply FOC, one must have a clear understanding of the meanings and physical implications of AC motor parameters.

- For permanent magnet synchronous motors (PMSMs), in SI units, if the back electromotive force (EMF) is trapezoidal and DC excitation control (i.e., 120-degree conduction) is used, then $K_e = K_T$. However, if the back EMF is sinusoidal and sinusoidal excitation (i.e., FOC control) is used, the correct back EMF constant K_e must be derived using the method presented in this section.
- All motor parameter derivations in this section are based on SI standard units. If the parameters in the motor datasheet are not in SI units, they must be converted to SI units before being applied in a vector control system.

5.6 Butterworth Filter Design

Filters are widely used in control systems to eliminate noise in feedback signals or to suppress resonance phenomena [4]. However, the requirements for filters in control systems differ from those in communication systems. In the field of communications, key performance indicators typically focus on signal distortion or attenuation. In contrast, control systems require filters that introduce minimal phase lag at the gain crossover frequency while effectively attenuating high-frequency signals.

Among various types of filters used in control systems, low-pass filters are the most common. The Butterworth filter is frequently employed because, for a given order, it offers the steepest attenuation without introducing peaking in the Bode plot. Additionally, it introduces minimal phase lag in the low-frequency range. Therefore, this section will introduce the design of Butterworth low-pass filters [4].

- **General Formula for Odd-Order Low-Pass Butterworth Filters**

For an odd-order low-pass Butterworth filter, the general form is given as follows [4]:

$$T(s) = \left(\frac{\omega_N}{s+\omega_N}\right)\prod_{i=1}^{\frac{M-1}{2}}\left(\frac{\omega_N^2}{s^2+2\cos(\theta_i)\omega_N s+\omega_N^2}\right), \theta_i = \mathrm{i} \times 180/M \qquad (5.6.1)$$

where M is the order of the filter.

For example, if you want to design a third-order Butterworth low-pass filter with a cutoff frequency of 1000 (rad/s), then $M = 3$ and $\omega_N = 1000$. The transfer function of this third-order Butterworth low-pass filter can be written as

$$T(s) = \left(\frac{1000}{s+1000}\right)\left(\frac{1000^2}{s^2+2\times\cos(\theta_1)\times 1000s+1000^2}\right) \qquad (5.6.1)$$

where $\theta_1 = 1 \times \frac{180}{3} = 60°$.

- **General Formula for Even-Order Low-Pass Butterworth Filters**

For an even-order low-pass Butterworth filter, the general form is given as follows [4]:

$$T(s) = \prod_{i=1}^{\frac{M}{2}} \left(\frac{\omega_N^2}{s^2 + 2\cos(\theta_i)\omega_N s + \omega_N^2} \right), \theta_i = (\mathrm{i} - 0.5) \times 180/M \tag{5.6.2}$$

where M is the order of the filter.

For example, if you want to design a second-order Butterworth low-pass filter with a cutoff frequency of 1000 (rad/s), then $M = 2$ and $\omega_N = 1000$. The transfer function of this second -order Butterworth low-pass filter can be written as

$$T(s) = \left(\frac{1000^2}{s^2 + 2 \times \cos(\theta_1) \times 1000s + 1000^2} \right) \tag{5.6.3}$$

where $\theta_1 = (1 - 0.5) \times \frac{180}{2} = 45°$.

From Eq. (5.6.3), we observe that the transfer function of a second-order Butterworth low-pass filter is identical to the standard second-order system transfer function with a damping ratio of 0.707 ($\zeta = \cos(45°) = 0.707$).

- **MATLAB Simulation**

Step 1:

Next, consider the following three low-pass filters:

1. **First-order low-pass filter**

$$\frac{\omega_N}{s + \omega_N} \quad \text{where} \quad \omega_N = 1000$$

2. **Second-order Butterworth low-pass filter**

$$\frac{\omega_N^2}{s^2 + 2\cos(\theta_1)\omega_N s + \omega_N^2} \quad \text{where} \quad \omega_N = 1000 \quad \text{and} \quad \theta_1 = 45°$$

3. **Third-order Butterworth low-pass filter**

$$\left(\frac{\omega_N}{s + \omega_N} \right) \left(\frac{\omega_N^2}{s^2 + 2\cos(\theta_1)\omega_N s + \omega_N^2} \right) \quad \text{where} \quad \omega_N = 1000 \quad \text{and} \quad \theta_1 = 60°$$

Step 2:

You can run the example script **m5_6_1** to plot the Bode diagrams of the three low-pass filters listed above. The resulting magnitude and phase curves are shown in Fig. 5.43.

MATLAB Example Script: m5_6_1.m

```
wn=1000;
tf_1lpf = tf(wn,[1 wn]);
tf_2bwlpf = tf(wn^2,[1 2*cos(45*pi/180)*wn wn^2]);
tf_3bwlpf =tf(wn,[1 wn])*tf(wn^2,[1 2*cos(60*pi/180)*wn wn^2]);
h=bodeoptions; h.PhaseMatching='on';
h.Title.FontSize = 14;
h.XLabel.FontSize = 14;
h.YLabel.FontSize = 14;
h.TickLabel.FontSize = 14;
bodeplot(tf_1lpf,'-b',tf_2bwlpf,'-.r',tf_3bwlpf,'.g',{1,100000},h);
legend('1_order_LPF','2_order_BWLPF', '3_order_BWLPF');
h = findobj(gcf,'type','line');
set(h,'linewidth',2); grid on;
```

Step 3:

As shown in Fig. 5.43, the **third-order Butterworth filter** has the steepest magnitude curve. Although it provides the best high-frequency attenuation among the three filters, it also introduces the **largest phase lag**. This is expected, because a third-order transfer function has three poles, which necessarily produce more phase lag than the second- or first-order filters.

By contrast, the **first-order low-pass filter** offers the poorest high-frequency attenuation, but—because its transfer function has only a single pole—its maximum phase lag is limited to **90°**, making it the smallest among the three filters. This minimal phase delay is the main reason first-order low-pass filters are so widely used in control systems.

The **second-order Butterworth filter** provides performance that falls between the first- and third-order cases: its high-frequency attenuation and phase lag are moderate, offering a compromise between signal filtering and phase delay.

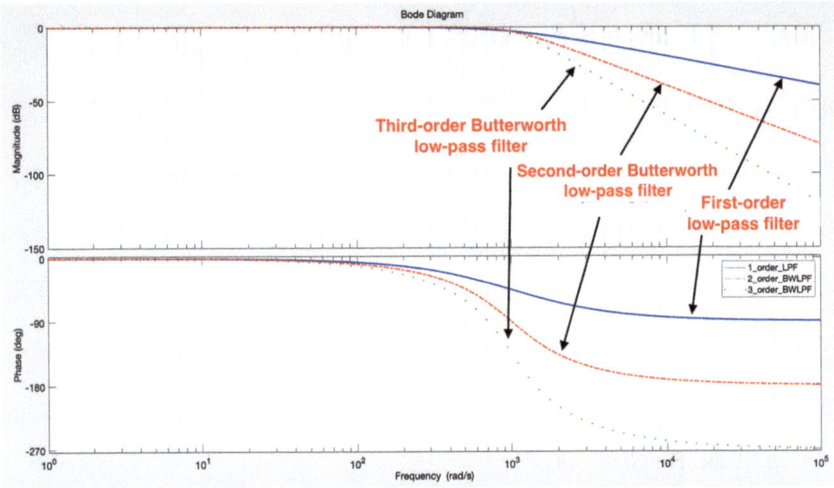

Fig. 5.43 Bode plots of first-, second-, and third-order butterworth low-pass filters

Summary of This Section:

- The **order of a filter** equals the number of poles in its transfer function. The higher the order, the **greater the phase lag** and the stronger its impact on control-system stability.
- The transfer function of a **second-order Butterworth low-pass filter** is **identical** to the standard second-order transfer function with a **damping ratio of 0.707**.
- Besides Butterworth filters, other classical prototypes—such as **Chebyshev** and **Bessel** filters—exist. However, they are used less frequently in control-system design, so they are not discussed further in this book.

5.7 Notch Filter Design

In control systems, in addition to low-pass filters, notch filters are also frequently used. Unlike a band-reject (band-stop) filter, which suppresses a range of frequencies, a notch filter is designed to attenuate a single, specific frequency. It is most commonly applied in motion-control systems for vibration suppression [4].

A typical second-order notch filter can be expressed as

$$T(s) = \frac{s^2 + \omega_N^2}{s^2 + 2\zeta\omega_N s + \omega_N^2} \tag{5.7.1}$$

where ω_N is the notch frequency to be rejected, ζ is the damping ratio that sets the notch sharpness and phase characteristics.

By substituting $s = j\omega_N$ into Eq. (5.7.1), the numerator becomes zero, indicating that the notch filter effectively suppresses signals at the specific frequency ω_N.

- **For frequencies much higher than ω_N:**
  In Eq. (5.7.1), the s^2 terms dominate both the numerator and the denominator, so the transfer function approximates to 1.
- **For frequencies much lower than ω_N:**
  Equation (5.7.1) reduces to $\frac{\omega_N^2}{\omega_N^2} = 1$.

These approximations show that the filter passes low- and high-frequency components while strongly attenuating the single, specified frequency ω_N—the hallmark behavior of a notch filter.

In general, a notch filter is designed by adjusting its damping ratio ζ. A smaller ζ produces a sharper (narrower) notch, while ζ is typically chosen to be less than 1.

You can open the example script **m5_7_1** to plot and compare the Bode diagrams of notch filters with damping ratios $\zeta = 0.2$ and $\zeta = 0.8$. The resulting magnitude and phase curves are shown in Fig. 5.44.

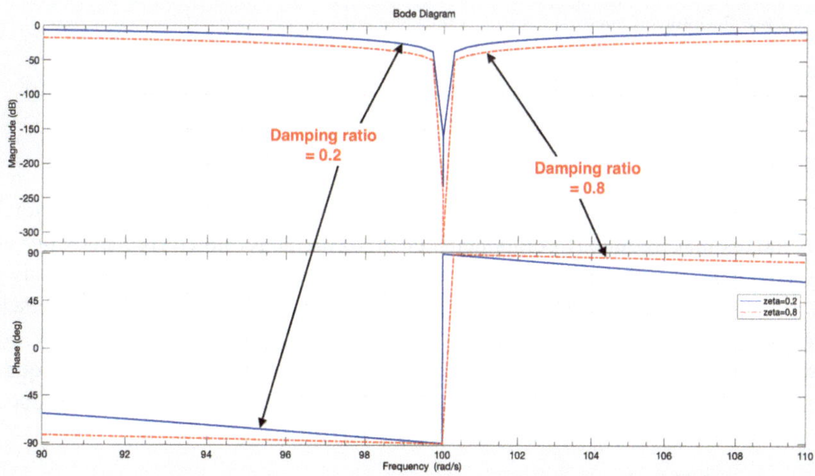

Fig. 5.44 Bode plots of notch filters with different damping ratios

MATLAB Example Script: m5_7_1.m

```
wn=100;
zeta1=0.2; zeta2=0.8;
tf_2notch_1 = tf([1 0 wn^2],[1 2*zeta1*wn wn^2]);
tf_2notch_2 = tf([1 0 wn^2],[1 2*zeta2*wn wn^2]);
h=bodeoptions; h.PhaseMatching='on';
h.Title.FontSize = 14;
h.XLabel.FontSize = 14;
h.YLabel.FontSize = 14;
h.TickLabel.FontSize = 14;
bodeplot(tf_2notch_1,'-b',tf_2notch_2,'-.r',{90,110},h);
legend('zeta=0.2','zeta=0.8');
h = findobj(gcf,'type','line');
set(h,'linewidth',2); grid on;
```

As shown in Fig. 5.44, the smaller the damping ratio ζ, the sharper the notch. This means that frequency components close to the target notch frequency ω_N are less affected—i.e., the filter is more selective—although the attenuation exactly at ω_N is weaker than that provided by a notch with a higher damping ratio. Conversely, a higher damping ratio offers stronger attenuation at the precise frequency ω_N; however, because the notch is broader, it also affects a wider range of nearby frequencies. In addition, the higher the damping ratio, the larger the phase lag introduced for signals below the notch frequency.

Summary of This Section:

- **Raising the order of a notch filter is generally unnecessary**.

The ability of a notch filter to reject the specific frequency ω_N stems from the numerator becoming zero at that frequency; adding additional poles does not enhance this fundamental property.

- **A second-order notch filter is sufficient for most control applications.**

If multiple discrete frequencies must be suppressed, simply connect several second-order notch filters in parallel (one for each target frequency).

- **When designing a notch filter for a control system, both attenuation and phase lag must be balanced.**

Stronger attenuation (higher damping ratio) broadens the notch and introduces more phase delay, whereas a sharper notch (lower damping ratio) is more selective but provides less attenuation exactly at the notch frequency.

5.8 Digital Filter Design Workflow

In Sects. 5.6 and 5.7 we introduced design methods for Butterworth low-pass filters and second-order notch filters, respectively. The designs presented there are expressed as **continuous-time (Laplace-domain) transfer functions**. In practice, however, if you want to implement these filters on a computer, the continuous-time transfer functions must be **converted to digital form** [6].

The main differences between **digital** and **analog** filters are:

- **Digital filters** can be implemented in an MCU or FPGA, and their parameters can be modified easily at virtually no cost.
- **Analog filters** must be built from discrete electronic components. Once the hardware is fabricated, modifications are difficult—or very expensive. In addition, the inherent tolerances of electronic components can reduce filter accuracy, and component values may drift over time or with environmental changes, further affecting filter performance.

Figure 5.45 illustrates a typical digital-filter design flow [6].

In the following pages, I will use a first-order low-pass filter as an example and guide you step-by-step—from **STEP 1** through **STEP 5**—to walk through the entire process.

First, let us consider an input signal Vin defined as follows:

$$V_{in} = \sin(2\pi \times 2 \times t) + \sin(2\pi \times 50 \times t) \tag{5.8.1}$$

It consists of a 2 (Hz) sine wave superimposed with a 50 (Hz) sine wave (see the upper portion of Fig. 5.46). Our goal is to design a low-pass filter that removes the 50 (Hz) component, leaving only the 2 (Hz) sine wave (see the lower portion of Fig. 5.46).

Step 1 (Designing the analog filter):

First, we take the general form of a first-order low-pass filter.

Selecting a cutoff frequency of $f_c = 5$(Hz) gives $\omega_c = 2\pi \times 5 = 31.42$ (rad/s).

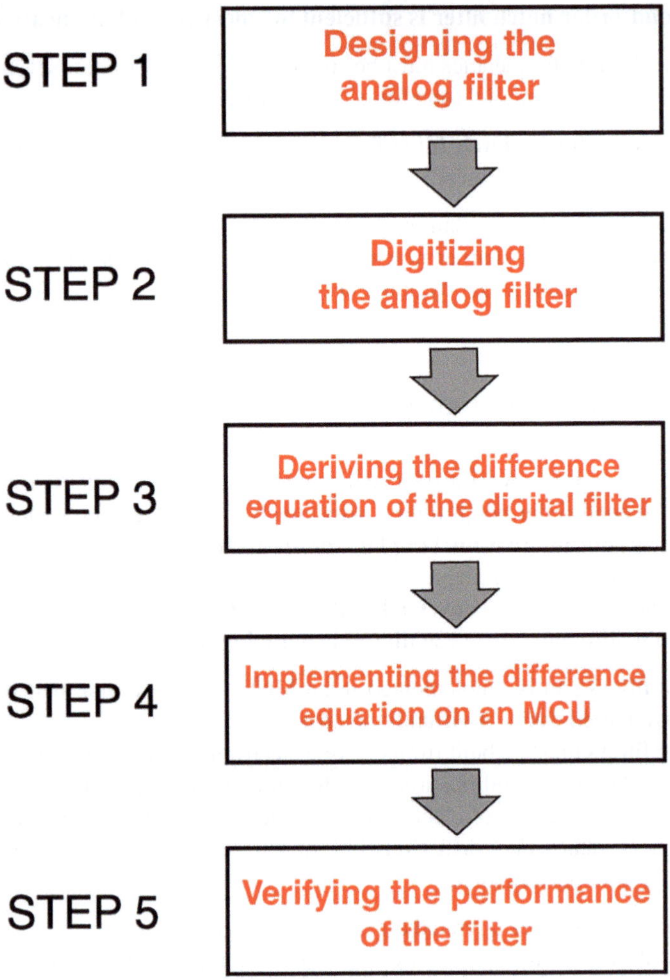

Fig. 5.45 Typical digital filter design process

Therefore, the analog transfer function of the designed first-order low-pass filter is

$$Analog\ LPF = \frac{31.42}{s + 31.42} \tag{5.8.2}$$

We use MATLAB to plot its Bode diagram.

From Fig. 5.47 we observe that the 2 Hz component is attenuated by − 0.649 dB. Converted to a linear scale, the gain is

$$10^{\frac{-0.649}{20}} = 0.928 \tag{5.8.3}$$

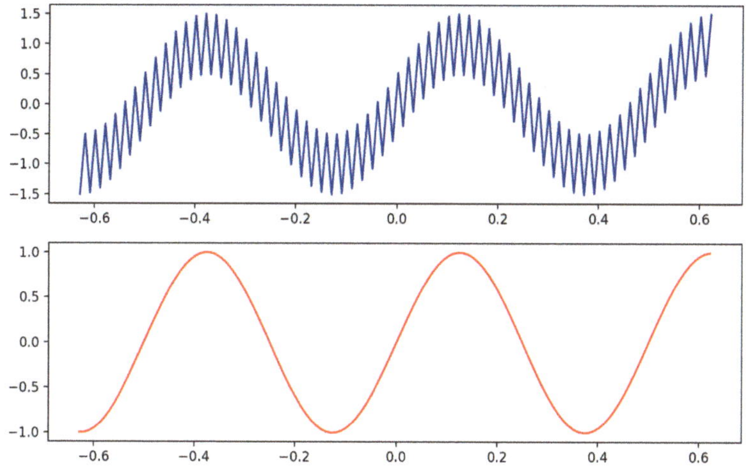

Fig. 5.46 Input signal waveforms (Upper: original signal with high-frequency noise; lower: original signal)

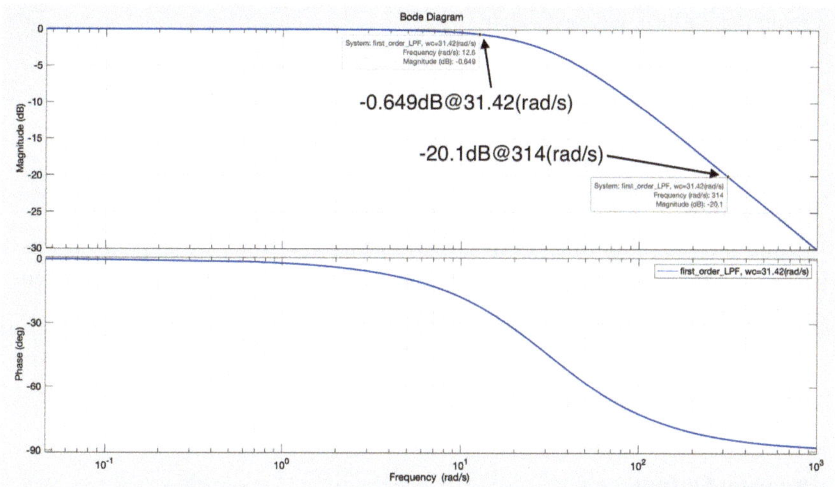

Fig. 5.47 Bode plot of the analog low-pass filter, example script: m5_8_1.m

Therefore, an attenuation of –0.649 dB means that the 2 Hz signal is reduced to 0.928 times its original amplitude.

Next, the attenuation for the 50 Hz signal is calculated as

$$10^{\frac{-20.1}{20}} = 0.0989 \tag{5.8.4}$$

Therefore, the 50 Hz signal will be reduced to approximately 0.0989 of its original amplitude.

Assuming this filter meets the application requirements, the next step is to digitizing the analog filter.

Step 2 (Digitizing the analog filter):

Assuming we digitize this filter using the bilinear (Tustin) transform and plan to implement it on a microcontroller with a sampling time of 1 (ms), we can convert the analog filter to a discrete one with the following MATLAB code.

MATLAB Example Script: m5_8_2.m

```
wc=2*pi*5;

Ts=0.001;

tf_lpf = tf(wc, [1 wc]);

dtf_lpf = c2d(tf_lpf, Ts, 'tustin');
```

Explanation

Setting the sampling time Ts to 1 (ms) means that when we implement this filter on a microcontroller, the filter calculations must be executed at a fixed interval of 1 (ms).

After running the program, you will obtain the following results:

Program Output

```
Dtf_lpf =

0.01547 z + 0.01547

-------------------

z - 0.9691

Sample time: 0.001 seconds

Discrete-time transfer function.
```

Therefore, we obtain the following Z-domain transfer function for the digital filter.

$$\frac{0.01547z + 0.01547}{z - 0.9691} \tag{5.8.5}$$

We need to express the digital filter in the form shown in Eq. (5.8.6).

$$H(z) = \frac{Y(z)}{X(z)} = \frac{b_0 + b_1 z^{-1} + b_2 z^{-2} + \dots}{1 + a_1 z^{-1} + a_2 z^{-2} + \dots} \tag{5.8.6}$$

Therefore, divide both the numerator and the denominator of Eq. (5.8.5) by the highest power of z; this yields

$$\frac{0.01547z + 0.01547}{z - 0.9691} = \frac{0.01547 + 0.01547 \times z^{-1}}{1 - 0.9691 \times z^{-1}} \tag{5.8.7}$$

Step 3 (Deriving the difference equation of the digital filter):

Once the digital filter has been expressed in the form of Eq. (5.8.6), it can be easily converted into the corresponding difference equation.

$$\begin{aligned} y[n] = &-a_1 y[n-1] - a_2 y[n-2] + \dots \\ &+ b_0 x[n] + b_1 x[n-1] + b_2 x[n-2] + \dots \end{aligned} \tag{5.8.8}$$

Therefore, based on Eqs. (5.8.7) and (5.8.8), we obtain the following difference equation:

$$y[n] = 0.9691 \times y[n-1] + 0.01547 \times x[n] + 0.01547 \times x[n-1] \tag{5.8.9}$$

At this point, we have derived the digital filter's difference equation. We can now implement this equation on a microcontroller; for demonstration purposes, I will use an **Arduino** as the microcontroller platform.

Step 4 (Implementing the difference equation on an MCU):

Next, we will use an **Arduino Uno R3** board to implement our first-order digital low-pass filter. Open the Arduino IDE on your computer and upload the following code to the Arduino Uno.

Arduino Code

```
Float xn1 = 0; // = x[n-1]
    float yn1 = 0; // = y[n-1]
    void setup() {
            Serial.begin(115200); //Baud rate =115200bps
    }
    void loop() {
            float t = micros()/1.0e6; //retrieve the current time in seconds
            //input signal
            float xn = sin(2*PI*2*t) + sin(2*PI*50*t);
            //difference equation
            float yn = 0.9691*yn1 + 0.01547*xn + 0.01547*xn1;
              //store the current x value so it can be used as x[n-1] in the next
calculation
            xn1 = xn;
              //store the current y value so it can be used as y[n-1] in the next
calculation
            yn1 = yn;
            Serial.print(xn);
            Serial.print(" ");
            Serial.println(yn);
            // Delaying for 1 ms allows the loop() function to run at a fixed 1 ms
interval
            delay(1);
    }
```

Step 5 (Verifying the performance of the filter):

- **Verify using an analog signal**

If the sketch uploads to the Arduino Uno successfully, open the **Serial Plotter** in the Arduino IDE (go to **Tools → Serial Plotter**). In the lower-left corner, set the **Baud rate** to **115200**. After the plotter starts, you should see a display similar to Fig. 3.19:

- The **blue waveform** is the analog input we created—a 2 (Hz) sine wave with a 50 (Hz) sine wave superimposed.
- The **red waveform** is the signal after passing through the low-pass filter.

From the red trace you can see that the 50 (Hz) component has been greatly attenuated, although a small portion of the 50 (Hz) signal still rides on top of the 2 (Hz) waveform. This residual ripple appears because a first-order low-pass filter can attenuate the 50 (Hz) component by roughly 90%, leaving about 10% of it in the output (Fig. 5.48).

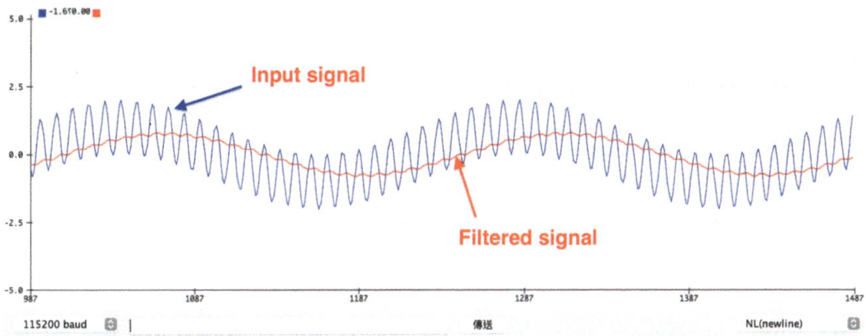

Fig. 5.48 Signal filtering simulation with Arduino

- **Verify using a real signal (generated by a signal generator)**

Next, we will verify the filter with a real physical signal. We will use an **NI myDAQ** as the signal generator to produce the input signal, read V_{in} through **Arduino pin A0**, and display the filtered result via the serial plotter. Figure 5.49 shows the hardware wiring diagram.

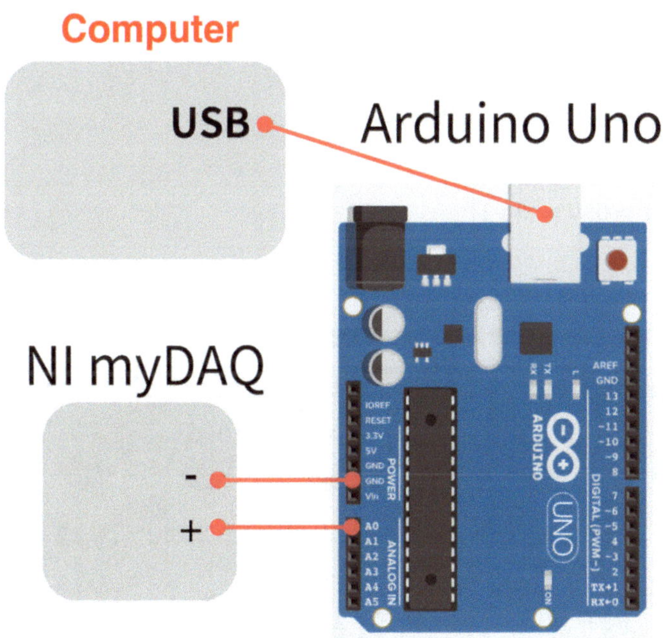

Fig. 5.49 Hardware wiring diagram

Explanation

The **NI myDAQ** is a portable data-acquisition and measurement device from National Instruments (NI). Using the software bundled with the myDAQ, the unit can function as a signal generator, oscilloscope, spectrum analyzer, multimeter, and more—making it extremely versatile. For details, refer to:

https://www.ni.com/zh-tw/shop/engineering-education/portable-student-devices/mydaq/what-is-mydaq.html

We will use the **A0 pin** on the Arduino Uno to read the analog signal. Because the Arduino Uno's analog input can measure only 0–5 V and converts that range into a 10-bit integer (0–1023), we must add a **2 V DC offset** to the input signal (this does not affect the filtering result) so that the entire waveform stays above 0 V.

Thus, the input voltage is modified to

$$V_{in} = \sin(2\pi \times 2 \times t) + \sin(2\pi \times 50 \times t) + 2$$

We generate this voltage with the **NI myDAQ** and view the resulting waveform on an oscilloscope (see Fig. 5.50).

Next, please upload the following code to the Arduino Uno.

Arduino Code

```
Int sensorPin = A0; // use Pin A0

int sensorValue = 0;

float xn1 = 0; // = x[n-1]
```

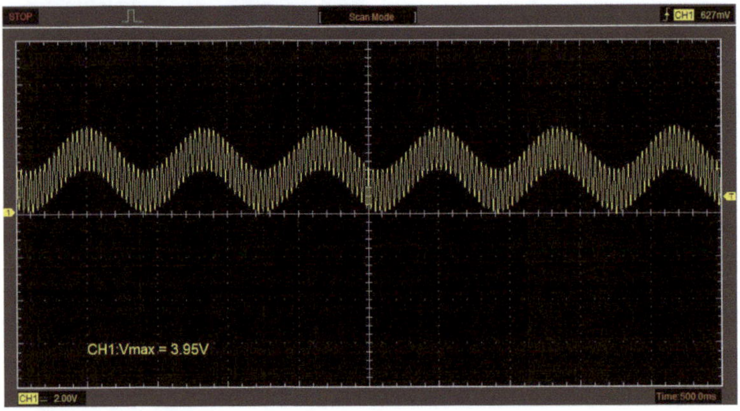

Fig. 5.50 Input signal generated by NI myDAQ

```
float yn1 = 0; // = y[n-1]
void setup() {
        Serial.begin(115200); //Baud rate=115200bps
  }
  void loop() {
        //Read analog signal from Pin A0
        sensorValue = analogRead(sensorPin);
        float xn = sensorValue;
        //Difference equation
        float yn = 0.9691*yn1 + 0.01547*xn + 0.01547*xn1;
          //store the current x value so it can be used as x[n-1] in the next
calculation
        xn1 = xn;
          //store the current y value so it can be used as y[n-1] in the next
calculation
        yn1 = yn;
        //Send back the input signal
        Serial.print(sensorValue);
        Serial.print(" ");
        //Send back the filtered result
        Serial.println(yn);
        // Delaying for 1 ms allows the loop() function to run at a fixed 1 ms
interval
        delay(1);
  }
```

If the sketch uploads to the Arduino Uno successfully, open the **Serial Plotter** in the Arduino IDE (Tools → Serial Plotter). You should see a display similar to Fig. 5.48: the **blue waveform** shows the digitized values of the analog voltage read by the Arduino, while the **red waveform** is the filtered output. As you can see, the result matches the analog test—the filter output still contains a certain proportion of the 50 Hz high-frequency component (Fig. 5.51).

- **Comparison of the Bode plots for the analog and digital filters**

Next, open the example script **m5_8_3**. When you run it, the script will plot the Bode diagrams of both the analog filter and its digital counterpart, as shown in Fig. 5.52.

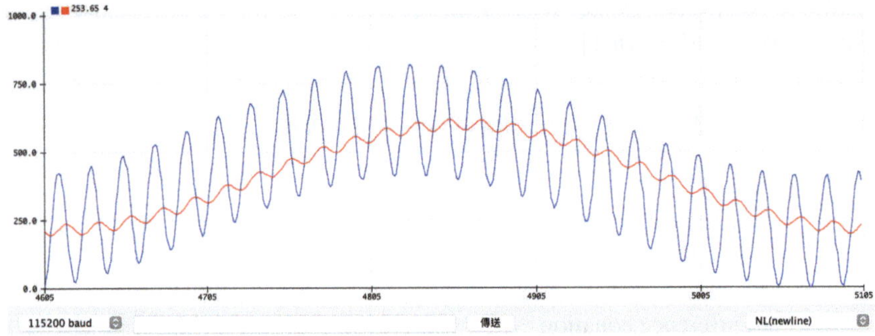

Fig. 5.51 Filtering results plotted by the serial plotter in Arduino IDE

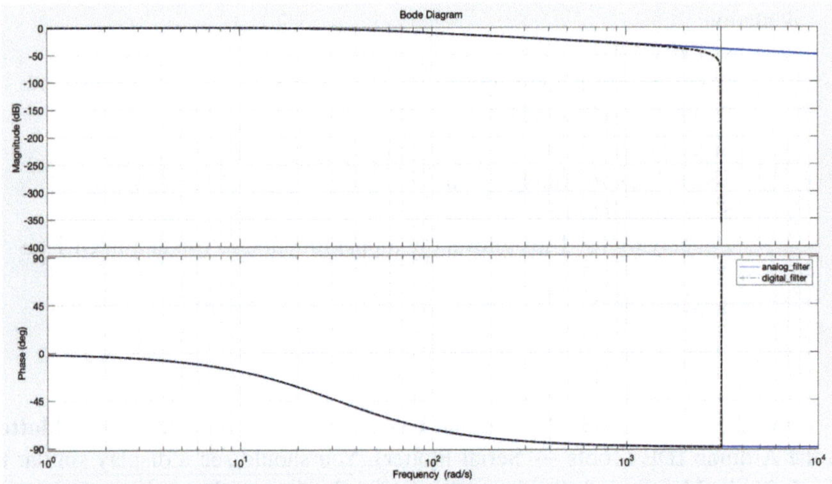

Fig. 5.52 Bode plot comparison of analog and digital filters

MATLAB Example Script: m5_8_3.m

```
wc=2*pi*5;

Ts=0.001;

tf_lpf = tf(wc, [1 wc]);

dtf_lpf = c2d(tf_lpf, Ts, 'tustin');

h=bodeoptions; h.PhaseMatching='on';

h.Title.FontSize = 14;

h.XLabel.FontSize = 14;

h.YLabel.FontSize = 14;

h.TickLabel.FontSize = 14;

bodeplot(tf_lpf,'-b',dtf_lpf,'-.k',{0,10000},h);

legend('analog_filter','digital_filter');

h = findobj(gcf,'type','line');

set(h,'linewidth',2); grid on;
```

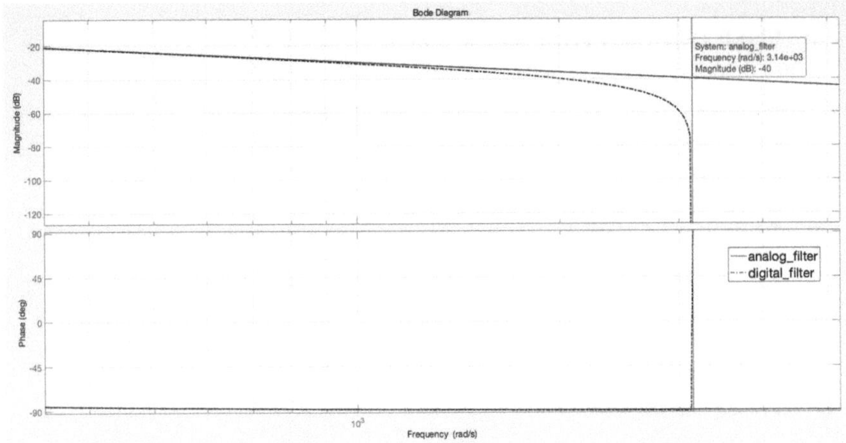

Fig. 5.53 Bode plot comparison of analog and digital filters (Zoomed near the Nyquist frequency)

From Fig. 5.52 it can be seen that the digital filter's Bode plot matches the analog filter very closely over a wide range of frequencies. However, as the frequency approaches the Nyquist frequency of 3141 (rad/s) (half the sampling rate), the distortion in the digital filter becomes increasingly pronounced, as illustrated in Fig. 5.53.

Therefore, it is important to understand that a digitized filter is merely a mathematical approximation of its analog counterpart. Consequently, the Bode plots of the two will be similar but never identical. By increasing the sampling rate, the digital filter can match the analog filter over a wider frequency range.

Summary of This Section:

- This section focused on explaining the digital-filter design workflow; we did not optimize the filter's attenuation or phase lag. In practical applications, you must choose an appropriate filter order to meet the required attenuation and evaluate the impact of the resulting phase delay.
- Although the examples here used an Arduino to validate the digital filter, you can also perform simulation and verification in SIMULINK.
- The design flow presented is not the only approach. Some engineers prefer to design a digital filter directly and skip the analog-prototype step. However, beginning with an analog filter provides clearer physical intuition, which is why the author recommends this method.

5.9 Speed Estimation via Encoder Pulse and the MT Method

In motor speed control systems, optical encoders or magnetic encoders are commonly used to calculate motor speed. Whether using incremental encoders or absolute encoders makes no difference for speed control purposes. The fundamental principle is that the controller measures the motor speed by analyzing the pulse signals returned by the encoder.

The higher the required bandwidth of the speed control loop, the higher the PPR (pulses per revolution) of the encoder must be. However, this also makes the system more sensitive to mechanical vibrations. To accurately measure speed from pulse signals, a commonly used approach is the MT method. As the name implies, the MT method combines the M method and the T method.

The **M method** calculates the number of encoder pulses within a fixed time interval. It is suitable for medium to high-speed ranges, where the encoder can return a sufficient number of pulses within the fixed interval for accurate speed measurement. However, at low speeds, the number of pulses within a fixed interval may be too few, leading to significant speed measurement errors and delays.

To compensate for this, the **T method** is used. The T method involves using a high-frequency pulse train, generated by the controller, to measure the time interval between encoder pulses. This greatly improves speed measurement accuracy at low speeds.

The **MT method** combines the advantages of both the M and T methods to achieve more accurate speed estimation across a wider speed range. The implementation of the M method, T method, and MT method will be explained in detail in the following sections [2].

- **M method**

Figure 5.54 illustrates the concept of the M-method. Here, T_s represents the speed sampling period, and β denotes the mechanical angle (in radians) corresponding to the interval between encoder pulses. The measured motor speed N_m can then be expressed as:

$$N_m = \frac{60}{2\pi}\left(\frac{m\beta}{T_s}\right) = \frac{60}{2\pi}\left(\frac{m}{T_s}\right)\left(\frac{2\pi}{PPR}\right) = \frac{60m}{T_s \times PPR}(\text{rpm}) \tag{5.9.1}$$

where $\beta = \frac{2\pi}{PPR}$.

The M-method yields larger speed measurement errors at low speeds. For example, suppose the encoder has a PPR (pulses per revolution) of 4000, and the speed sampling period is $T_s = 0.001$ seconds. If only one pulse is received during one sampling period (i.e., $m = 1$), then the lowest measurable speed using the M-method is:

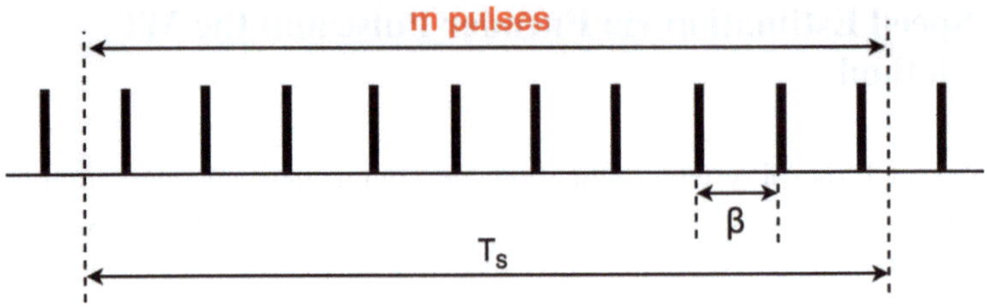

Fig. 5.54 Pulse timing diagram of the M-method

$$N_m = \frac{60 \times 1}{0.001 \times 4000} = 15(\text{rpm})$$

If the rotational speed falls below 15 RPM, it is possible that not even a single pulse will be received within one speed sampling period. As a result, the measured speed will exhibit both error and delay, and this error and delay will increase as the speed decreases.

- **T method**

Figure 5.55 illustrates the T-method. As shown in the figure, the computer inserts a high-speed pulse sequence between the encoder pulses. Here, T_h represents the period of the high-speed pulse sequence, and β represents the mechanical angle (in radians) corresponding to the interval between encoder pulses. The motor speed measured using the T-method, N_T, can be expressed as:

$$N_T = \frac{60}{2\pi}\left(\frac{\beta}{mT_h}\right) = \frac{60}{2\pi}\left(\frac{f_h}{m}\right)\left(\frac{2\pi}{PPR}\right) = \frac{60 f_h}{m \times PPR} \quad (\text{rpm}) \tag{5.9.2}$$

Here, $\beta = \frac{2\pi}{PPR}$, and f_h is the frequency of the high-speed pulse sequence, i.e., $f_h = 1/T_h$.

- **MT method**

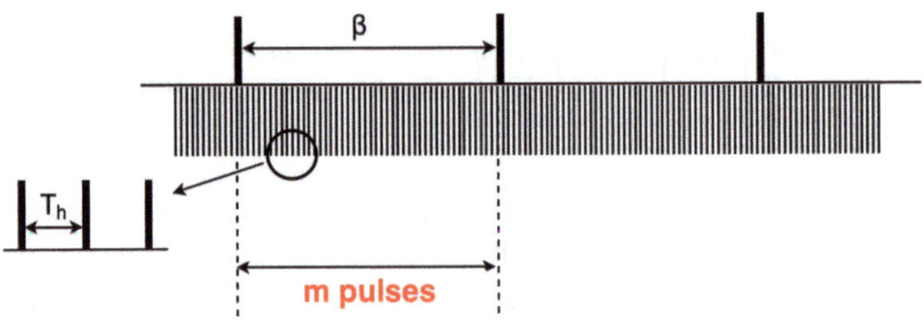

Fig. 5.55 Pulse timing diagram of the T-method

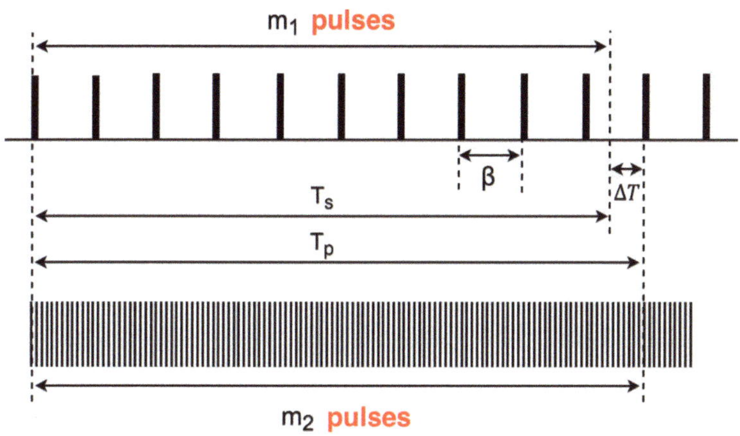

Fig. 5.56 Pulse timing diagram of the MT-method

Finally, we can combine the advantages of both the M-method and the T-method to form the MT-method, achieving more accurate speed measurement performance, as shown in Fig. 5.56. Here, T_s is the speed sampling period, and T_p is the computation period of the T method. There is a time difference ΔT between them, where $\Delta T = T_p - T_s$. The speed measurement formula using the MT method can be expressed as:

$$N_{MT} = \frac{60m_1\beta}{2\pi(T_s + \Delta T)} = \frac{60m_1}{(T_s + \Delta T) \times PPR} = \frac{60m_1}{(T_h m_2) \times PPR} \text{ (rpm)} \quad (5.9.3)$$

Therefore, at medium to high speeds, the performance of the MT-method approaches that of the M-method, i.e.,

$$N_{MT} = \frac{60m_1}{(T_s + \Delta T) \times PPR} \text{ (rpm)} \quad (5.9.4)$$

However, at very low speeds, as shown in Fig. 5.57, the MT-method approaches the T-method, i.e.,

$$N_{MT} = \frac{60m_1}{(T_h m_2) \times PPR} \text{ (rpm)} \quad (5.9.5)$$

Although at very low speeds the value of m_1 is very small, the T method can improve speed measurement resolution at low speeds through the denominator term $T_h m_2$. Therefore, by using the MT-method, the ratio between the M- and T-methods can be automatically adjusted based on the motor speed, significantly reducing speed measurement error and delay at low speeds.

However, at extremely low speeds, the time difference ΔT increases significantly, causing a mismatch between the nominal speed sampling period T_s and the actual

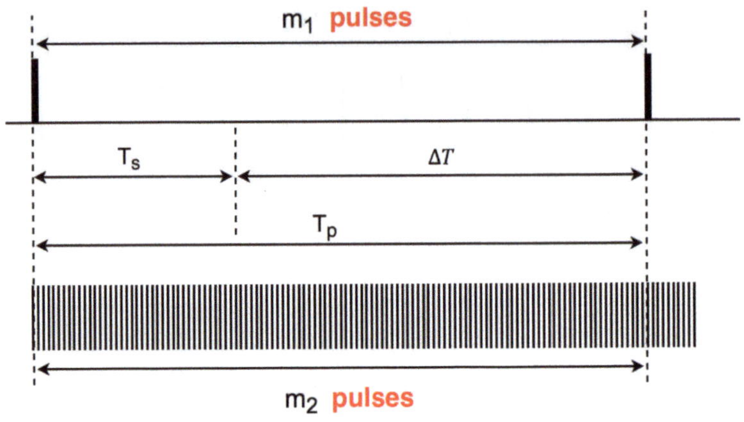

Fig. 5.57 MT-method pulse timing diagram at very low speed

speed sampling period T_p, which results in a noticeable increase in sampling delay. This in turn limits the bandwidth of the speed control loop.

Summary of This Section:

- By using the MT-method, the ratio between the M- and T-methods can be automatically adjusted based on motor speed, significantly reducing speed measurement error and delay at low speeds. However, at extremely low speeds, the time difference ΔT increases significantly, resulting in a mismatch between the nominal speed sampling period T_s and the actual speed sampling period T_p, which causes a noticeable increase in sampling delay and thereby limits the bandwidth of the speed control loop.
- The MT-method is a commonly used technique for motor speed measurement based on encoder pulses. Alternatively, a Luenberger observer can be used in conjunction with encoder pulse signals to estimate motor speed. For more details, please refer to Sect. 5.11 of this book.

5.10 The Nature and Design of the Classical Luenberger Observer

In general, using sensors incurs additional costs. In some harsh environments, sensors may even be impossible to install. Furthermore, sensors are prone to failure, which can reduce the reliability of the system. In certain scenarios, observers can be used to replace or enhance the role of sensors in control systems. In fact, observers can sometimes provide better accuracy and phase characteristics than physical sensors.

In AC motor control systems, the Luenberger observer architecture is widely used [1–4]. For example, in sensorless control techniques for permanent magnet synchronous motors (PMSMs), the back-EMF observer is essentially a Luenberger observer [3]. Since back-EMF is difficult to measure directly with sensors, a Luenberger observer in combination with current sensors is often used to estimate the back-EMF, thereby enabling sensorless speed control of the motor.

Another typical application is when low-cost sensors are used. In such cases, the Luenberger observer can effectively eliminate the phase lag introduced by the sensors and improve the performance and stability of the control loop.

In this section, we will introduce the classical Luenberger observer architecture, teach you how to design a Luenberger observer compensator, and finally demonstrate how it can effectively mitigate phase lag introduced by sensors through MATLAB/ SIMULINK simulations and verification.

- **Classical Luenberger Observer Architecture**

Figure 5.58 illustrates a typical Luenberger observer architecture. The dashed box at the top represents the actual physical system in operation, which consists of the plant and the sensor. The dashed box at the bottom depicts the Luenberger observer. In essence, the Luenberger observer includes a digital simulation module of the physical system—namely, the plant model $G_{p\_est}(s)$ *and the sensor model* $G_{s\_est}(s)$. This digital simulation module runs on a microcontroller with the goal of making the observer output closely match—or even exactly replicate—the output of the actual physical system.

To achieve this, the observer's internal models $G_{p\_est}(s)$ *and* $G_{s\_est}(s)$ must closely approximate the real plant and sensor. When the observer's output equals the real system output, the observer error $E_o(s)$ becomes zero, indicating that the observed state $C_o(s)$ is identical to the actual state $C(s)$ of the physical system.

In practice, however, due to modeling inaccuracies, noise, and disturbances, there will inevitably be some observer error $E_o(s)$. Therefore, an observer compensator $G_{co}(s)$ is introduced to form a closed-loop control system that drives the observer error

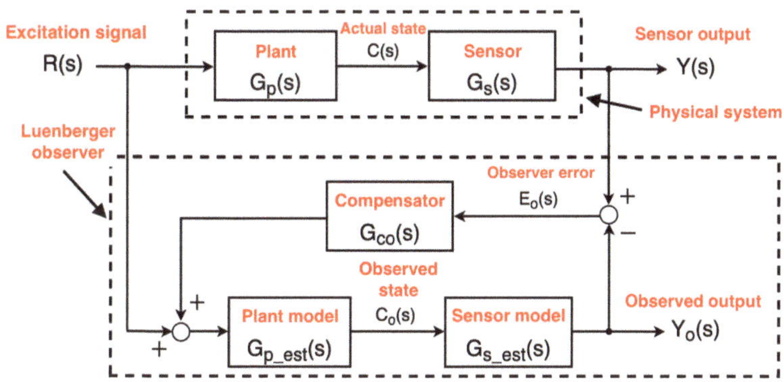

Fig. 5.58 Block diagram of a typical luenberger observer

$E_o(s)$ toward zero. This constitutes the core operating principle of the Luenberger observer.

Ideally, the sensor transfer function $G_s(s) = 1$, but in practice, sensors exhibit low-pass filter characteristics and introduce phase lag. They are typically modeled using first-order or second-order low-pass filter models. As for the plant transfer function $G_p(s)$, it is commonly represented as an integrator with gain. For example, in a motor speed control loop, the mechanical plant is modeled as $\frac{1}{Js}$, which is a single integrator. In a motor position control loop, the mechanical plant is modeled as $\frac{1}{Js^2}$, which is a double integrator.

When the plant is a single integrator, a PI-type observer compensator $G_{co}(s)$ is sufficient to stabilize the observer. However, if the plant is a double integrator, a PID-type observer compensator $G_{co}(s)$ is required for stability. This is because a double integrator inherently introduces 180 degrees of phase lag, making it difficult to stabilize. A PI-type compensator alone cannot provide the necessary positive phase margin (PM). A PID-type compensator is needed to generate sufficient phase margin and ensure the observer remains stable.

Figure 5.59 shows a typical control system architecture using a Luenberger observer. As illustrated, the Luenberger observer provides an accurate estimated state $C_o(s)$ for the main control loop, which is then used as the feedback signal for closed-loop control.

We can also represent the Luenberger observer from Fig. 5.58 in the form shown in Fig. 5.60, which offers a more intuitive understanding. In this representation, the actual sensor output $Y(s)$ becomes the input to the control loop, while the estimated sensor output $Y_o(s)$ becomes the output of the loop. The excitation signal $R(s)$ to the plant is now interpreted as a feedforward input to the loop.

This structure makes it clear that designing the observer compensator is fundamentally the same as designing the PI controller in a feedback control loop. Therefore, the controller design principles introduced in Chap. 2 can also be applied directly to the design of the observer compensator.

Using Mason's Rule, we can derive the Laplace transform of the estimated state $C_o(s)$ in Fig. 5.60 as follows:

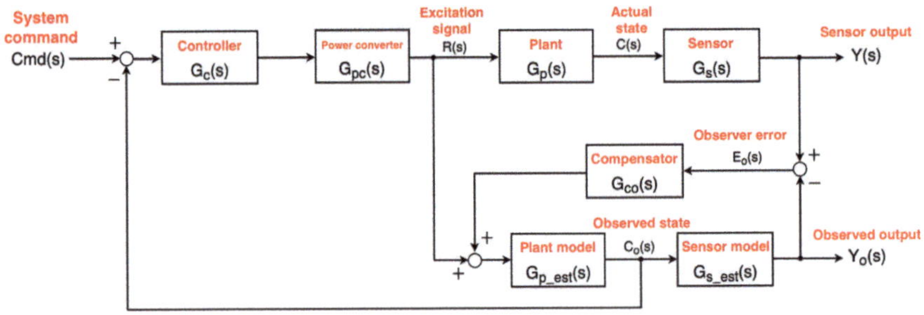

Fig. 5.59 Block diagram of a typical control system using a Luenberger observer

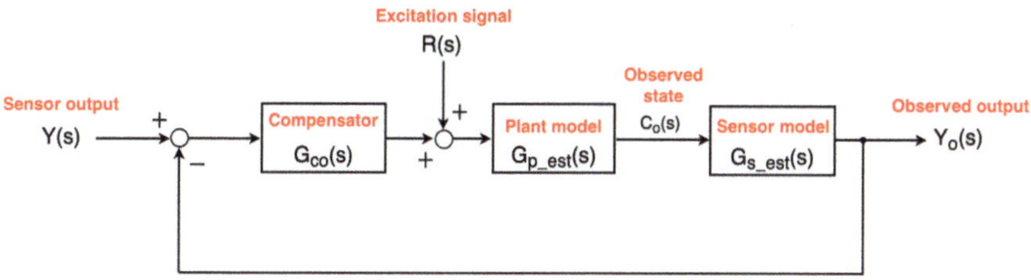

Fig. 5.60 Another representation of the Luenberger observer structure

$$C_o(s) = Y(s)\frac{G_{p\_est}(s)G_{co}(s)}{1 + G_{p\_est}(s)G_{co}(s)G_{s\_est}(s)} + R(s)\frac{G_{p\_est}(s)}{1 + G_{p\_est}(s)G_{co}(s)G_{s\_est}(s)} \tag{5.10.1}$$

- **Using the Luenberger Observer to Eliminate Sensor Phase Lag**

Step 1:

Next, we use the permanent magnet synchronous motor (PMSM) speed control loop from Sect. 2.3.1 as an example.

The motor's moment of inertia parameter, $J = 0.00016$, is the same as in Sect. 2.3.1. Assuming the speed control loop bandwidth is designed as 942 (rad/s) (note: 942 (rad/s) corresponds to 150 Hz), the PI controller parameters for the speed loop, designed using the example script m2_3_2, are as follows:

$$K_{p\_\omega} = J\omega_{sc} = 0.00016 \times 942 = 0.15$$

$$K_{i\_\omega} = K_{p\_\omega} \times \frac{\omega_{sc}}{5} = 0.15 \times \frac{942}{5} = 28.42$$

The constructed SIMULINK control loop is shown in Fig. 5.61. After running the simulation, the speed response waveform can be obtained as shown in Fig. 5.62.

Step 2:

In STEP 1, when designing the speed control loop, we assumed the speed sensor's transfer function to be unit gain, i.e., $G_s(s) = 1$, without considering any phase delay

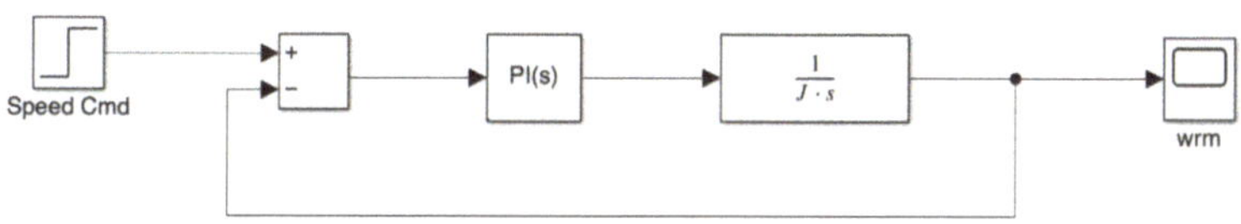

Fig. 5.61 Ideal motor speed control loop, example model: mdl_Luenberger_1.slx

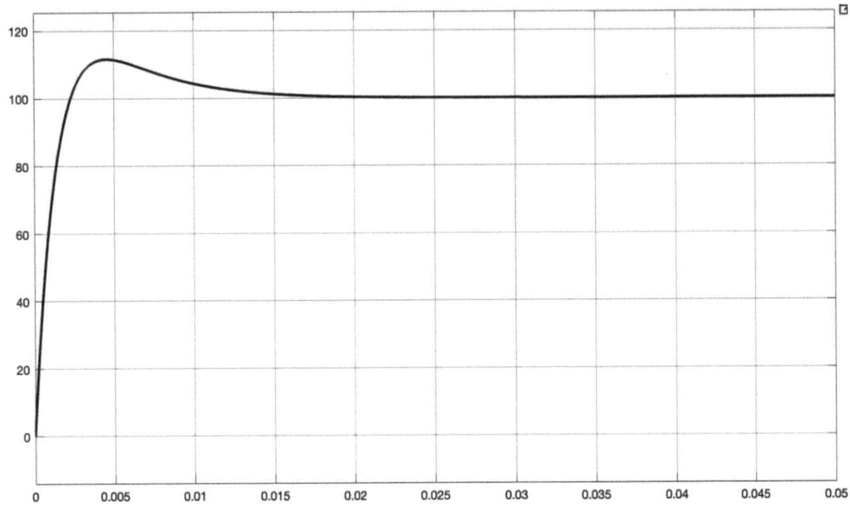

Fig. 5.62 Speed response waveform

caused by the speed sensor—this represents an ideal case, which is unattainable in reality.

In practice, lower-cost sensors might be used to reduce expenses, and such sensors typically have lower bandwidth, resulting in greater phase delay.

Here, we assume the speed sensor has a bandwidth of only 628 (rad/s) and can be modeled as a first-order low-pass filter. Therefore, we modify the SIMULINK block diagram in Fig. 5.61 as shown in Fig. 5.63.

After running the system simulation, the speed response waveform is obtained as shown in Fig. 5.64.

From the speed response waveform shown in Fig. 5.64, we can see that after adding the sensor, the transient response exhibits significant oscillations. This is caused by the phase lag introduced by the speed sensor, which erodes the phase margin (PM) of the control loop. As a result, the system's stability is greatly reduced, negatively affecting its damping characteristics.

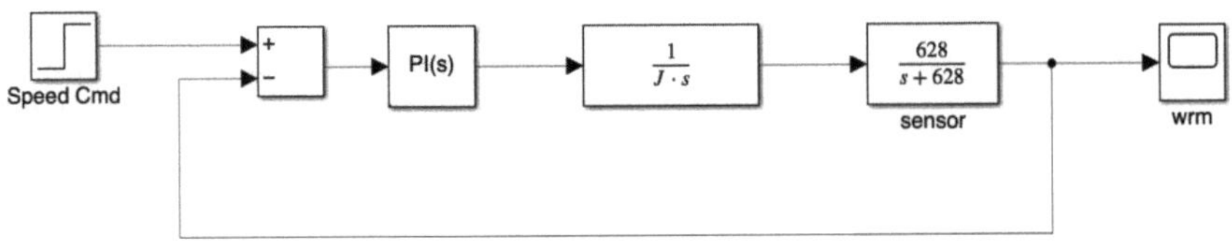

Fig. 5.63 Motor speed control loop with a non-ideal speed sensor, example model: mdl_Luenberger_1.slx

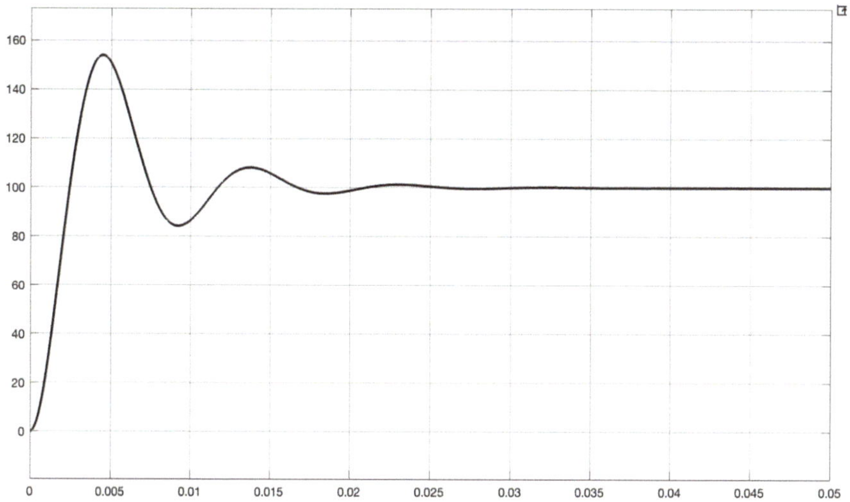

Fig. 5.64 Speed response waveform

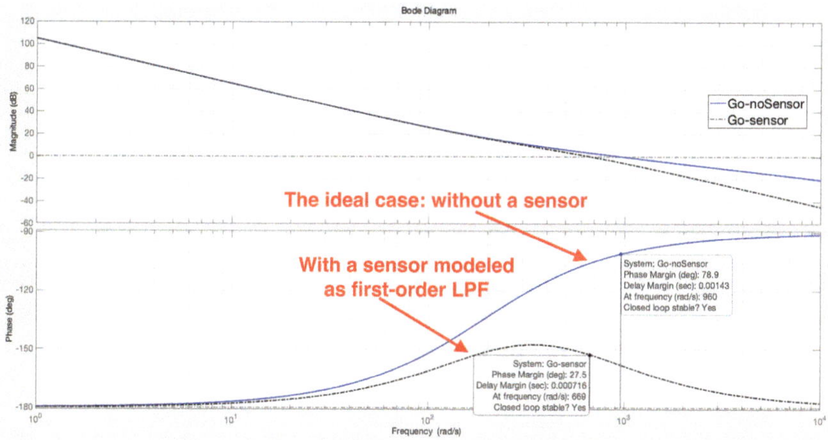

Fig. 5.65 Bode plots of the open-loop system considering a non-ideal speed sensor and of the ideal open-loop system

You can use the example program m5_10_1 to plot the open-loop Bode diagram of the system. As shown in Fig. 5.65, the diagram displays the open-loop Bode plot without the sensor and the plot with the sensor included. From the figure, we observe that before adding the sensor, the system's phase margin (PM) was 78.9 degrees. After including the speed sensor, due to the additional phase lag, the phase margin drops significantly to 27.5 degrees.

MATLAB Example Script: m5_10_1.m

```
wsc=942; J=0.00016;
Kp_w=J*wsc; Ki_w=Kp_w*wsc/5;
tf_pi=tf([Kp_w Ki_w],[1 0]);
tf_sensor = tf(628,[1 628]);
tf_plant=tf(1,[J 0]);
Go_noSensor = tf_pi*tf_plant;
Go_sensor = tf_pi*tf_sensor*tf_plant;
h=bodeoptions; h.PhaseMatching='on';
h.Title.FontSize = 14;
h.XLabel.FontSize = 14;
h.YLabel.FontSize = 14;
h.TickLabel.FontSize = 14;
bodeplot(Go_noSensor,'-b',Go_sensor,'-.k',{1,10000},h);
legend('Go-noSensor','Go-sensor')
h = findobj(gcf,'type','line');
set(h,'linewidth',2);
grid on;
```

Step 2:

Next, we add a Luenberger observer to address this issue. The Luenberger observer is incorporated into the system shown in Fig. 5.63, resulting in the modified system shown in Fig. 5.66. As illustrated, the main speed control loop no longer uses the sensor output as the speed feedback signal; instead, it uses the observed state $C_o(s)$ from the observer.

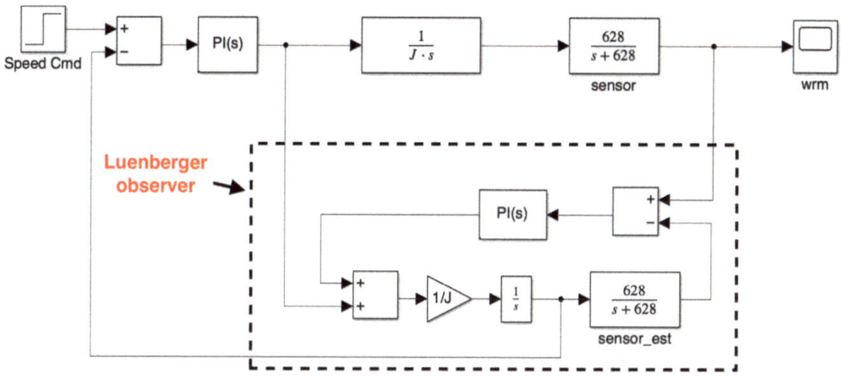

Fig. 5.66 Motor speed control loop with a luenberger observer, example model: mdl_Luenberger_1.slx

By using the observed state $C_o(s)$, the system can eliminate the phase delay effect introduced by the speed sensor.

Next, we need to design the observer compensator $G_{co}(s)$. Since the plant is a single integrator system, a PI controller is sufficient for this purpose. You can refer to the control architecture in Fig. 5.66 and use the PI controller design method that accounts for time delay effects, as discussed in Sect. 2.3.2, to design the observer compensator $G_{co}(s)$.

First, set the excitation signal $R(s)$ to zero. The open-loop transfer function of the Luenberger observer, $G_{L\_open}(s)$, *can then be expressed as:*

$$G_{L\_open}(s) = \left(K_{p\_co} + \frac{K_{i\_co}}{s}\right) \times e^{-sT_{LPF}} \times \frac{1}{Js} \tag{5.10.2}$$

where K_{p_co} *and* K_{i_co} *are the proportional and integral gains of the observer compensator* $G_{co}(s)$, T_{LPF} is the time delay introduced by the speed sensor, given by $T_{LPF} = \frac{1}{628} = 0.0016$ seconds, and the moment of inertia is $J = 0.00016$.

Next, using Eqs. (2.3.19) to (2.3.21) to design the PI controller parameters, we can choose $\alpha = 2$. According to Eq. (2.3.20), the closed-loop bandwidth of the observer ω_{Lc} can then be designed as:

$$\omega_{Lc} = \frac{1}{\alpha T_{LPF}} = 312.5 \quad (\text{rad/s}) \tag{5.10.3}$$

The corresponding PI controller parameters are as follows:

$$K_{p\_co} = \frac{J}{\alpha T_{LPF}} = 0.05, \quad K_{i\_co} = \frac{K_{p\_co}}{\alpha^2 T_{LPF}} = 7.8125 \tag{5.10.4}$$

Using the method above, the observer closed-loop bandwidth ω_{Lc} can be designed as 312.5 (rad/s). However, please note that the observer loop also includes a feedforward input $R(s)$. This feedforward term $R(s)$ can effectively enhance the observer's response speed and suppress transient overshoot (if this is unclear, you may refer back to the feedforward compensation discussion in Chap. 2).

Next, input the PI controller parameters from Eq. (5.10.4) into the system shown in Fig. 5.66, and run the SIMULINK simulation.

Step 3:

Figure 5.67 shows the speed response waveforms under three different conditions:

(1) the ideal case without a sensor (Fig. 5.62),
(2) with a sensor included (Fig. 5.64), and
(3) using the Luenberger observer.

From the waveforms, it can be observed that the inherent phase lag of the sensor erodes the system's phase margin and causes oscillations. In contrast, although the response with the Luenberger observer is slightly slower, it significantly mitigates the oscillation caused by the sensor's phase delay.

Summary of this section:

- The control plant used in this section's example is the motor speed loop mechanical plant $\frac{1}{Js}$, which is a single integrator. Therefore, a PI controller is sufficient to stabilize the observer loop. In the next section, I will introduce another classic application of the Luenberger observer: estimating motor speed using position signals from an optical encoder. In that case, the plant is a double integrator

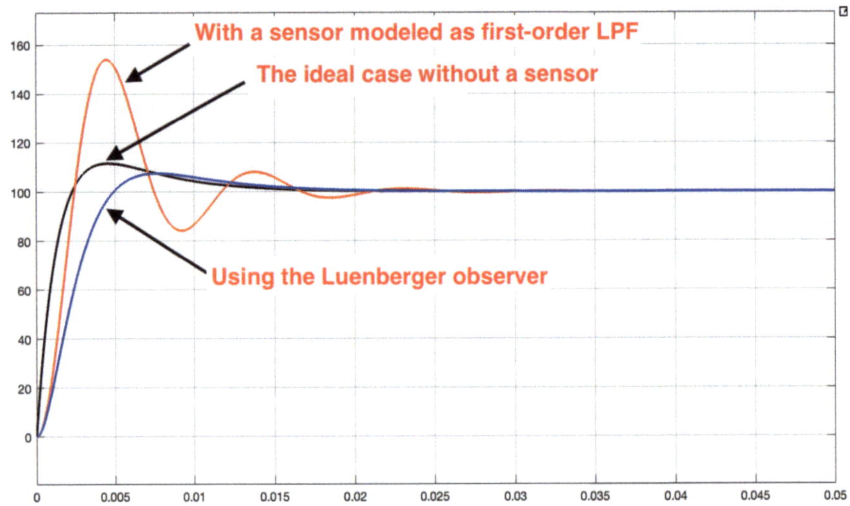

Fig. 5.67 Comparison of motor speed responses: with Luenberger observer versus the two previous cases

system, so a PID-type observer compensator will be required to stabilize the observer loop.

- This section adopts the method from Sect. 2.3.2 to design the observer compensator. This method enables effective bandwidth design for the observer compensator. Furthermore, the observer loop includes a feedforward input $R(s)$, which can significantly improve the observer's response speed and suppress transient overshoot.

5.11 Luenberger Observer Design for Speed and Disturbance Estimation

In Sect. 5.10, we explored the general structure of the Luenberger observer and the method for designing its compensator. In this section, I will introduce a classic application of the Luenberger observer in motor control: real-time estimation of motor speed using encoder position signals [2, 4]. We will employ state-space pole placement to design the Luenberger observer. Since this method is developed based on the mechanical equations of the motor, it is highly versatile and can be applied not only to permanent magnet synchronous motors (PMSMs) but also to other motion systems equipped with encoders, such as induction motors or switched reluctance motors.

- **Derivation of the State Observer Equations**

First, consider the state-space equations of a linear time-invariant (LTI) system as follows:

$$\dot{\boldsymbol{x}} = \boldsymbol{A}\boldsymbol{x} + \boldsymbol{B}\boldsymbol{u} \tag{5.11.1}$$

where $\boldsymbol{x}$ represents the system state vector, $\boldsymbol{A}$ is the system matrix, $\boldsymbol{B}$ is the input matrix, and $\boldsymbol{u}$ represents the system input vector.

Assuming the system states are observable, the state observer for this linear time-invariant system can be expressed as:

$$\dot{\hat{\boldsymbol{x}}} = \boldsymbol{A}\hat{\boldsymbol{x}} + \boldsymbol{B}\boldsymbol{u} \tag{5.11.2}$$

Here, $\hat{\boldsymbol{x}}$ represents the estimated value of the system state $\boldsymbol{x}$. If the initial state $\boldsymbol{x}(0)$ of the system is known precisely and the matrices $\boldsymbol{A}$ and $\boldsymbol{B}$ are completely accurate, then the correct system state $\hat{\boldsymbol{x}}$ can be estimated at any time using Eq. (5.11.2). However, in reality, this is difficult to achieve because the system's initial state $\boldsymbol{x}(0)$ is often unknown. Even if the initial state is known, there may still be inaccuracies in the $\boldsymbol{A}$ and $\boldsymbol{B}$ matrices. As a result, the estimated state $\hat{\boldsymbol{x}}$ will deviate from the actual state $\boldsymbol{x}$, and thus, we define the estimation error $\tilde{\boldsymbol{x}}$.

$$\tilde{\boldsymbol{x}} = \boldsymbol{x} - \hat{\boldsymbol{x}} \tag{5.11.3}$$

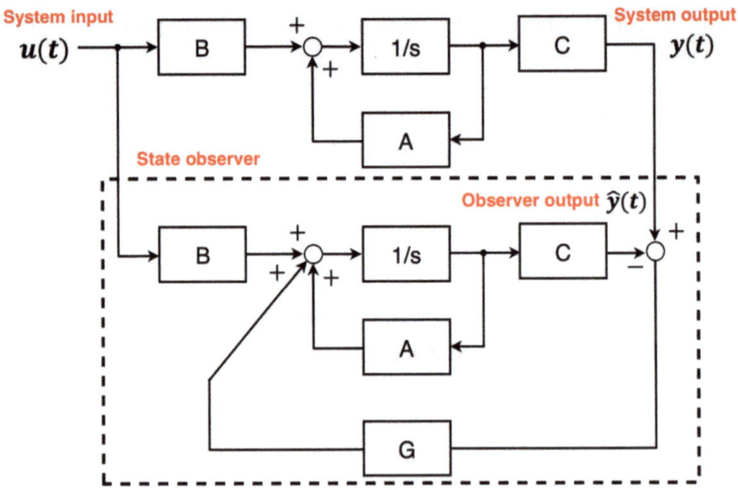

Fig. 5.68 Block diagram of a typical Luenberger observer system

Using Eqs. (5.11.1) and (5.11.2), we can obtain:

$$\dot{\tilde{\boldsymbol{x}}} = \boldsymbol{A}\tilde{\boldsymbol{x}} \tag{5.11.4}$$

If the system is stable, the estimation error $\tilde{\boldsymbol{x}}$ will converge to zero over time. This implies that all eigenvalues of matrix A lie in the left half of the complex plane. The rate at which the error converges depends on the magnitude of these eigenvalues. We can determine the speed of convergence by designing the state feedback matrix accordingly. Thus, the observer can be designed as shown in Fig. 5.68.

As shown in Fig. 5.68, the dashed box represents the state observer. Without loss of generality, we can introduce an output matrix $\boldsymbol{C}$, which maps the system state to the system output. The matrix $\boldsymbol{G}$ is the observer gain matrix. By designing the matrix $\boldsymbol{G}$, the observer can be tuned to achieve the desired performance.

The state-space equation of the observer with the observer gain matrix $\boldsymbol{G}$ included is given by:

$$\dot{\hat{\boldsymbol{x}}} = \boldsymbol{A}\hat{\boldsymbol{x}} + \boldsymbol{B}\boldsymbol{u} + \boldsymbol{G}(\boldsymbol{y} - \boldsymbol{C}\hat{\boldsymbol{x}}) \tag{5.11.5}$$

Subtracting Eq. (5.11.5) from Eq. (5.11.1), we obtain:

$$\dot{\tilde{\boldsymbol{x}}} = (\boldsymbol{A} - \boldsymbol{G}\boldsymbol{C})\tilde{\boldsymbol{x}} \tag{5.11.6}$$

The characteristic equation of Eq. (5.11.6) can be obtained from Eq. (5.11.7).

$$\det[sI - (\boldsymbol{A} - \boldsymbol{G}\boldsymbol{C})] = 0 \tag{5.11.7}$$

Here, det[*] represents the determinant of the matrix *.

Therefore, by designing the state feedback gain matrix $\boldsymbol{G}$, we can make the estimation error $\tilde{\boldsymbol{x}}$ converge to zero at the desired speed. In practice, even if there are reasonable errors in the $\boldsymbol{A}$, $\boldsymbol{B}$, and $\boldsymbol{C}$ matrices, a properly designed feedback gain matrix $\boldsymbol{G}$ can still ensure that the estimation error $\tilde{\boldsymbol{x}}$ converges to zero.

- **Design of a State Observer Based on the Motor Mechanical Model**

Next, we can apply the observer model described above to a motor speed control system. Consider the motor mechanical equation as follows:

$$T_e = J\frac{d\omega_{rm}}{dt} + B\omega_{rm} + T_L \tag{5.11.8}$$

We choose the system state as $\boldsymbol{x} = \begin{bmatrix} \theta_{rm} & \omega_{rm} & T_L \end{bmatrix}^T$, where θ_{rm} is the mechanical angle of the motor, which can be measured using an encoder. Thus, the motor mechanical equation can be converted into a state-space equation as follows:

$$\frac{d}{dt}\begin{bmatrix} \theta_{rm} \\ \omega_{rm} \\ T_L \end{bmatrix} = \begin{bmatrix} 0 & 1 & 0 \\ 0 & -\frac{B}{J} & -\frac{1}{J} \\ 0 & 0 & 0 \end{bmatrix}\begin{bmatrix} \theta_{rm} \\ \omega_{rm} \\ T_L \end{bmatrix} + \begin{bmatrix} 0 \\ \frac{1}{J} \\ 0 \end{bmatrix} T_e \tag{5.11.9}$$

Here, we set $\frac{dT_L}{dt} = 0$, assuming that the variation of the load torque with respect to time is much slower than the variations of angle and speed. Based on the state-space equation in (5.11.9), we can design the corresponding state observer as shown in Eq. (5.11.10).

$$\begin{aligned}\frac{d}{dt}\begin{bmatrix} \hat{\theta}_{rm} \\ \hat{\omega}_{rm} \\ \hat{T}_L \end{bmatrix} &= \begin{bmatrix} 0 & 1 & 0 \\ 0 & -\frac{\hat{B}}{\hat{J}} & -\frac{1}{\hat{J}} \\ 0 & 0 & 0 \end{bmatrix}\begin{bmatrix} \hat{\theta}_{rm} \\ \hat{\omega}_{rm} \\ \hat{T}_L \end{bmatrix} + \begin{bmatrix} 0 \\ \frac{1}{\hat{J}} \\ 0 \end{bmatrix} T_e \\ &+ \begin{bmatrix} g_1 \\ g_2 \\ g_3 \end{bmatrix}\left(\theta_{rm} - \begin{bmatrix} 1 & 0 & 0 \end{bmatrix}\begin{bmatrix} \hat{\theta}_{rm} \\ \hat{\omega}_{rm} \\ \hat{T}_L \end{bmatrix}\right)\end{aligned} \tag{5.11.10}$$

Here, the state feedback gain matrix is $\boldsymbol{G} = \begin{bmatrix} g_1 & g_2 & g_3 \end{bmatrix}^T$, and the output matrix is $\boldsymbol{C} = \begin{bmatrix} 1 & 0 & 0 \end{bmatrix}$. We can use the pole placement method to design the state feedback gain matrix $\boldsymbol{G}$. First, based on Eqs. (5.11.7) and (5.11.10), we can write the system's characteristic equation as follows:

$$\det[sI - (\boldsymbol{A} - \boldsymbol{GC})] = s^3 + \frac{g_1\hat{J} + \hat{B}}{\hat{J}}s^2 + \frac{g_2\hat{J} + g_1\hat{B}}{\hat{J}}s - \frac{g_3}{\hat{J}} = 0 \tag{5.11.11}$$

Next, the characteristic equation can be designed as a triple root form, that is:

$$\delta(s) = (s - \beta)^3 = s^3 - 3\beta s^2 + 3\beta^2 s - \beta^3 = 0 \tag{5.11.12}$$

Assuming friction is neglected, i.e., $\hat{B} = 0$, and using Eqs. (5.11.11) and (5.11.12), the state feedback gain values can be obtained as follows:

$$\begin{aligned} g_1 &= -3\beta \\ g_2 &= 3\beta^2 \\ g_3 &= \hat{J}\beta^3 \end{aligned} \tag{5.11.13}$$

In general, the observer's bandwidth should be significantly greater than that of the control loop that uses the estimated state. When designing the observer using the pole placement method, the location of the poles has a substantial impact on the observer's performance. If the poles are placed to the left of the imaginary axis (i.e., negative real part), the observer will be stable, which is the desired condition during observer design. The following summarizes the effects of placing poles farther from the imaginary axis:

- **Response Speed**: The farther the poles are from the imaginary axis (i.e., the larger their absolute values), the faster the observer's response will be, meaning higher bandwidth. In other words, the observer output can more quickly track changes in observed inputs or states.
- **Noise Sensitivity**: Placing poles farther from the imaginary axis increases the observer's sensitivity to input or measurement noise. Thus, while aiming for fast response, the noise rejection capability must also be considered.
- **Stability Margin**: Poles placed farther from the imaginary axis provide a greater stability margin, making it more likely that the observer remains stable under variations in system parameters.
- **Overshooting**: Placing poles too far left may cause excessive overshooting in the observer, which could be undesirable for some applications.

In this context, we design the state feedback gain matrix using a triple-root characteristic equation. The parameter β determines the observer's bandwidth and performance. If β is too large, it may lead to oscillation or instability. If β is too small, the observer may become too susceptible to noise.

To simplify the design of the observer's bandwidth, the characteristic equation can be configured in the form of a third-order Butterworth filter. This allows the observer's bandwidth to be directly determined by the cutoff frequency of the Butterworth filter.

- **Design of the Feedback Gain Matrix Based on the Butterworth Filter**

Consider the general form of a third-order Butterworth low-pass filter.

$$\frac{\omega_N}{s + \omega_N}\left(\frac{\omega_N^2}{s^2 + \omega_N s + \omega_N^2}\right) \tag{5.11.14}$$

where ω_N is the cut-off frequency of the filte.

The characteristic equation of Eq. (5.11.14) can be expressed as:

$$s^3 + 2\omega_N s^2 + 2\omega_N^2 s + \omega_N^3 \tag{5.11.15}$$

Assuming the friction is neglected, i.e., $\hat{B} = 0$, and using Eqs. (5.11.11) and (5.11.15), the state feedback gain values can be designed as follows:

$$\begin{aligned} g_1 &= 2\omega_N \\ g_2 &= 2\omega_N^2 \\ g_3 &= -\hat{J}\omega_N^3 \end{aligned} \tag{5.11.16}$$

Therefore, using Eq. (5.11.16), the pole placement problem of the observer can be transformed into a bandwidth design problem. In other words, the observer's bandwidth can be determined by the cutoff frequency of the Butterworth filter.

- **Luenberger Observer Structure for State Estimation**

We can expand and rearrange the right-hand side of Eq. (5.11.10) as follows:

$$\frac{d}{dt}\begin{bmatrix} \hat{\theta}_{rm} \\ \hat{\omega}_{rm} \\ \hat{T}_L \end{bmatrix} = \begin{bmatrix} \hat{\omega}_{rm} + g_1(\theta_{rm} - \hat{\theta}_{rm}) \\ -\frac{\hat{B}}{\hat{J}}\hat{\omega}_{rm} - \frac{1}{\hat{J}}\hat{T}_L + \frac{1}{\hat{J}}T_e + g_2(\theta_{rm} - \hat{\theta}_{rm}) \\ g_3(\theta_{rm} - \hat{\theta}_{rm}) \end{bmatrix} \tag{5.11.17}$$

Equation (5.11.17) can be transformed into the block diagram shown in Fig. 5.69.

The control block diagram in Fig. 5.69 is quite similar to the Luenberger control block diagram shown in Fig. 5.60. Therefore, we have successfully transformed the state-space-based observer into the general form of a Luenberger observer. As shown in the figure, the gain g_3 can be interpreted as the integral gain, the gain $g_2 \times \hat{J}$ as the proportional gain, and the feedforward gain g_1 can be translated into a derivative gain $g_1 \times \hat{J}$. Thus, the control block diagram in Fig. 5.69 can be equivalently represented by the one in Fig. 5.70.

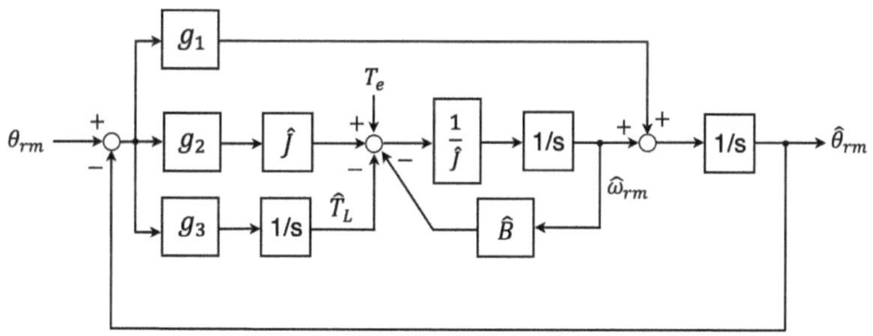

Fig. 5.69 Luenberger observer system block diagram

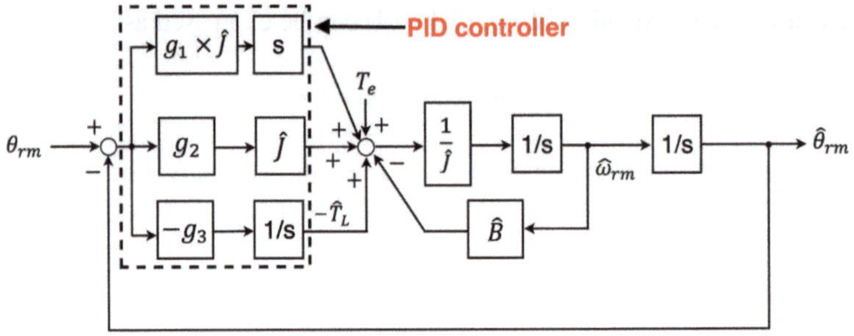

Fig. 5.70 Equivalent transformation of the Luenberger observer system block diagram

From Fig. 5.70, we can observe that when the friction coefficient is neglected (i.e., $\hat{B} = 0$), the plant becomes a double integrator system. Therefore, to stabilize the control loop, a PID controller is required for compensation. The resulting PID controller parameters for the observer are as follows:

- **Proportional gain of the observer**: $K_p = g_2 \times \hat{J}$
- **Integral gain of the observer**: $K_i = -g_3$ (5.11.18)
- **Derivative gain of the observer**: $K_d = g_1 \times \hat{J}$

In practice, to avoid the amplification of noise caused by the derivative operation, implementation is typically based on the structure in Fig. 5.69. Using Eq. (5.11.16), the PI controller parameters and feedforward gain g_1 for the observer can be designed as follows:

- **Proportional gain of the observer**: $K_p = 2\omega_N^2 \times \hat{J}$
- **Integral gain of the observer**: $K_i = \hat{J}\omega_N^3$ (5.11.19)
- **Feedforward gain**: $g_1 = 2\omega_N$

- **MATLAB/SIMULINK Simulation**

Step 1:

We continue using the motor speed control loop from Sect. 5.10 for simulation verification. Here, we still neglect the friction coefficient, i.e., $\hat{B} = 0$, and assume that motor angle information is available through the encoder. Since the bandwidth of the motor speed loop in Sect. 5.10 was designed to be 942 (rad/s), and the observer bandwidth should be greater than the control loop bandwidth (typically, the observer bandwidth should be 2 to 5 times that of the control loop bandwidth), we choose the cutoff frequency of the third-order Butterworth filter to be $\omega_N = 3 \times 942 = 2826$ (rad/s).

Thus, the observer's PID controller parameters are designed as follows:

$$\begin{aligned} &\bullet\ \textbf{Proportional gain}: K_p = 2\omega_N^2 \times \hat{J} = 2555.6 \\ &\bullet\ \textbf{Integral gain}: K_i = \hat{J}\omega_N^3 = 3.61e06 \\ &\bullet\ \textbf{Feedforward gain}: g_1 = 2\omega_N = 5652 \end{aligned} \tag{5.11.20}$$

You may use the example program m5_11_1 to list the observer's characteristic roots. In this example, using the characteristic equation of a third-order Butterworth low-pass filter, the characteristic roots are:

$$-2826, -1413 + \mathrm{j}2447.4, \text{ and } -1413 - \mathrm{j}2447.4.$$

MATLAB Example script: m5_11_1.m

```
Wn=2826;
roots([1 2*wn 2*wn^2 wn^3]).
```

Step 2:

Please open the example program mdl_Luenberger_2.slx. Once opened, the model should appear as shown in Fig. 5.71. The compensator parameters designed in Eq. (5.11.20) have already been updated in the simulation's PID controller module.

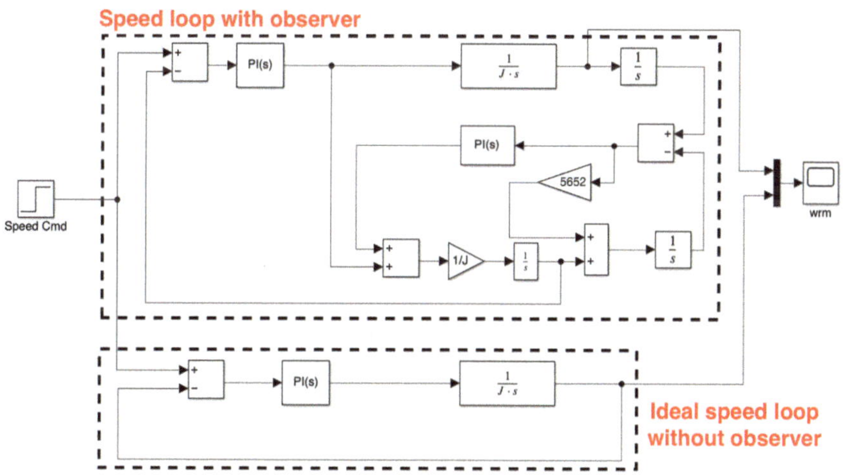

Fig. 5.71 SIMULINK simulation of the speed control loop using a Luenberger observer, example model: mdl_Luenberger_2.slx

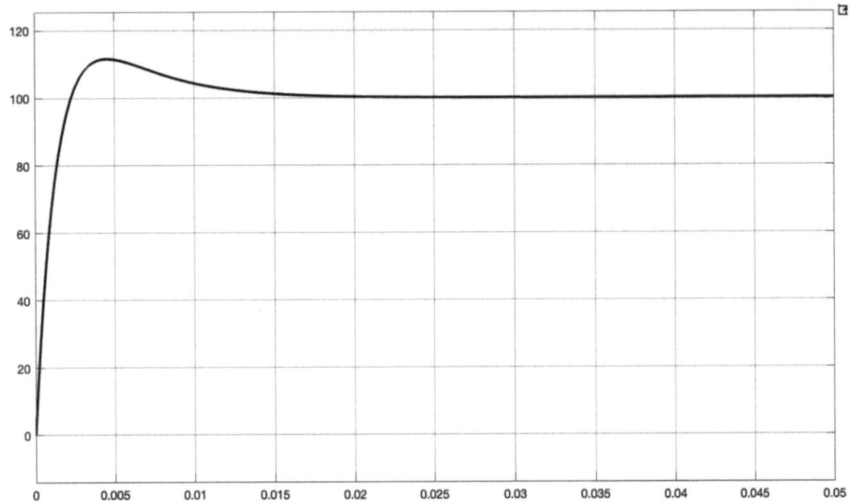

Fig. 5.72 Speed response waveforms

In the figure, the upper part represents the motor speed control system using the Luenberger observer, while the lower part shows the ideal speed control loop. Both control loops use the same PI controller parameters for speed regulation. After running the simulation, the speed response waveforms of the two systems can be obtained, as illustrated in Fig. 5.72.

Step 3:

From the speed response waveform in Fig. 5.72, we can see that the speed responses of the two systems overlap almost perfectly. This demonstrates that, in theory, a Luenberger speed observer based on encoder signals can provide performance nearly equivalent to that of an ideal speed sensor.

However, this is under ideal conditions—in the simulation system, the motor inertia used in the Luenberger observer matches the actual system exactly. In reality, there will inevitably be parameter mismatches between the observer model and the true physical system.

Step 4:

If we intentionally set the feedforward gain g_1 in the observer compensator to zero and run the simulation again, you will notice that the system eventually becomes unstable and diverges. This outcome proves that a double integrator plant must be compensated using a PID controller to provide sufficient stability margin.

(Note: The feedforward gain g_1 *is equivalent to the derivative gain in a PID controller).*

Step 5:

Next, we will conduct a disturbance rejection test for the observer. Referring to Fig. 5.70, we can see that the output of the integrator is equivalent to the estimated load torque $\hat{T}_L$. Therefore, this Luenberger observer can also be used as a load torque estimator.

Please open the example file mdl_Luenberger_3.slx. Upon opening, you will see the setup shown in Fig. 5.73. In the figure, the upper portion depicts the speed control loop with disturbance estimation. Since the output of the integrator is the negative estimated disturbance $-\hat{T}_L$, it must be multiplied by -1 before being added to the output of the speed PI controller. The lower portion shows the speed control loop without disturbance estimation.

After running the simulation, you will obtain a comparison of speed response waveforms, as shown in Fig. 5.74.

As shown in Fig. 5.74, when an impulsive load of 0.5 (Nm) is applied at 0.1 s, the speed control loop with disturbance estimation clearly demonstrates better disturbance rejection compared to the one without disturbance estimation.

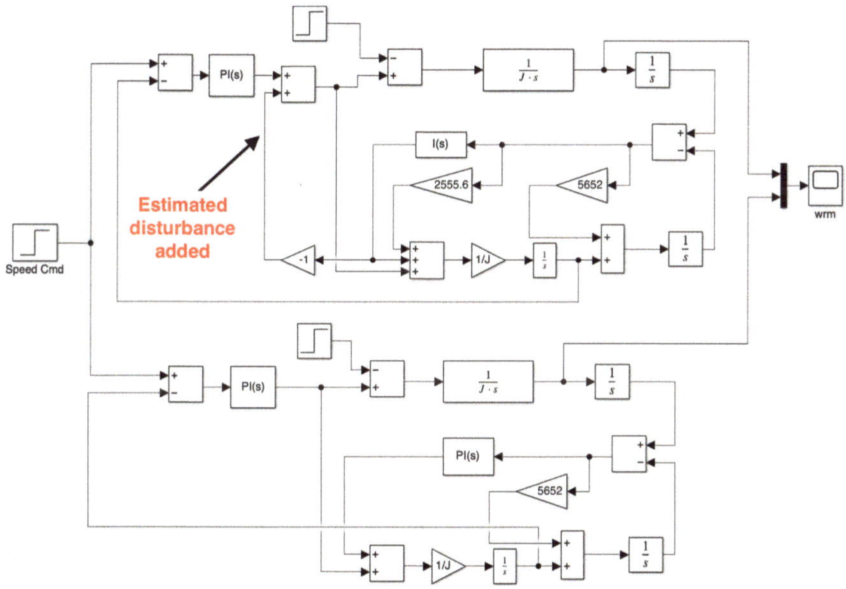

Fig. 5.73 SIMULINK simulation of the disturbance rejection performance of the speed control loop using a Luenberger observer, example model: mdl_Luenberger_3.slx

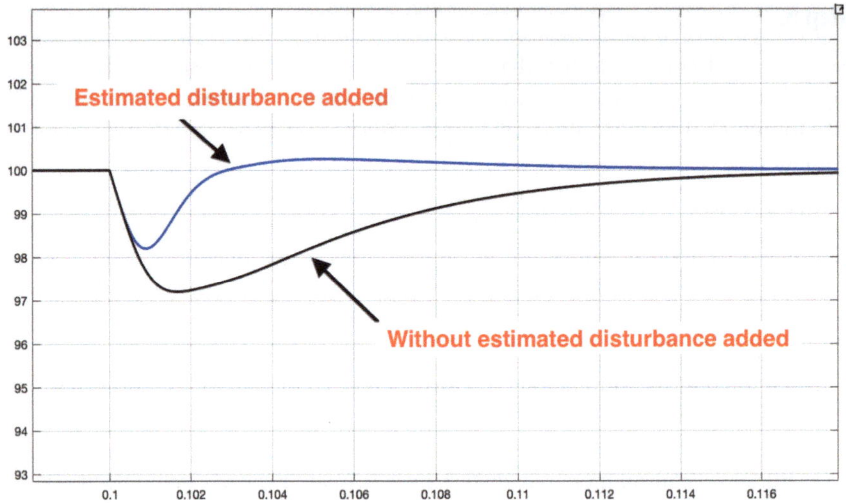

Fig. 5.74 Speed response waveforms

This confirms that the integrator output of the Luenberger observer can effectively function as a disturbance estimator for the system, thereby significantly enhancing the system's disturbance rejection capability.

Step 6:

Since the disturbance estimation signal in Fig. 5.73 is derived from the integrator output, its response speed may be relatively slow. As an alternative, you can create a separate Luenberger observer specifically for disturbance estimation, as shown in the SIMULINK block diagram of Fig. 5.75.

As illustrated, this dedicated disturbance observer still takes the actual motor position as its input, but the output of its PI controller is used as the estimated disturbance, which is then fed into the output control signal of the speed loop's PI controller. The controller parameters in this disturbance observer are identical to those used in the speed observer.

After running the SIMULINK simulation, you can obtain the speed response waveform shown in Fig. 5.75 (Fig. 5.76).

Step 7:

As shown in Fig. 5.77, the independently operating disturbance observer performs significantly better. When a shock load of 0.5 (Nm) is applied at 0.1 s, the maximum speed drop is less than 1 (rad/s). Compared with Fig. 5.74, where disturbance estimation relies solely on the integrator output of the speed observer, the independent disturbance observer improves disturbance rejection by nearly double, making this method highly worth considering.

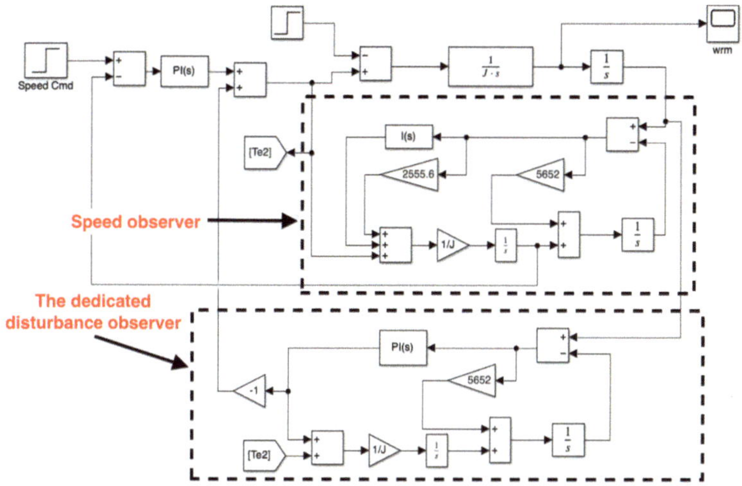

Fig. 5.75 SIMULINK simulation of the disturbance rejection performance of the speed control loop using a dedicated Luenberger observer as a disturbance observer, example model: mdl_Luenberger_4.slx

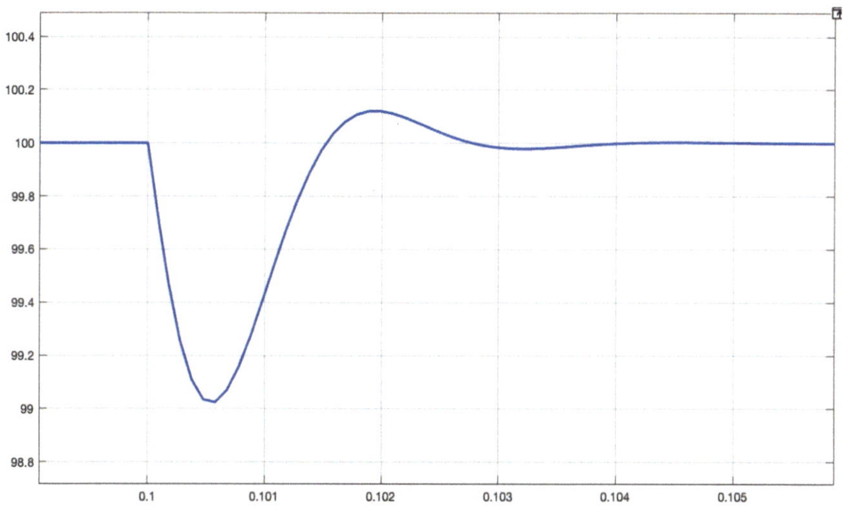

Fig. 5.76 Speed response waveform

Another advantage of using a dedicated disturbance observer is the ability to tune its controller parameters independently, enabling optimization of the disturbance estimation and further enhancing the system's robustness against external disturbances.

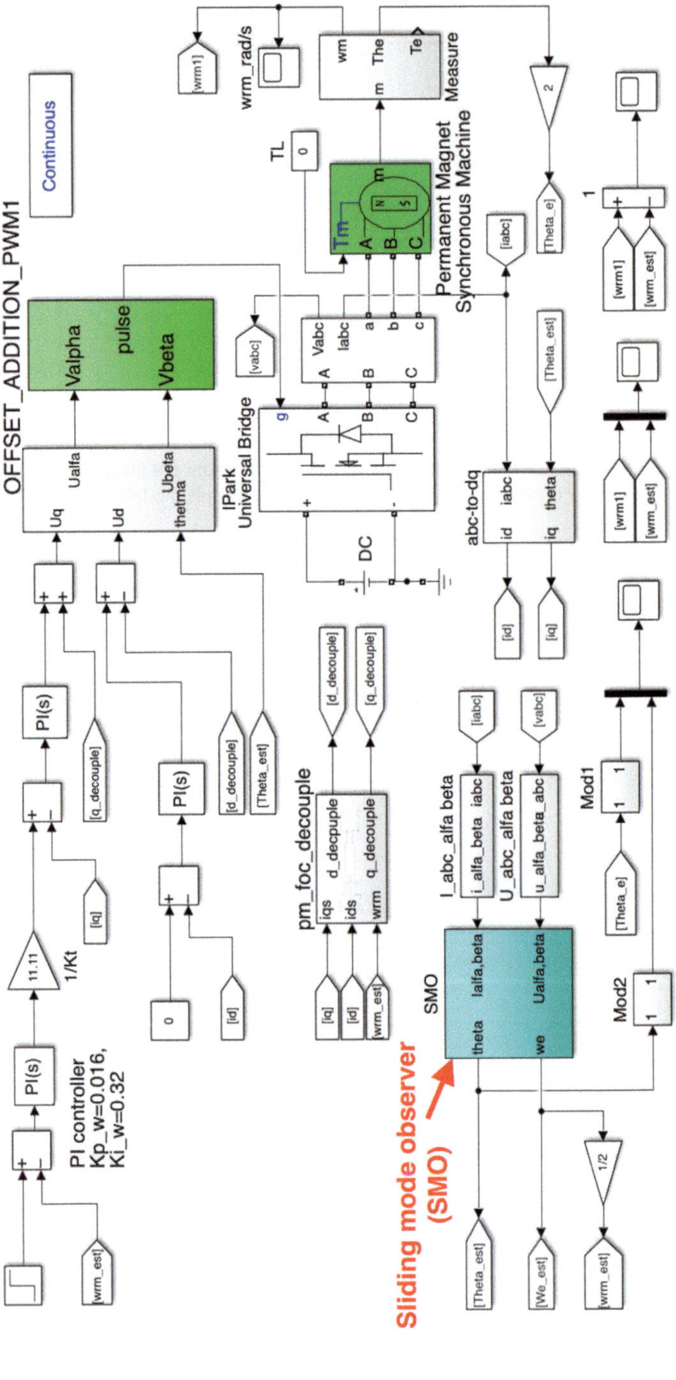

Fig. 5.77 Speed-sensorless simulation system of a permanent magnet synchronous motor using a sliding mode observer, example model: mdl_SMO_spm_foc_inv.slx

Summary of This Section:

- Observer Bandwidth Consideration:
  In general, the bandwidth of the observer should be significantly greater than that of the control loop utilizing the estimated states. The location of the poles greatly influences the observer's performance:
  - Placing poles farther left on the imaginary axis results in faster response (higher bandwidth) and better stability, but reduced noise immunity.
  - Placing poles closer to the imaginary axis yields slower response and lower stability, but better resistance to noise.
- Compensator Selection Based on Plant Type:
  - If the controlled plant is a single integrator system, such as $\frac{1}{Js}$, a PI controller is sufficient to stabilize the observer loop.
  - If the controlled plant is a double integrator system, such as $\frac{1}{Js^2}$, a PID controller is required. This is because a double integrator introduces an inherent 180° phase lag, making it difficult to stabilize using only PI compensation. A PID structure is necessary to ensure sufficient phase margin (PM) for stability.
- Feedforward Input T_e in the Observer:
  The observer loop includes a feedforward path driven by the torque input T_e, which:
  - Speeds up the observer's dynamic response
  - Suppresses transient overshoot

This section demonstrates that by properly designing the observer structure and its compensator, and incorporating feedforward and independent disturbance estimation, one can achieve robust and high-performance motor control.

5.12 Sensorless PMSM Control Using Sliding Mode Observers

In this chapter, we introduce a Sliding Mode Observer (SMO) for sensorless speed control of Permanent Magnet Synchronous Motors (PMSMs) [7]. This method is a model-based observer designed to estimate the rotor electrical angle θ_r and electrical angular velocity ω_r of the motor.

The SMO primarily estimates the back electromotive force (back-EMF) of the PMSM, from which the rotor electrical angle θ_e and electrical angular frequency ω_e can be derived. Compared to traditional PI controller-based methods, Sliding Mode Control (SMC) offers the advantages of fewer parameters to tune, guaranteed system stability, and strong robustness to parameter variations and disturbances.

- **Design of Sliding Mode Observer for PMSM**

Consider the mathematical model of a Surface-mounted Permanent Magnet Synchronous Motor (SPM-PMSM) in the two-axis stationary (α-β) reference frame [3, 4] as follows:

$$\frac{d}{dt}\begin{bmatrix} i_{s\alpha} \\ i_{s\beta} \end{bmatrix} = \begin{bmatrix} -\frac{R_s}{L_s} & 0 \\ 0 & -\frac{R_s}{L_s} \end{bmatrix}\begin{bmatrix} i_{s\alpha} \\ i_{s\beta} \end{bmatrix} + \frac{1}{L_s}\begin{bmatrix} v_{s\alpha} - e_{s\alpha} \\ v_{s\beta} - e_{s\beta} \end{bmatrix} \tag{5.12.1}$$

where $v_{s\alpha}$ and $v_{s\beta}$ are the stator voltages of the motor in the two-axis stationary (α-β) reference frame, $i_{s\alpha}$ and $i_{s\beta}$ are the stator currents in the (α-β) frame, $e_{s\alpha}$ and $e_{s\beta}$ are the back electromotive forces (back-EMF) in the (α-β) frame, and L_s is the stator inductance of the motor windings.

The stator back electromotive forces $e_{s\alpha}$ and $e_{s\beta}$ of the motor in the two-axis stationary (α-β) reference frame can be expressed as follows [3, 4]:

$$e_{s\alpha} = -\lambda_f \omega_r \sin(\theta_r) \tag{5.12.2}$$

$$e_{s\beta} = \lambda_f \omega_r \cos(\theta_r) \tag{5.12.3}$$

From Eqs. (5.12.2) and (5.12.3), we can derive Eqs. (5.1.4) and (5.1.5) to obtain the rotor position and speed information, respectively.

$$\theta_r = \tan^{-1}\frac{-e_{s\alpha}}{e_{s\beta}} \tag{5.12.4}$$

$$\omega_r = \frac{\sqrt{e_{s\alpha}^2 + e_{s\beta}^2}}{\lambda_f} \tag{5.12.5}$$

Therefore, the back-EMF information in the stationary α-β reference frame is crucial for estimating the rotor position and speed. To estimate the motor's back-EMF, we need to construct an observer model as follows:

$$\frac{d}{dt}\begin{bmatrix} \hat{i}_{s\alpha} \\ \hat{i}_{s\beta} \end{bmatrix} = \begin{bmatrix} -\frac{R_s}{L_s} & 0 \\ 0 & -\frac{R_s}{L_s} \end{bmatrix}\begin{bmatrix} \hat{i}_{s\alpha} \\ \hat{i}_{s\beta} \end{bmatrix} + \frac{1}{L_s}\begin{bmatrix} v_{s\alpha} - z_\alpha \\ v_{s\beta} - z_\beta \end{bmatrix} \tag{5.12.6}$$

where $\hat{i}_{s\alpha}$ *and* $\hat{i}_{s\beta}$ are the estimated stator currents of the motor, and z_α and z_β are the sliding mode control efforts (SMO control efforts).

In the observer, the stator current is the only state that needs to be considered and can be obtained through measurement. Therefore, the sliding surface $s_{\alpha\beta}$ can be defined based on the stator current state.

$$\boldsymbol{s}_{\alpha\beta} = \begin{bmatrix} s_\alpha \\ s_\beta \end{bmatrix} = \tilde{i}_{\alpha\beta} = \hat{i}_{\alpha\beta} - i_{\alpha\beta} = \begin{bmatrix} \hat{i}_{s\alpha} - i_{s\alpha} \\ \hat{i}_{s\beta} - i_{s\beta} \end{bmatrix} \tag{5.12.7}$$

where $\hat{i}_{\alpha\beta} = \left[\hat{i}_{s\alpha}\ \hat{i}_{s\beta}\right]^T$, $i_{\alpha\beta} = \left[i_{s\alpha}\ i_{s\beta}\right]^T$, and $\tilde{i}_{\alpha\beta}$ *represents the error between the estimated and actual stator currents. To force the system state to reach the sliding surface, the sliding mode control effort can be designed as:*

$$z_{\alpha\beta} = K_{\alpha\beta}\mathrm{sgn}(\tilde{i}_{\alpha\beta}) \tag{5.12.8}$$

where the sliding mode gain is defined as $K_{\alpha\beta} = \left[K_\alpha\ K_\beta\right]^T$.

Using Eqs. (5.12.1) and (5.12.6), we can derive the stator current error equation as follows:

$$\frac{d}{dt}\begin{bmatrix} \tilde{i}_{s\alpha} \\ \tilde{i}_{s\beta} \end{bmatrix} = \begin{bmatrix} -\frac{R_s}{L_s} & 0 \\ 0 & -\frac{R_s}{L_s} \end{bmatrix}\begin{bmatrix} \tilde{i}_{s\alpha} \\ \tilde{i}_{s\beta} \end{bmatrix} + \frac{1}{L_s}\begin{bmatrix} e_{s\alpha} - z_\alpha \\ e_{s\beta} - z_\beta \end{bmatrix} \tag{5.12.9}$$

Next, we perform a stability analysis of this sliding mode observer. First, define the Lyapunov function as follows:

$$\boldsymbol{V}_{\alpha\beta} = \frac{1}{2}\boldsymbol{s}_{\alpha\beta}^T\boldsymbol{s}_{\alpha\beta} = \frac{1}{2}\left(s_\alpha^2 + s_\beta^2\right) \tag{5.12.10}$$

From Eq. (5.12.10), we know that $\boldsymbol{V}_{\alpha\beta}$ is always positive, and it can be considered as the energy function of the estimation error. Therefore, for the system to be **asymptotically stable**, it must satisfy the condition $\frac{d\boldsymbol{V}_{\alpha\beta}}{dt} < 0$ whenever $\boldsymbol{V}_{\alpha\beta} > 0$.

The derivative $d\boldsymbol{V}_{\alpha\beta}/dt$ can be expressed as:

$$\frac{d\boldsymbol{V}_{\alpha\beta}}{dt} = \left(s_\alpha\frac{ds_\alpha}{dt} + s_\beta\frac{ds_\beta}{dt}\right) \tag{5.12.11}$$

Substituting Eq. (5.12.7) into Eq. (5.12.11), we obtain:

$$\begin{aligned} \frac{d\boldsymbol{V}_{\alpha\beta}}{dt} &= \tilde{i}_{s\alpha}\cdot\frac{d\tilde{i}_{s\alpha}}{dt} + \tilde{i}_{s\beta}\cdot\frac{d\tilde{i}_{s\beta}}{dt} \\ &= \tilde{i}_{s\alpha}\left(-\frac{R_s}{L_s}\tilde{i}_{s\alpha} + \frac{1}{L_s}(e_{s\alpha} - z_\alpha)\right) + \tilde{i}_{s\beta}\left(-\frac{R_s}{L_s}\tilde{i}_{s\beta} + \frac{1}{L_s}(e_{s\beta} - z_\beta)\right) \end{aligned} \tag{5.12.12}$$

To satisfy $\frac{d\boldsymbol{V}_{\alpha\beta}}{dt} < 0$, Eq. (5.12.12) can be decomposed into the following two parts:

$$-\frac{R_s}{L_s}\left(\tilde{i}_{s\alpha}^2 + \tilde{i}_{s\beta}^2\right) < 0$$

$$\frac{\tilde{i}_{s\alpha}}{L_s}\left(e_{s\alpha} - K_\alpha \text{sgn}\left(\tilde{i}_{s\alpha}\right)\right) + \frac{\tilde{i}_{s\beta}}{L_s}\left(e_{s\beta} - K_\beta \text{sgn}\left(\tilde{i}_{s\beta}\right)\right) < 0 \tag{5.12.13}$$

From Eq. (5.12.13), it can be seen that the sliding mode gain $K_{\alpha\beta}$ must satisfy the following condition:

$$\begin{aligned} K_\alpha &> \max(|e_{s\alpha}|) \\ K_\beta &> \max\left(\left|e_{s\beta}\right|\right) \end{aligned} \tag{5.12.14}$$

Based on the above analysis, if the sliding mode gain $K_{\alpha\beta}$ satisfies the condition given in Eq. (5.12.14), then the system states will reach the sliding surface defined by Eq. (5.12.7). Once the system states reach the sliding surface, they will remain on it.

When the system states reach the sliding surface, i.e., when $s_{\alpha\beta} = 0$, according to Eq. (5.12.9), we have:

$$\begin{aligned} e_{s\alpha} &= z_\alpha = K_\alpha \text{sgn}\left(\tilde{i}_{s\alpha}\right) \\ e_{s\beta} &= z_\beta = K_\beta \text{sgn}\left(\tilde{i}_{s\beta}\right) \end{aligned} \tag{5.12.15}$$

Therefore, we can conclude that once the system states reach the sliding surface, the sliding mode control signal $z_{\alpha\beta}$ becomes equal to the motor back-EMF $e_{\alpha\beta} = \left[\, e_{s\alpha} \; e_{s\beta} \,\right]^T$. However, since the actual sliding mode control signal is a discontinuous high-frequency switching signal, a low-pass filter is required in practical applications to filter it, i.e.,

$$\hat{e}_{\alpha\beta} = \frac{\omega_{cut-off}}{s + \omega_{cut-off}} z_{\alpha\beta} \tag{5.12.16}$$

where $\hat{e}_{\alpha\beta}$ *is the estimated back-EMF after filtering. The estimated back-EMF* $\hat{e}_{\alpha\beta}$ can be used to obtain the estimated rotor position and speed of the permanent magnet synchronous motor (PMSM) via Eqs. (5.12.4) and (5.12.5), as follows:

$$\hat{\theta}_r = -\tan^{-1}\frac{\hat{e}_{s\alpha}}{\hat{e}_{s\beta}} \tag{5.12.17}$$

$$\hat{\omega}_r = \frac{\sqrt{\hat{e}_{s\alpha}^2 + \hat{e}_{s\beta}^2}}{\lambda_f} \tag{5.12.18}$$

Since the use of a low-pass filter introduces rotor angle estimation errors due to phase delay, a compensation term ε can be introduced to correct the phase delay caused by the low-pass filter.

$$\varepsilon = \tan^{-1}\frac{\hat{\omega}_r}{\omega_{cut-off}} \tag{5.12.19}$$

Therefore, Eq. (5.12.17) can be modified as follows:

$$\hat{\theta}_r = -\tan^{-1}\frac{\hat{e}_{s\alpha}}{\hat{e}_{s\beta}} + \varepsilon = -\tan^{-1}\frac{\hat{e}_{s\alpha}}{\hat{e}_{s\beta}} + \tan^{-1}\frac{\hat{\omega}_r}{\omega_{cut-off}} \quad (5.12.20)$$

The above completes the design process of the Sliding Mode Observer (SMO) for the permanent magnet synchronous motor. Next, we will use SIMULINK to perform system simulation.

Tips

To simplify the derivation of the formulas, this section deliberately uses a Surface Permanent Magnet (SPM) synchronous motor for the design of the Sliding Mode Observer (SMO). However, the SMO can also be applied to Interior Permanent Magnet (IPM) synchronous motors. The mathematical model of the IPM motor in the two-axis stationary reference frame (α-β) can be expressed as [7]:

$$\frac{d}{dt}\begin{bmatrix} i_{s\alpha} \\ i_{s\beta} \end{bmatrix} = \begin{bmatrix} -\frac{R_s}{L_d} & \frac{\omega_r L_\Delta}{L_d} \\ -\frac{\omega_r L_\Delta}{L_d} & -\frac{R_s}{L_d} \end{bmatrix}\begin{bmatrix} i_{s\alpha} \\ i_{s\beta} \end{bmatrix} + \frac{1}{L_d}\left(\begin{bmatrix} v_{s\alpha} \\ v_{s\beta} \end{bmatrix} - e_{s\alpha\beta}\right)$$

where

$$L_\Delta = L_q - L_d,\ e_{s\alpha\beta} = E_{ex}\begin{bmatrix} -\sin(\theta_r) \\ \cos(\theta_r) \end{bmatrix},$$

$$E_{ex} = (L_d - L_q)\left(\omega_r i_d - \frac{di_q}{dt}\right) + \omega_r \lambda_f$$

(*Note: i_d and i_q are the stator d-axis and q-axis currents, respectively*).

You may use the above IPM motor mathematical model to replace equation (5.12.1) in order to design a Sliding Mode Observer (SMO) suitable for IPM motors.

- **MATLAB/SIMULINK Simulation**

Step 1:

Since this section uses a Surface Permanent Magnet (SPM) synchronous motor for the design and derivation, simulation should also be based on SPM motor parameters. To achieve this, the author simply modifies the IPM motor parameters in Table 2.3 by setting the stator q-axis inductance L_q to 5.7 (mH), making $L_d = L_q$, while keeping the other parameters unchanged. As a result, the IPM motor parameters in Table 2.3 become SPM motor parameters, as shown in Table 5.3.

Table 5.3 SPM parameters and bandwidth design specifications

Parameter	Symbol	Value	Unit
Stator resistance	R_s	1.2	Ω
Stator d-axis inductance	L_d	5.7	mH
Stator q-axis inductance	L_q	5.7	mH
Rotor flux linkage	λ_f	0.03	Wb
Number of poles	P	4	–
Moment of inertia	J	0.00016	kg m^2
Friction coefficient (Note: Here, the friction coefficient B is set to zero.)	B	0	N m s/rad
Bandwidth of d- and q-axis current control loops	ω_d, ω_q	1000	rad/s
Speed loop bandwidth	ω_s	100	rad/s

Please start by running the MATLAB example script spm_params.m to load the motor parameters for this section.

Step 2:

Next, we will use SIMULINK's built-in Simscape motor and inverter modules for simulation. Please open the example model file mdl_SMO_spm_foc_inv.slx. Once opened, it will appear as shown in Fig. 5.77.

The dashed box in Fig. 5.77 represents the Sliding Mode Observer (SMO). Double-clicking it reveals the block diagram shown in Fig. 5.78. As shown, the SMO consists of two main blocks: the SMO block and the Arctan_function block.

- The SMO block primarily implements the computations from Eqs. (5.12.9) and (5.12.16).
- The Arctan_function block performs the operations described in Eqs. (5.12.17) through (5.12.20) to calculate the estimated rotor position $\hat{\theta}_r$ and speed $\hat{\omega}_r$.

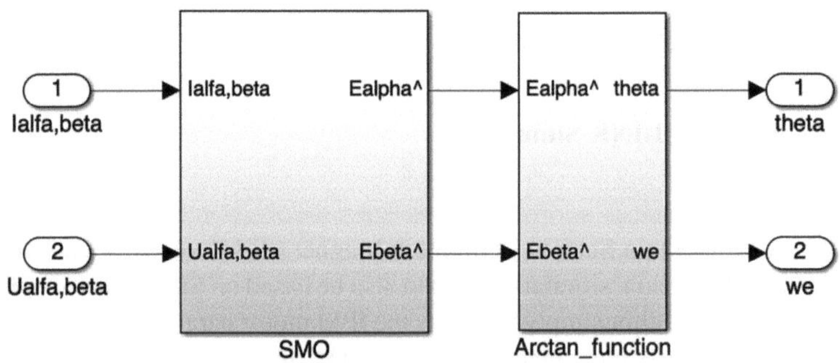

Fig. 5.78 Composition of the sliding mode observer

Step 3:

Double-clicking the SMO block reveals the system diagram shown in Fig. 5.79.

- The upper section of the diagram is responsible for estimating the back-EMF $\hat{e}_\alpha$,
- while the lower section handles the estimation of $\hat{e}_\beta$.

In this setup, a low-pass filter with a bandwidth of 2000 (rad/s) is used to filter the sliding mode control signals $z_{\alpha\beta}$.

Since the current control loop bandwidth has been designed to be 1000 (rad/s), and typically the observer's bandwidth should be higher than that of the loop using its estimated values, a bandwidth of 2000 (rad/s) for the observer is considered sufficient for simulation purposes. In real-world applications, the observer bandwidth can be increased further depending on system requirements.

Step 4:

From Fig. 5.79, we can also see that the sliding mode gain $K_{\alpha\beta}$ used in the SMO is 24.

How do we determine the sliding mode gain?

We need to refer to Eq. (5.12.14), which provides the necessary condition for the stability of the SMO observer. According to this equation, the sliding mode gain $K_{\alpha\beta}$ must be greater than the maximum back-EMF. We can go back to the motor specifications in Table 5.3. The motor's torque constant K_T can be calculated as follows:

$$K_T = \frac{3}{2}\frac{P}{2}\lambda_f = 0.09 \quad (\text{Nm/A})$$

Since in SI units the motor torque constant K_T is equal to the back-EMF constant K_e, where K_e has units of V/(rad/s), we can use this to estimate the maximum back-EMF.

In this example, the maximum speed command is 100 rad/s, which corresponds to a maximum back-EMF of approximately 9 V. Therefore, selecting a sliding mode gain $K_{\alpha\beta}K_{\alpha\beta} = 24$ ensures that the SMO (Sliding Mode Observer) can operate stably for speeds up to around **260 rad/s**, well above the commanded maximum, providing a sufficient safety margin for robustness.

Since the motor is driven with sinusoidal control (FOC), the motor's back-EMF constant is given by $K_e = \frac{2}{3}K_T$ (please refer to Sect. 5.5 of this book). The back-EMF constant K_e has units of V/(rad/s). In this example, the maximum speed command is 100 (rad/s), so the maximum back-EMF of the motor is 6 (V). Therefore, the sliding mode gain $K_{\alpha\beta}$ is set to 24 to ensure that the SMO (Sliding Mode Observer) can operate stably at speeds up to 400 (rad/s).

Step 5:

Next, open the Arctan_function block to view the system diagram shown in Fig. 5.80. As illustrated, the Arctan_function block uses Eq. (5.12.17) to calculate the estimated

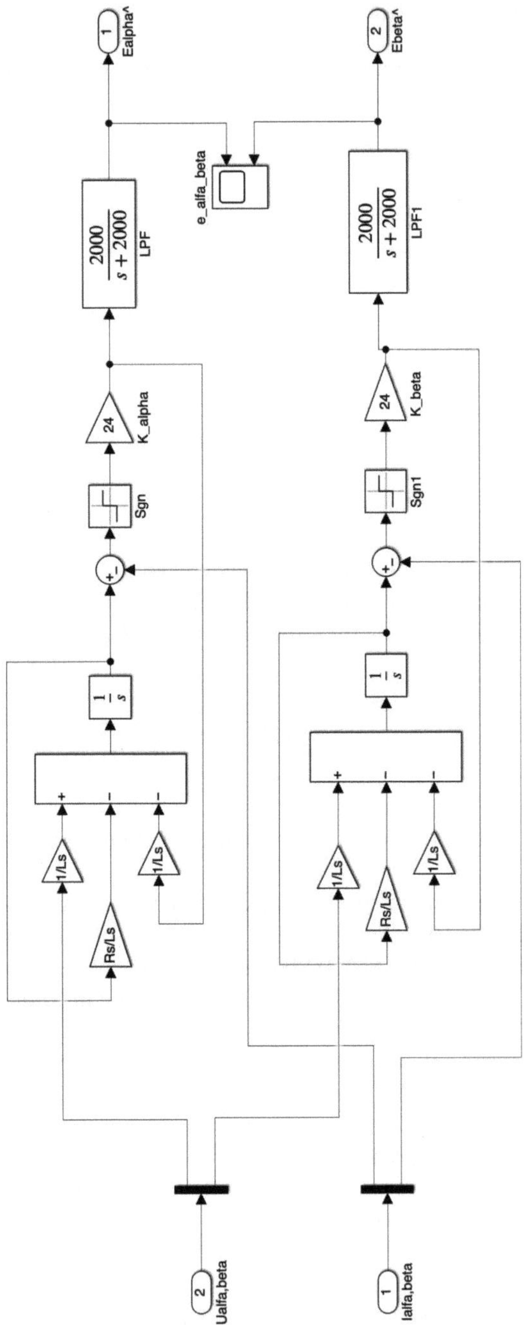

Fig. 5.79 Composition of the SMO block

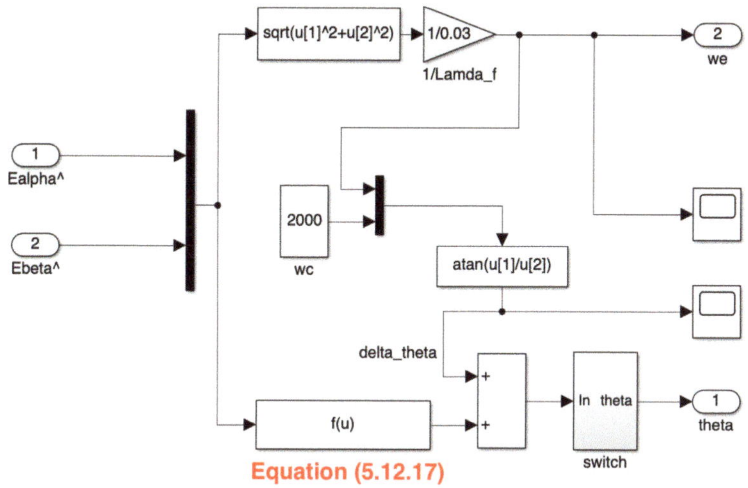

Fig. 5.80 Composition of the arctan function block

rotor electrical angle $\hat{\theta}_r$. The variable delta_theta represents the compensation term ε, which is used to compensate for the phase delay introduced by the low-pass filter. Meanwhile, Eq. (5.12.18) is used to calculate the estimated rotor electrical angular speed $\hat{\omega}_r$.

Step 6:

By running the SIMULINK system simulation of this example program, you can obtain the speed response waveform shown in Fig. 5.81 and the estimated angle waveform shown in Fig. 5.82. Both waveform plots simultaneously display the actual motor values and the values estimated by the SMO. As shown in the figures, the sliding mode observer (SMO) operates effectively—the estimated angle and speed closely match the actual motor values.

Step 7:

Figure 5.83 shows the estimated back-EMF waveforms: the upper plot represents the estimated back-EMF $\hat{e}_\alpha$, *and the lower plot represents* $\hat{e}_\beta$. From the waveforms, it is evident that the low-pass filter has effectively eliminated the high-frequency switching signals.

Step 8:

In this example program, the mechanical speed command ω_{rm}^* is 100 (rad/s), and the motor has 4 poles. Therefore, the synchronous speed $\hat{\omega}_r$ is 200 (rad/s). We can calculate the phase delay caused by the low-pass filter as follows:

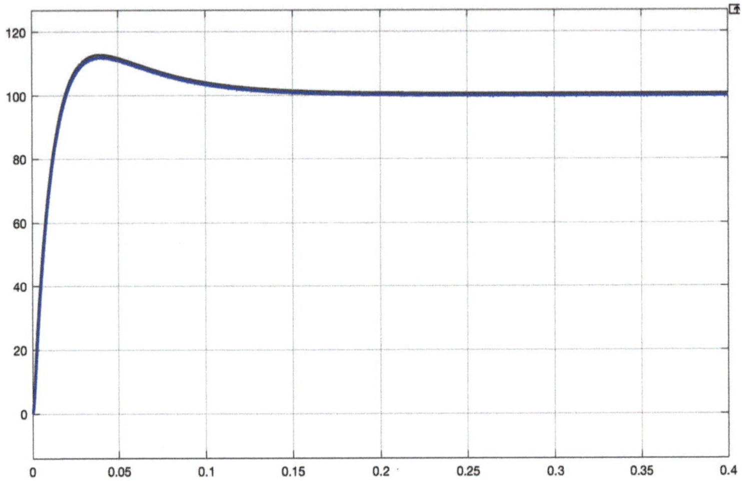

Fig. 5.81 Speed response waveform

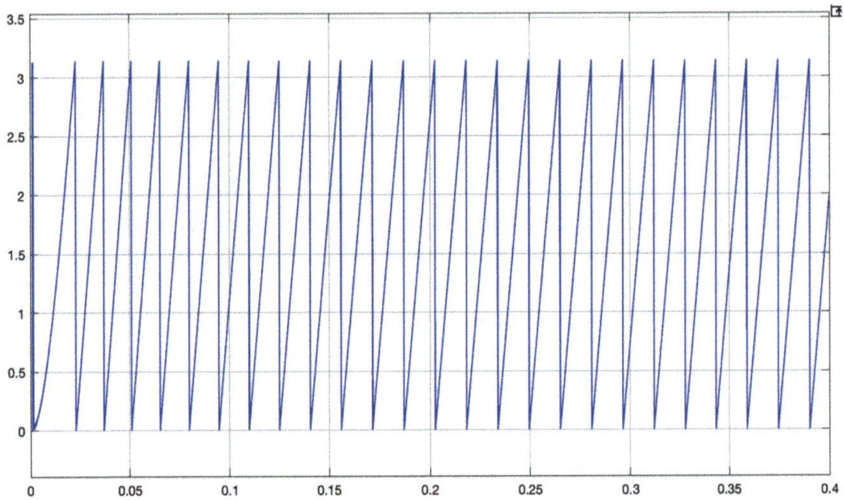

Fig. 5.82 Estimated angle waveform

$$\angle\frac{\omega_{cut-off}}{s+\omega_{cut-off}} = \angle\frac{1}{\frac{s}{\omega_{cut-off}}+1} = \angle\frac{1}{j\frac{\omega}{\omega_{cut-off}}+1} \approx \angle-\tan^{-1}\left(\frac{\omega}{\omega_{cut-off}}\right) \tag{5.12.21}$$

When $\omega = 200$ and $\omega_{cut-off} = 2000$ are substituted into Eq. (5.12.21), we obtain:

$$\angle-\tan^{-1}\left(\frac{\omega}{\omega_{cut-off}}\right) = \angle-\tan^{-1}\left(\frac{200}{2000}\right) = \angle-5.7^{\circ} \tag{5.12.22}$$

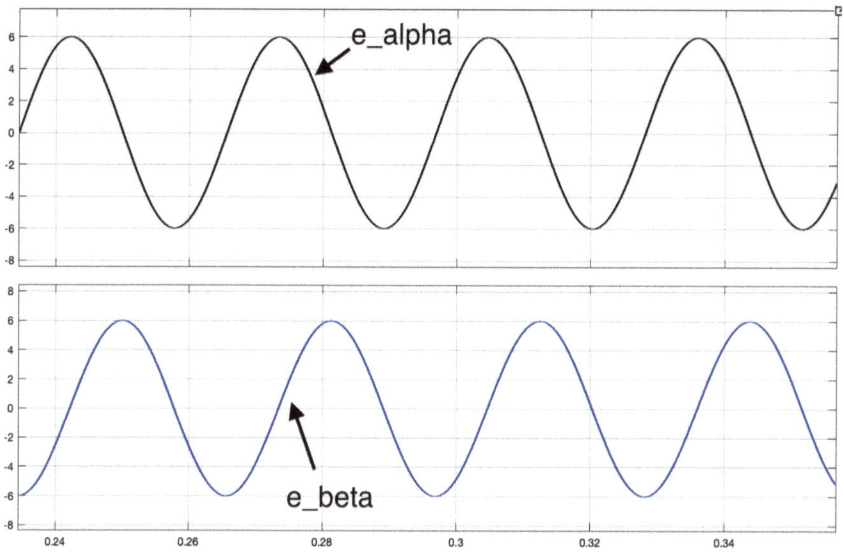

Fig. 5.83 Back-EMF waveforms

From Eq. (5.12.22), it can be seen that the phase delay caused by the low-pass filter is − 5.7°. Therefore, we need to use Eq. (5.12.19), i.e., the phase compensation term $\varepsilon = \tan^{-1}\frac{\hat{\omega}_r}{\omega_{cut-off}} = 5.7°$, to compensate for the phase delay.

Figure 5.84 shows the waveform of the compensation term ε. As can be seen from the figure, the steady-state phase compensation value is 5.7°, which matches the calculated result.

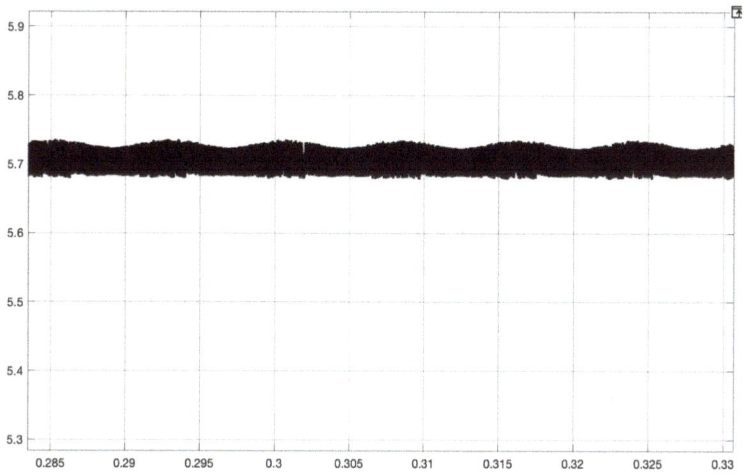

Fig. 5.84 Phase compensation waveform

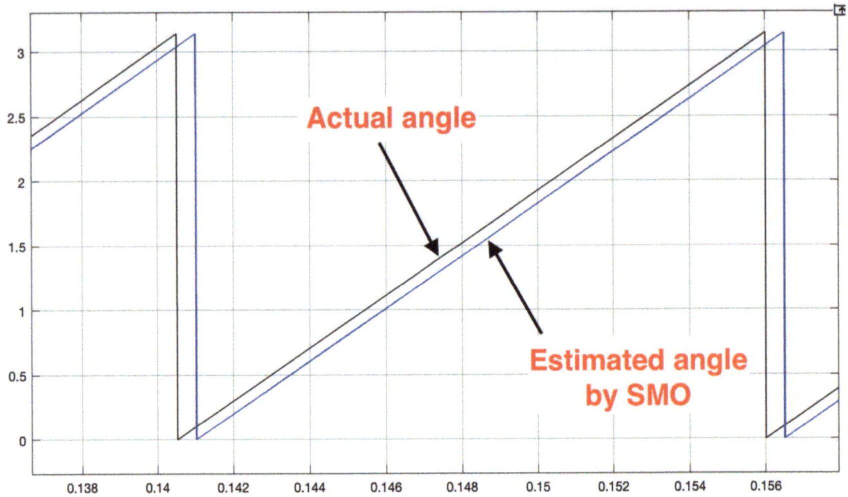

Fig. 5.85 Actual rotor angle and estimated rotor angle

Step 9:

If the phase compensation term ε is not applied, and the SIMULINK simulation is executed again, the waveforms of the actual rotor angle and the estimated rotor angle can be observed, as shown in Fig. 5.85. It can be seen that an error occurs between the SMO-estimated angle and the actual rotor angle. Moreover, it is evident that the SMO-estimated angle lags behind the actual rotor angle. This angular lag is caused by the low-pass filter.

Furthermore, this phase delay increases with the rotor speed, making it even more significant at higher speeds. Therefore, in practical applications, phase delay compensation is essential to ensure accurate estimation performance.

Summary of This Section:

- The sliding mode observer (SMO) introduced in this section is widely used in sensorless control of permanent magnet synchronous motors (PMSMs). Since only two parameters need to be tuned—the sliding mode gain $K_{\alpha\beta}$ and the low-pass filter cutoff frequency $\omega_{cut-off}$—and both can be determined through design, this algorithm is highly versatile and practical.
- To simplify the mathematical derivation in this section, we specifically used the surface-mounted permanent magnet (SPM) PMSM model for observer design. However, the same SMO structure can also be applied to interior permanent magnet (IPM) PMSMs using the same design and derivation approach.

References

1. Chang-Hwan Liu, AC Motor Control: Principles of Vector Control and Direct Torque Control, Taipei: Dong Hua Book Co., 2001.
2. Sul, Seung-Ki, Control of Electric Machine Drive Systems, Wiley-IEEE Press, 2011.
3. Chih-Chun Yeh, AC Motor Control and Simulation Techniques: Mastering Core Algorithms of EV and Inverter Technology, Taipei: Wu-Nan Book Inc., 2023.
4. George Ellis, Control System Design Guide: Using Your Computer to Understand and Diagnose Feedback Controllers, Butterworth-Heinemann, 2016.
5 J. Bocker, S. Beineke, and A. Bahr, "On the control bandwidth of servo drives," in Proc. Eur. Conf. Power Electron. Appl., pp. 1–10, 2009.
6. Chih-Chun Yeh, The Self-Training of IoT Experts, Taiwan: Drmaster. Press Co. LTD., 2023.
7. Ying Zuo, Chunyan Lai and K. Lakshmi Varaha Iyer, "A Review of Sliding Mode Observer Based Sensorless Control of PMSM Drive," IEEE Trans. On Power Electronics, Vol. 38, No. 9, pp. 11352–11366, Sept., 2023.

Appendix
A Refresher on Classical Control Theory

While modern digital control techniques dominate today's motor drive applications, a solid grasp of classical control theory remains essential for effective design, analysis, and troubleshooting. Appendix revisits the foundational principles of classical control, providing a concise yet thorough review that connects theory directly to practical motor control scenarios.

The appendix begins with the core concepts of stability, establishing the criteria for determining whether a system will perform predictably under various conditions. It then distinguishes between Laplace transforms and transfer functions, clarifying their roles in system modeling and analysis. The discussion extends into the digital domain with sampling theory, Z-transforms, and zero-order hold effects, bridging the gap between continuous-time and discrete-time control.

Readers will explore the implementation of P, I, and D controllers in both the s- and z-domains, followed by an in-depth look at standard second-order systems and their dynamic characteristics. A steady-state error analysis, illustrated through a speed control case study, reinforces the connection between theoretical analysis and real-world performance.

The appendix also addresses more advanced topics, such as the behavior of non-minimum phase systems, and introduces pole placement methods—combining theoretical foundations with MATLAB-based examples. Finally, it presents linearization techniques for handling nonlinear systems, enabling more accurate modeling and control design.

By revisiting these classical concepts in a targeted and application-driven manner, Appendix equips readers with the analytical tools needed to interpret system behavior, design effective controllers, and ensure robust performance in both classical and modern motor control contexts.

C-C. Yeh, *AC Motor Control Loop Design*, Power Systems,
https://doi.org/10.1007/978-981-95-3261-2

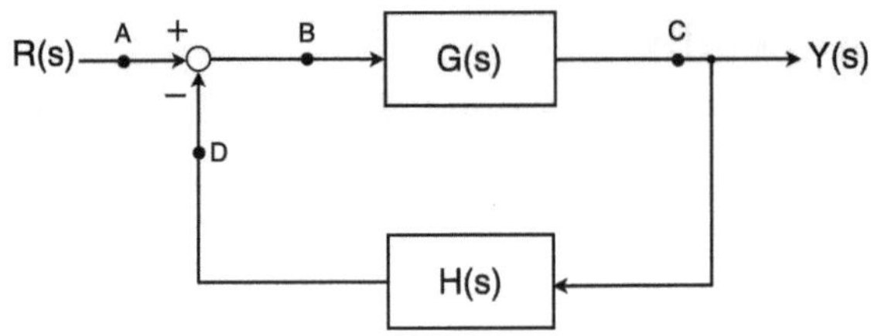

Fig. A.1 Block diagram of a typical closed-loop negative feedback control system

A.1 Core Concepts of Stability in Classical Control

For control systems, **system stability** is the most fundamental requirement determining whether the system is usable at all. If a system is unstable, it has no practical value. However, if the system is stable, we still need to evaluate **how stable** it is and whether it meets specified performance criteria.

In **modern control theory**, system stability is determined by the eigenvalues of the **system matrix** (i.e., the **A matrix**). A system is stable if all eigenvalues have negative real parts (i.e., they lie in the **left-half complex plane**). The further these eigenvalues are from the imaginary axis, the higher the degree of stability. This also means that the system states or state estimation errors converge to zero more quickly over time.

In contrast, **classical control theory** typically deals with **single-input single-output (SISO)** negative feedback systems and interprets stability in a different way. In classical analysis, system stability is often evaluated using the **phase margin (PM)** and **gain margin (GM)** as metrics [1]. The definition of instability in this context is also very precise.

Before discussing stability in more detail, it's important to clarify a concept that often confuses learners: the difference between the **open-loop transfer function** and the **closed-loop transfer function**. Understanding this distinction is crucial for a clear grasp of system stability.

- **Difference Between Open-Loop and Closed-Loop Transfer Functions**

Figure A.1 Shows a typical negative feedback control system, where $G(s)$ is the forward path transfer function, $H(s)$ is the feedback path transfer function, $R(s)$ is the system input, and $Y(s)$ is the system output.

To fully characterize the relationship between the system input $R(s)$ and output $Y(s)$, the **closed-loop transfer function** $G_{close}(s)$ must be derived, which is given by:

$$G_{close}(s) = \frac{Y(s)}{R(s)} = \frac{G(s)}{1 + G(s)H(s)} \tag{A.1.1}$$

Equation (A.1.1) also has its own magnitude and phase characteristics. When we use the closed-loop transfer function, we are examining the relationship between the magnitude and phase of the input $R(s)$ and the output $Y(s)$. Additionally, when it comes to designing or measuring the system's bandwidth, the closed-loop transfer function is what we use.

When we use the closed-loop transfer function, we treat the system as a whole, but we cannot see the internal details. However, to analyze the stability of the closed-loop system, we need to understand the magnitude and phase relationship between the input and feedback signal at the feedback node—that is, between points A and D in Fig. A.1. Therefore, the open-loop transfer function is defined as follows:

$$G_{open}(s) = G(s)H(s) \tag{A.1.2}$$

The physical meaning of the open-loop transfer function $G(s)H(s)$ is that it represents the gain and phase changes experienced by a signal as it travels from point A through points B and C, and then returns to point D within the control loop. Therefore, the open-loop transfer function helps us observe the magnitude and phase relationship between the system input and the feedback signal.

But why do we need to observe the magnitude and phase between the input and the feedback signal? To answer this question, we first need to explore the phenomenon of instability in classical negative feedback systems.

- **The Phenomenon of Instability in Classical Negative Feedback Systems**

As shown in the classical negative feedback system of Fig. A.1, there are two necessary conditions for the system to become unstable:

1. **The feedback signal has a phase delay of 180°** relative to the input signal.
2. **The magnitude of the open-loop gain** $G(s) \times H(s)$ **is greater than or equal to 1.**

If both conditions are satisfied simultaneously, the system will become unstable.

Let us first describe this instability phenomenon in words:

Assume the input $R(s)$ is a standard sinusoidal signal. When this input passes through point A to point B and generates an output at point C, the output is then fed back to point D, where it is subtracted from the original input signal $R(s)$.

Now, suppose the signal passing through the open-loop transfer function $G(s) \times H(s)$ experiences **no attenuation**—thus satisfying the second instability condition—and also undergoes a **180° phase delay**, as stated in the first condition.

For a sinusoidal input, a 180° phase delay causes the feedback signal to be exactly **out of phase** with the input. When this delayed feedback is subtracted at the summing junction, it is effectively equivalent to **adding a signal that is in-phase** with the original input (since subtracting a negative is the same as adding a positive).

As a result, the **negative feedback loop turns into a positive feedback loop**, and the input continues to be reinforced by its own in-phase feedback signal. This leads to an **amplification loop**, causing the system output to **diverge** over time and thus leading to **system instability**.

The above provides a textual description of instability. Next, we can verify this behavior using **SIMULINK**. First, please open the example model file **mdl_A_1_1.slx**. Once opened, the layout will appear as shown in Fig. A.2.

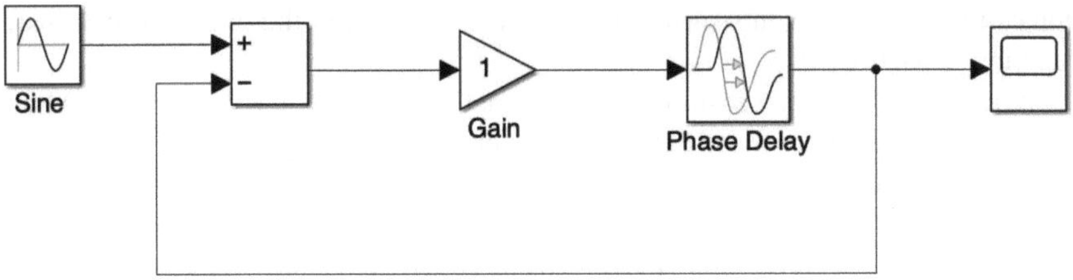

Fig. A.2 SIMULINK simulation of a typical negative feedback control system with phase delay, example model: mdl_A_1_1.slx

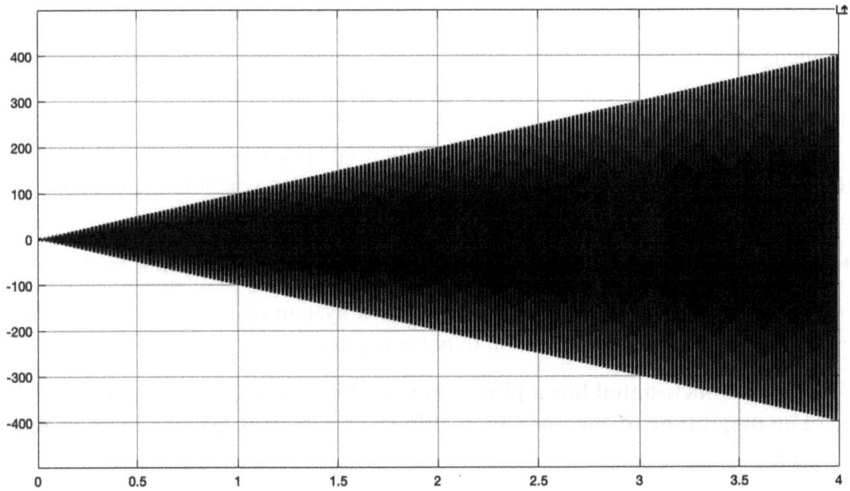

Fig. A.3 Output response waveform at the gain of 1

As shown in the figure, the input used in the example simulation is as follows:

$$R(s) = \sin(2 \times \pi \times 50) \tag{A.1.3}$$

In the program, the **Phase Delay** block is the built-in *Transport Delay* module in SIMULINK. The delay time of the Phase Delay block has been set to **(1/50)*0.5**, which corresponds exactly to a **180° phase delay**.

Next, set the **Gain** to **1**, then run the SIMULINK simulation. The output waveform can be obtained as shown in Fig. A.3.

As shown in Fig. A.3, when the loop gain is set to 1 and the phase delay is 180°, the input continuously adds to an in-phase version of itself, causing the output response to diverge.

Next, set the Gain to 0.9 and run the SIMULINK simulation again. The output waveform obtained is shown in Fig. A.4. As illustrated, the system output exhibits

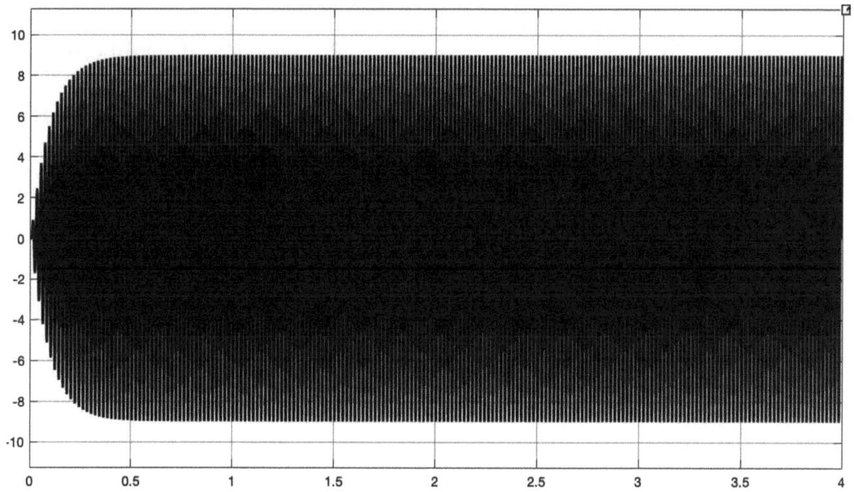

Fig. A.4 Output response waveform at the gain of 0.9

bounded oscillation **but does not diver**ge. This demonstrates that although the output oscillates, the system does not enter an unstable state.

- **Further Exploration of the Open-Loop Transfer Function $G(s)H(s)$**

From the above analysis and simulation verification, we can conclude that a negative feedback system becomes unstable only when the following two necessary conditions are met:

1. The feedback signal has a phase delay of 180° relative to the input signal.
2. The magnitude of the open-loop transfer function $G(s) \times H(s)$ is greater than or equal to 1.

The phase delay of the feedback signal in the first condition corresponds to the phase lag of the open-loop transfer function G(s)H(s). This explains why, in classical control theory, in addition to the closed-loop transfer function, it is also essential to define the system's open-loop transfer function. By analyzing the magnitude and phase characteristics of the open-loop transfer function, not only can we determine whether the system is stable, but we can also evaluate how stable it is using two key stability indicators: **Phase Margin (PM)** and **Gain Margin (GM)**. Their physical meanings are as follows:

- **Phase Margin (PM)**: When the magnitude of the open-loop transfer function $G(s)H(s)$ is equal to 1, the **phase margin** is the amount by which the phase φ is above $-180°$. The system is stable only if the phase margin is positive. The mathematical definition of phase margin is:

$$\text{PM} = \angle G_{open}(j\omega_{gc}) - (-180°) \tag{A.1.4}$$

where ω_{gc} is the frequency at which the magnitude of the open-loop transfer function $G(s)H(s)$ is equal to 1. This frequency is also known as the **gain crossover frequency**.

- **Gain Margin (GM)**: When the phase of the open-loop transfer function $G(s)H(s)$ is $-180°$, the **gain margin** represents how far the gain is from 1 (or 0 dB). The system is stable only if the gain margin is positive. The mathematical definition of gain margin is:

$$\text{GM} = \frac{1}{\left|G_{open}(j\omega_{pc})\right|} \tag{A.1.5}$$

where ω_{pc} is the frequency at which the phase of the open-loop transfer function $G(s)H(s)$ is -180°, also known as the **phase crossover frequency**.

Equation (A.1.5) can also be expressed in decibels (dB) as:

$$\text{GM in dB} = 20\log(\text{GM}) = 20\log\left(\frac{1}{\left|G_{open}(j\omega_{pc})\right|}\right) \tag{A.1.6}$$

From the definitions of the two stability indices above, it becomes clear that they represent two sides of the same coin as the two previously mentioned instability conditions. These stability indices essentially convey how far the system is from the critical instability condition—namely, when the open-loop transfer function $G(s)H(s)$ has a **gain of 1** and a **phase of $-180°$** at the same time. The greater the distance from this condition, the more stable the system is.

- **Understanding Stability Using Bode Plots**

Next, we use the Bode plot to help us understand the stability margins. Assume that the forward path transfer function in Fig. A.1 is given by $G(s) = \frac{1}{s^3+2s^2+s}$, and $H(s) = 1$.

We can then compute the closed-loop transfer function of the system as follows:

$$G_{close}(s) = \frac{G(s)}{1 + G(s)H(s)} = \frac{1}{s^3 + 2s^2 + s + 1} \tag{A.1.7}$$

The characteristic equation is $s^3 + 2s^2 + s + 1$. You can use MATLAB's roots command to find the roots of the characteristic equation as follows:

$$-1.7549, -0.1226 + 0.7449i, -0.1226 - 0.7449i$$

From the roots of the characteristic equation, we can determine that they are all located in the left-half plane, which means the system is stable. However, this does not tell us how stable the system is. Therefore, we need to plot the Bode diagram of the open-loop transfer function to help evaluate the system's stability.

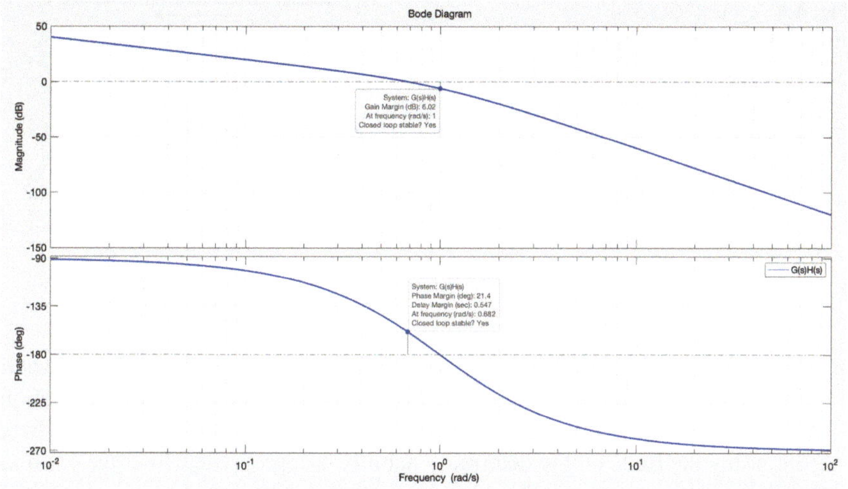

Fig. A.5 Bode plot of the open-loop transfer function $G(s)H(s)$

You can use MATLAB to plot the Bode diagram of the open-loop transfer function $G(s)H(s)$, as shown in Fig. A.5.

As shown in the figure, when the gain of $G(s)H(s)$ is 1, the corresponding gain crossover frequency $\omega_{gc} = 0.682$ (rad/s). At this point, the phase is 21.4° away from -180°, which means the system's phase margin (PM) is 21.4°.

When the phase of the open-loop transfer function $G(s)H(s)$ reaches -180°, the corresponding phase crossover frequency ω_{pc} is 1 (rad/s). At this point, the gain is 6.02 (dB) below 0 (dB), which means the system's gain margin (GM) is 6.02 (dB).

Summary of This Section:

- In modern control theory, the stability of a system is determined by the eigenvalues of the system matrix (i.e., the A matrix). A system is stable if all eigenvalues have negative real parts (i.e., they lie in the left half of the complex plane). The farther the eigenvalues are from the imaginary axis, the greater the system's stability, which also indicates a faster convergence rate of system states or state errors over time. However, in classical control theory, stability is typically analyzed for single-input single-output (SISO) negative feedback systems, where a different interpretation is used. Stability is often evaluated using *Phase Margin (PM)* and *Gain Margin (GM)*.
- When using the closed-loop transfer function, it is employed to study the magnitude and phase relationship between input and output or to assess the system bandwidth. In contrast, when analyzing system stability, the open-loop transfer function is used.

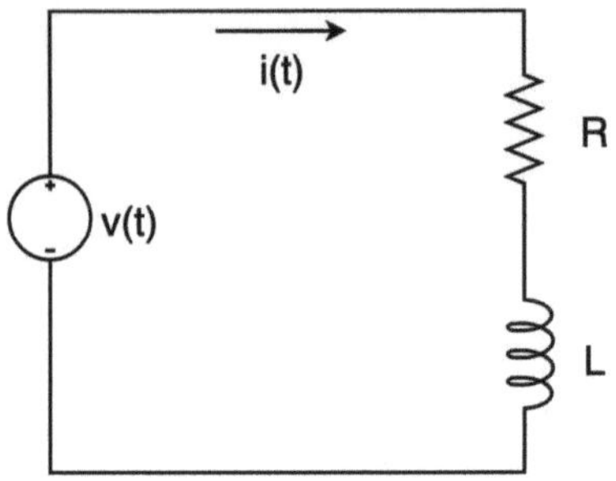

Fig. A.6 Typical RL series circuit

- For **minimum-phase systems** (i.e., systems with no zeros or poles in the right half-plane or on the imaginary axis, excluding the origin), there is a unique relationship between their frequency-domain magnitude and phase characteristics. Therefore, using the Bode plot to determine stability is appropriate. However, for **non-minimum-phase systems**, there is no unique correspondence between magnitude and phase characteristics. As a result, relying solely on the Bode plot to evaluate the stability of non-minimum-phase systems may lead to misjudgment. To accurately determine the stability of non-minimum-phase systems, use the *Nyquist Plot*. For a deeper understanding of the physical meaning of non-minimum-phase systems, refer to Sect. A.7 of this book.

A.2 Laplace Transforms Versus Transfer Functions

Classical control theory extensively uses **transfer functions** for the design and analysis of control systems. A transfer function represents the relationship between the input and output of a single-input single-output (SISO) linear system. Behind the operation of a transfer function lies the **differential equation** of the linear system. By applying the **Laplace transform**, the differential equation of the system can be converted into a transfer function [1, 2].

A control system can consist of multiple transfer functions. Thanks to the mathematical properties of the Laplace transform, transfer functions can be combined using simple arithmetic operations—addition, subtraction, multiplication, and division—without needing to deal with complex calculus operations. This significantly reduces the workload and computational effort required for control system design.

Therefore, the **Laplace transform** serves as a fundamental mathematical tool in classical control theory. But what is the difference between the Laplace transform and the transfer function? Can the two be considered equivalent? In this section, the author will clarify the subtle differences between the two concepts.

- **Ordinary Differential Equations and Laplace Transform**

Here, the author uses a simple series RL circuit as an example, as shown in Fig. A.6.

Where $v(t)$ is the input voltage, R is the circuit resistance, L is the circuit inductance, and $i(t)$ is the input current. This circuit system can be described by the following differential equation:

$$v(t) = Ri(t) + L\frac{di(t)}{dt} \tag{A.2.1}$$

To solve this differential equation, certain solution techniques for differential equations are required, which we will temporarily omit here. By applying such techniques, the general solution of Eq. (A.2.1) can be expressed as:

$$i(t) = \frac{v(t)}{R} + ce^{-(R/L)t} \tag{A.2.2}$$

where the constant c is related to the initial conditions of the circuit. Assuming the initial current is non-zero, i.e., $i(0) \neq 0$, then:

$$c = i(0) - \frac{v(0)}{R} = \frac{Ri(0) - v(0)}{R}$$

Assuming the input voltage $v(t)$ is a DC voltage E, i.e., $v(t) = E$, then:

$$c = \frac{Ri(0) - E}{R}$$

Thus, Eq. (A.2.2) can be rewritten as:

$$i(t) = \frac{E}{R} + \left[\frac{Ri(0) - E}{R}\right]e^{-(R/L)t} \tag{A.2.4}$$

Equation (A.2.4) is the particular solution of the differential equation. A particular solution refers to the specific solution obtained from a differential equation based on given conditions and the system's initial values.

The above shows the solution process for the differential equation. If we use the **Laplace transform** to solve it, the process becomes much simpler, as it replaces complex calculus operations with basic arithmetic.

First, the **general formula** for the Laplace transform of the n-th derivative of a function $f(t)$ is given by:

$$Laplace\{f^{(n)}\} = s^n L\{f\} - \sum_{i=1}^{n} s^{n-i} f^{(i-1)}(0) \tag{A.2.5}$$

Here, $Laplace\{*\}$ denotes the Laplace transform of a function, and s is the Laplace operator. Using Eq. (A.2.5), we can apply the Laplace transform to Eq. (A.2.1) and obtain:

$$V(s) = R \times I(s) + L[sI(s) - i(0)] \tag{A.2.6}$$

Here, $V(s)$ and $I(s)$ represent the Laplace transforms of $v(t)$ and $i(t)$, respectively. If we want to solve for $i(t)$, Eq. (A.2.6) can be rearranged as follows:

$$V(s) = (R + Ls) \times I(s) - L \times i(0) \tag{A.2.7}$$

Since the input voltage $v(t)$ is a DC voltage, i.e., $v(t) = E$, its Laplace transform is $V(s) = \frac{E}{s}$. Substituting this into Eq. (A.2.7), we get:

$$I(s) = \frac{E}{s} \times \frac{1}{(R + Ls)} + \frac{L}{(R + Ls)} \times i(0) \tag{A.2.8}$$

Taking the inverse Laplace transform of Eq. (A.2.8), we obtain:

$$i(t) = \frac{E}{R} + \left[\frac{Ri(0) - E}{R}\right] e^{-(R/L)t} \tag{A.2.9}$$

As shown in Eq. (A.2.9), the solution obtained using the Laplace transform is identical to the one derived using differential equation solving techniques, as in Eq. (A.2.4). Both methods take into account the system's initial conditions, specifically the initial current value $i(0)$.

- **Laplace Transform and Transfer Function**

Next, we will derive the transfer function of the system shown in Fig. A.6. In this system, the input is the voltage $v(t)$, and the output is the current $i(t)$. Therefore, we need to determine the transfer function from the input to the output. To obtain the transfer function, we again apply the Laplace transform to Eq. (A.2.1). The resulting Laplace-transformed equation is the same as Eq. (A.2.7), as shown below:

$$V(s) = (R + Ls) \times I(s) - L \times i(0) \tag{A.2.10}$$

To obtain the transfer function of the output $I(s)$ with respect to the input $V(s)$, the initial current $i(0)$ must be set to zero. Therefore, when $i(0) = 0$, the transfer function from input $V(s)$ to output $I(s)$ is:

$$\frac{I(s)}{V(s)} = \frac{1}{R + Ls} \tag{A.2.11}$$

From the above steps, we can see that the main difference between the Laplace transform and the transfer function is as follows:

When using the **Laplace transform** to solve a differential equation, the **initial conditions of the system can be taken into account**.

However, when using a **transfer function** to solve a differential equation, the **initial conditions must be set to zero**. If the initial values are not set to zero, the linear differential equation **cannot be converted into a transfer function**.

Summary of This Section:

- A **transfer function** represents the relationship between the input and output of a **linear single-input single-output (SISO) system**, and it is fundamentally based on the **differential equations** that describe the system's dynamics.
- The **main difference** between the **Laplace transform** and the **transfer function** is that the Laplace transform allows the **initial conditions** of the system to be included when solving differential equations. In contrast, when using a transfer function to solve a system, the **initial conditions must be set to zero**. If the initial values are not zero, the linear differential equation **cannot be converted into a transfer function**.

A.3 Sampling, Z-Transforms, and Zero-Order Hold in Digital Systems

In this book, we frequently use Laplace transform-based transfer functions for the design and simulation of motor control systems. In practice, to implement a controller based on the Laplace transform transfer function, analog components such as operational amplifiers, resistors, inductors, and capacitors must be used. However, analog control systems have drawbacks such as high cost and difficulty in modification.

As a result, the mainstream approach today is to use digital control systems, which means implementing Laplace-based controllers using computers [3]. To realize such controllers on a computer, it is necessary to approximate the continuous-time transfer function from the Laplace domain using discrete-time integration methods. This is precisely the origin of the Z-transform. Essentially, the Z-transform is an approximation of the Laplace transform based on numerical integration.

Moreover, since computers process data using a fixed sampling period, the shorter the sampling time, the better the Z-transform approximates the Laplace-domain transfer function. For input signals, the computer performs an analog-to-digital (A/D) conversion, which includes sampling and quantization. After the input signal is converted into a digital signal and the controller completes its calculations, the output can be converted back into an analog signal using a digital-to-analog (D/A) module. The most common D/A module is the zero-order hold (ZOH).

In the following section, we will provide an in-depth explanation of the key mechanisms in digital control systems—sampling, the Z-transform, and the zero-order hold (ZOH)—and clarify some common misconceptions.

- **Sampling of Signals**

A typical digital control system architecture is shown in Fig. A.7. In this system, the command input Cmd is subtracted from the sensor output signal to produce

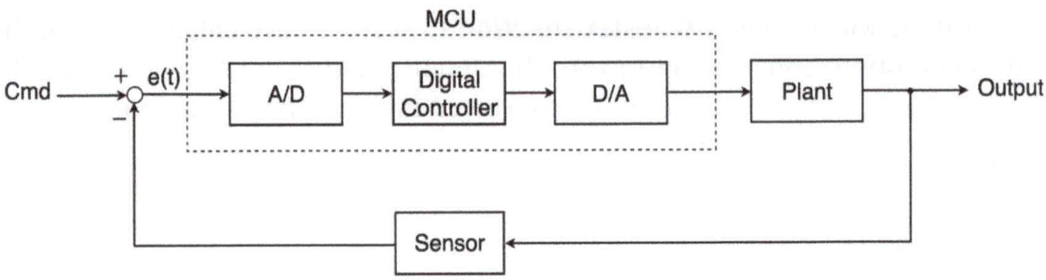

Fig. A.7 Typical digital control system block diagram

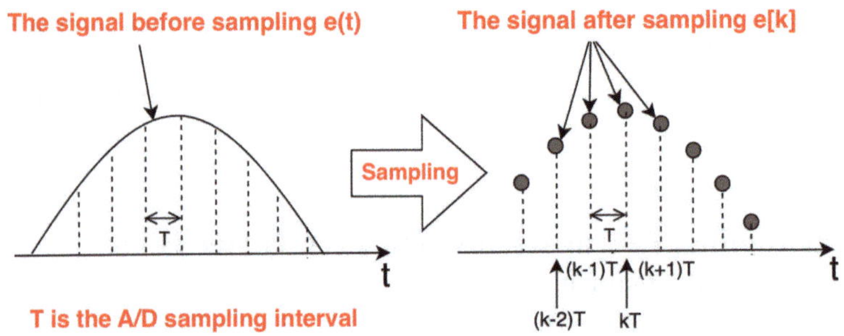

Fig. A.8 Illustration of signal sampling principle

the error signal $e(t)$. Before this error signal is processed by the digital controller, it undergoes analog-to-digital (A/D) conversion, which includes **sampling** and **quantization**.

Sampling refers to collecting the input signal $e(t)$ at uniform time intervals T. After sampling, the signal is no longer a continuous-time signal but becomes a discrete-time signal $e[k]$, as illustrated in Fig. A.8.

As shown in Fig. A.8, the continuous-time signal $e(t)$ before sampling and the discrete-time signal $e[k]$ after sampling are fundamentally different in nature. The discrete-time signal e[k] can be expressed using a mathematical series as follows:

$$e[k] = \sum_{k=0}^{\infty} e(kT)e^{-skT} \tag{A.3.1}$$

Here, e^{-skT} represents the Laplace transform of a delay of kT seconds. If we substitute $z = e^{sT}$ into Eq. (A.3.1), it becomes equivalent to performing a Z-transform on the continuous-time signal $e(t)$, that is:

$$Z\{e(t)\} = E(z) = \sum_{k=0}^{\infty} e(kT)z^{-k} \tag{A.3.2}$$

We can expand Eq. (A.3.2) as follows:

$$E(z) = e(0) + e(T)z^{-1} + e(2T)z^{-2} + e(3T)z^{-3} + \ldots \tag{A.3.3}$$

From the above derivation, we can understand that the essence of the Z-transform is a mathematical representation of a continuous-time signal after it has been discretized. The term z^{-1} corresponds to a time delay of T in the Laplace domain, that is, e^{-sT}.

If a digital system is to process analog signals, those signals must be sampled. Once a signal is sampled, it is no longer a continuous-time signal, and therefore the Laplace transform is no longer suitable to describe it. Instead, the Z-transform is used to represent the sampled signal and its computational behavior. Accordingly, any operations on discrete signals should also be described using the Z-transform, which will be elaborated in the following sections.

Tips

In fact, after a signal is sampled, it also undergoes quantization. Based on the resolution of the A/D module, each sampled signal point is converted into a corresponding digital value. Since this mechanism is relatively straightforward, it will not be discussed further here.

- **Sampling Delay of Signals**

Let's re-examine the difference between the signal before and after sampling. As shown in Fig. A.9, when the continuous-time signal is uniformly sampled with a sampling interval T, it forms the discrete-time signal $e[k]$. However, each discrete point $e[k]$ is considered to be valid throughout the entire sampling period.

In other words, the value $e(kT)$, which the computer uses for computation during the k-th sampling period, was actually sampled at the beginning of that period. A new input signal is not sampled until the start of the (k + 1)-th sampling period. Therefore, the signal processed by the computer, $e[k]$, lags behind the actual continuous-time signal $e(t)$.

So how much is this delay? Referring to Fig. A.9, if T is the sampling period, then the signal $e(t - T/2)$ can be regarded as the average waveform of $e[k]$. This suggests that the signal $e[k]$ used by the computer is delayed by $\frac{T}{2}$ compared to the actual continuous-time signal $e(t)$.

This is why, in Chap. 2, the sampling delay is equivalently modeled as a half-period delay, represented by the transfer function: $e^{-sT_{SH}} = e^{-s\times\frac{T_s}{2}}$ where T_s is the sampling period used in Chap. 2, corresponding to the sampling interval T used in this section.

- **The Essence of Z-Transform**

In Eqs. (A.3.2) and (A.3.3), we have already developed a preliminary understanding of the Z-transform. The essence of the Z-transform is that it provides a mathematical representation of a signal after discretization of a continuous-time signal.

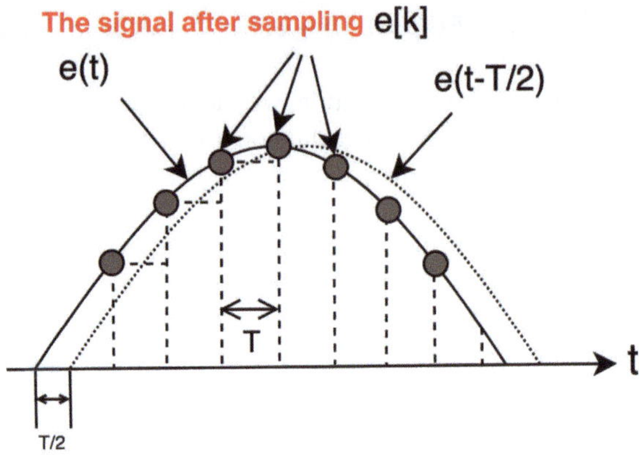

Fig. A.9 Illustration of original signal and sampled signal

However, Eqs. (A.3.2) and (A.3.3) represent the series form of the Z-transform. Generally, it is necessary to convert the series form into a closed-form expression for practical use. If the form of the input signal is known, one can refer to standard Z-transform tables to find the corresponding result. This will not be elaborated here, and readers can consult digital control textbooks for further details.

What the author wants to emphasize here is the practical approach in control system design: the standard practice is to first design a continuous-time transfer function based on Laplace transform methods, and then apply a Z-transform using the sampling time T to convert it into a discrete-time transfer function.

For example, suppose a designed PI controller has the following transfer function in the Laplace domain:

$$C(s) = K_p + \frac{K_i}{s} = 0.16 + \frac{0.73}{s} \tag{A.3.4}$$

Assuming a sampling time of $T = 0.001$ s and using the **bilinear transformation** for approximation, the bilinear transformation is given as:

$$s \cong \frac{2}{T}\left(\frac{z-1}{z+1}\right) \tag{A.3.5}$$

You can use the following MATLAB code to obtain the Z-transform form of Eq. (A.3.4):

```
tf_pi = tf([0.16 0.73],[1 0]);
tf_pi_z = c2d(tf_pi, 0.001, 'tustin');
```

The resulting Z-transform is as follows:

$$C(z) = \frac{0.1604z - 0.1596}{z - 1} \tag{A.3.6}$$

After obtaining the Z-transform in Eq. (A.3.6), we need to rearrange it into the following form:

$$H(z) = \frac{Y(z)}{X(z)} = \frac{b_0 + b_1 z^{-1} + b_2 z^{-2} + \ldots}{1 + a_1 z^{-1} + a_2 z^{-2} + \ldots} \tag{A.3.7}$$

Therefore, divide both the numerator and the denominator of Eq. (A.3.6) by the highest power of z.

$$C(z) = \frac{P(z)}{E(z)} = \frac{0.1604 - 0.1596z^{-1}}{1 - z^{-1}} \tag{A.3.8}$$

Assuming that the input to the PI controller is the error signal $E(z)$ and the output is $P(z)$, we can apply cross-multiplication to obtain:

$$P(z) - P(z)z^{-1} = 0.1604E(z) - 0.1596E(z)z^{-1} \tag{A.3.9}$$

Equation (A.3.9) can be converted into the following difference equation:

$$p[n] = p[n-1] + 0.1604 \times e[n] - 0.1596 \times e[n-1] \tag{A.3.10}$$

After obtaining the difference Eq. in (A.3.10), it can be directly implemented on a microcontroller as a digital PI controller. In this equation, $e[n]$ and $e[n-1]$ represent the sampled values of the error signal. Therefore, there is no need for a separate "sampling" transfer function—once a Laplace-domain transfer function is converted into its Z-domain form, the sampling behavior is inherently included, because the essence of the Z-transform is sampling.

Moreover, since a digital control system performs operations on sampled signals, the resulting control algorithm naturally takes the form of a difference equation like (A.3.10).

- **Zero-Order Hold (ZOH)**

During each sampling period, the computer calculates the output control signal $p[n]$ based on the controller's difference equation, such as Eq. (A.3.10). However, at this stage, $p[n]$ is a digital value and exists in a discrete state. To use this signal as the input control signal for the plant, it must be converted from a digital value to an analog signal $p(t)$ using a digital-to-analog (D/A) module. The most commonly used D/A conversion method is the zero-order hold (ZOH).

Figure A.10 illustrates how the output value $p[n]$ from the digital controller is converted into an analog signal $p(t)$ using a zero-order hold. Assuming the D/A

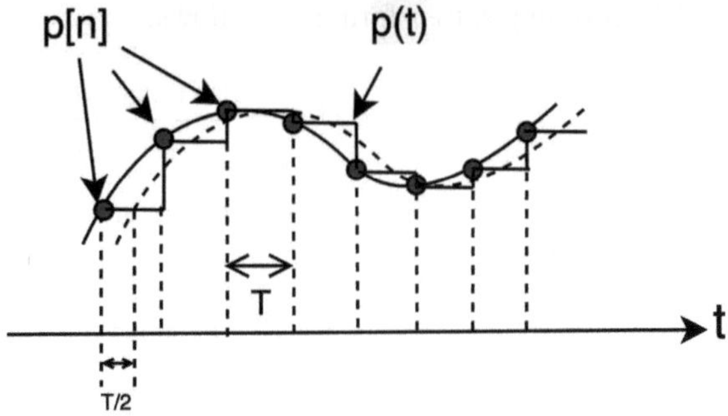

Fig. A.10 Illustration of reconstructing analog signal from digital signal via ZOH

module uses a sampling period of T, the resulting staircase signal $p(t)$ in the figure represents the analog signal produced by the ZOH.

In this section, we will not delve into the mathematical derivation of the zero-order hold in detail. Interested readers may refer to relevant textbooks on digital control. What the author would like to emphasize here is that the output delay characteristic of the analog signal $p(t)$ generated by the ZOH is similar to the delay effect caused by sampling. Specifically, the analog signal $p(t)$ lags behind the digital signal $p[n]$ by $\frac{T}{2}$.

This is why, as discussed in Chap. 2, the output delay T_o of the voltage command in the current control loop is modeled as one-half of the sampling period, that is, $T_o = \frac{T_s}{2}$, where T_s is the sampling period variable used in Chap. 2 and is the same as the sampling interval T in this section.

The transfer function T(z) of the Zero-Order Hold (ZOH) is given as follows:

$$T(s) = \frac{1 - e^{-Ts}}{s} = \frac{1 - z^{-1}}{s} = \frac{z-1}{z}\left(\frac{1}{s}\right) \tag{A.3.11}$$

where $z = e^{sT}$, in order to make its DC gain equal to 1, we multiply $T(z)$ by $\frac{1}{T}$, resulting in:

$$T(z) = \frac{z-1}{Tz}\left(\frac{1}{s}\right) \tag{A.3.12}$$

Equation (A.3.12) represents the transfer function of a Zero-Order Hold (ZOH), and it is also commonly used as the transfer function of a digital-to-analog (D/A) converter employing ZOH.

When considering sinusoidal steady-state, we can substitute $s = j\omega$ into Eq. (A.3.12), yielding:

$$T(z) = \frac{z-1}{Tz}\left(\frac{1}{s}\right) = \left(\frac{\sin(\omega T/2)}{\omega T/2}\right)\angle - \frac{\omega T}{2} \tag{A.3.13}$$

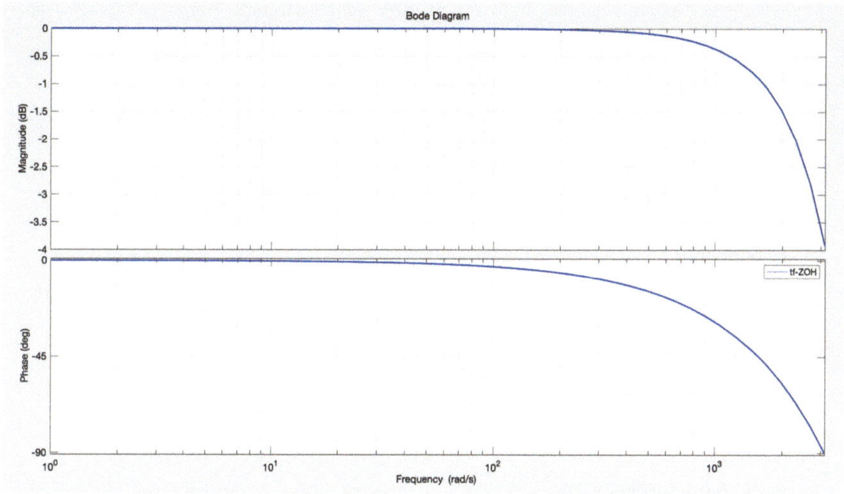

Fig. A.11 Bode plot of the zero-order hold (ZOH)

From Eq. (A.3.13), it can be seen that from the perspective of the transfer function, a Zero-Order Hold (ZOH) indeed causes a time delay of $\frac{T}{2}$.

You may open the example program mA_3_1 to plot the Bode diagram of Eq. (A.3.12), as shown in Fig. A.11, where the sampling time T is set to 0.001 s.

MATLAB Example Script: mA_3_1.m

```
s=tf('s');
T=0.001;
tf_ZOH=(1-exp(-T*s))/(T*s);
h=bodeoptions;
h.PhaseMatching='on';
h.Title.FontSize = 14;
h.XLabel.FontSize = 14;
h.YLabel.FontSize = 14;
h.TickLabel.FontSize = 14;
bodeplot(tf_ZOH,'-b',{1,2*pi*1000/2},h);
legend('tf-ZOH');
h = findobj(gcf,'type','line');
set(h,'linewidth',2); grid on;
```

As shown in Fig. A.11, when the signal frequency is much lower than half of the sampling frequency—that is, $\frac{2\times\pi\times1000}{2}$ (rad/s)—there is no significant attenuation in magnitude after the signal passes through the Zero-Order Hold (ZOH). However, as the signal frequency approaches half of the sampling frequency, the magnitude attenuation becomes more severe. This frequency, half of the sampling frequency, is known as the **Nyquist frequency**.

The phase exhibits a similar characteristic. As the signal frequency ω increases, the phase delay described by Eq. (A.3.13), which is $-\frac{\omega T}{2}$, also increases. The greater the phase delay, the more it undermines the stability of the control system.

Therefore, a key prerequisite for designing a well-performing digital control system is a thorough understanding of the relationship between the control loop bandwidth and the sampling frequency—namely, the fundamental principle revealed by the **Nyquist-Shannon sampling theorem**.

The relationship between the control loop bandwidth and the sampling frequency can be understood through the **Nyquist-Shannon sampling theorem**, a fundamental principle in signal processing. This theorem states that to perfectly reconstruct an original analog signal, the **sampling frequency** (in Hz) must be at least **twice the highest frequency component** present in the signal.

When the input signal frequency exceeds half of the sampling frequency, the sampling points are insufficient to accurately represent the signal. In such cases, **aliasing** occurs, where the observed signal no longer represents the original high-frequency input but appears as a lower-frequency distorted version.

In control systems, the **bandwidth** refers to the frequency range over which the system can effectively respond and control the input signal. A wide-bandwidth control system can respond quickly and track rapid changes in the input. However, in digital control systems, the controller must sample the input signal to generate control commands. This imposes a constraint on the achievable bandwidth. **According to the Nyquist theorem, the sampling frequency should be at least twice the control loop bandwidth** to ensure stability and performance. But in practical applications, it is typically recommended to set the **sampling frequency to 5 to 10 times the bandwidth**, or even higher. This ensures:

- Better signal reconstruction quality
- Reduced phase delay
- Improved system stability and responsiveness

Thus, choosing an appropriate sampling frequency is essential for designing high-performance digital control systems.

Summary of This Section:

- The primary difference between analog and digital control systems lies in the delays caused by sampling, computation, and digital-to-analog (D/A) conversion in digital control systems.
- Once a signal is sampled, it is no longer a continuous-time signal and thus cannot be accurately described using the Laplace transform. Instead, the Z-transform is used to describe the behavior and operations on sampled signals.
- The Bode plot of a Z-domain transfer function is only an approximation of that of its continuous-time (Laplace domain) counterpart. This is because the Z-transform is essentially an integral approximation of the Laplace transform. Therefore, it is important to examine the Bode plot of the Z-domain transfer function to evaluate the effects of magnitude attenuation and phase delay introduced by different sampling times, and to comprehensively assess the performance and stability of the digital control system.

A.4 Implementation of P, I, and D Controllers in the s- and z-Domains

In a typical motor control system, **P**, **I**, and **D** controllers are the most commonly used types of controllers, where:

- **P** stands for *Proportional* controller,
- **I** stands for *Integral* controller, and
- **D** stands for *Derivative* controller.

Each component has its own gain:

- The gain of the proportional controller is called the **proportional gain**, denoted as K_p,
- The gain of the integral controller is called the **integral gain**, denoted as K_i,
- The gain of the derivative controller is called the **derivative gain**, denoted as K_d.

The time-domain representation of a typical PID controller is given by:

$$PID\,output = K_p \times e(t) + K_i \times \int e(t)dt + K_d \times \frac{de(t)}{dt} \quad \text{(A.4.1)}$$

where $e(t)$ is the error signal, defined as the difference between the command (reference) and the feedback value.

By taking the **Laplace transform** of Eq. (A.4.1), the PID controller's Laplace-domain representation becomes:

$$PID\,output = \left(K_p + K_i \times \frac{1}{s} + K_d \times s\right) \times E(s) \quad \text{(A.4.2)}$$

where $E(s)$ is the Laplace transform of the error signal $e(t)$.

Next, the author will introduce the representations of P, I, and D controllers in both the S-domain and the Z-domain, and will use Bode plots to examine the differences caused by various Z-transform approximation methods [1, 2].

- **Proportional controller**

We can isolate the proportional controller term from Eq. (A.4.2) as follows:

$$P\,control\,output = K_p \times E(s) \quad \text{(A.4.3)}$$

As shown in Eq. (A.4.3), the proportional controller works by directly multiplying the error signal by a proportional gain K_p. If we convert Eq. (A.4.3) into the Z-domain, we get:

$$P\,control\,output = K_p \times E(z) \quad \text{(A.4.4)}$$

By converting Eq. (A.4.4) into a difference equation, we obtain:

$$p[n] = K_p \times e[n] \quad \text{(A.4.5)}$$

where $p[n]$ is the output of the proportional controller and $e[n]$ is the input, we can see from Eq. (A.4.5) that the behavior of the proportional controller is quite simple:

in each sampling period, it multiplies the input signal by the proportional gain K_p to produce the controller output. Therefore, the operation of a proportional controller is essentially the same in both the S-domain and the Z-domain.

- **Integral controller**

  We can isolate the integral controller term from Eq. (A.4.2) as follows:

$$I\ control\ output = K_i \times \frac{1}{s} \times E(s) \tag{A.4.6}$$

As shown in Eq. (A.4.6), the integral controller operates by integrating the error signal and then multiplying it by an integral gain K_i. To approximate the integral operation in Eq. (A.4.6), two common approximation methods are introduced. The first method is Euler's integration approximation, which can be expressed as follows: [2]

$$c[n] = c[n-1] + T \times r[n] \tag{A.4.7}$$

where $c[n]$ is the integrator output at the n-th sampling period, $c[n-1]$ is the output at the (n-1)-th sampling period, $r[n]$ is the integrator input at the n-th sampling period, and T is the sampling period.

Applying the Z-transform to Eq. (A.4.7), we get:

$$C(z) = C(z) \times z^{-1} + TR(z) \tag{A.4.8}$$

From Eq. (A.4.8), the Z-transform of the integrator using **Euler's integration** method is:

$$\frac{C(z)}{R(z)} = \frac{Tz}{z-1} \tag{A.4.9}$$

You can open the example script mA_4_1 to plot the Bode diagram of Eq. (A.4.9) and compare it with the Bode plot of the integrator in the S-domain, as shown in Fig. A.12.

MATLAB Example Script: mA_4_1.m

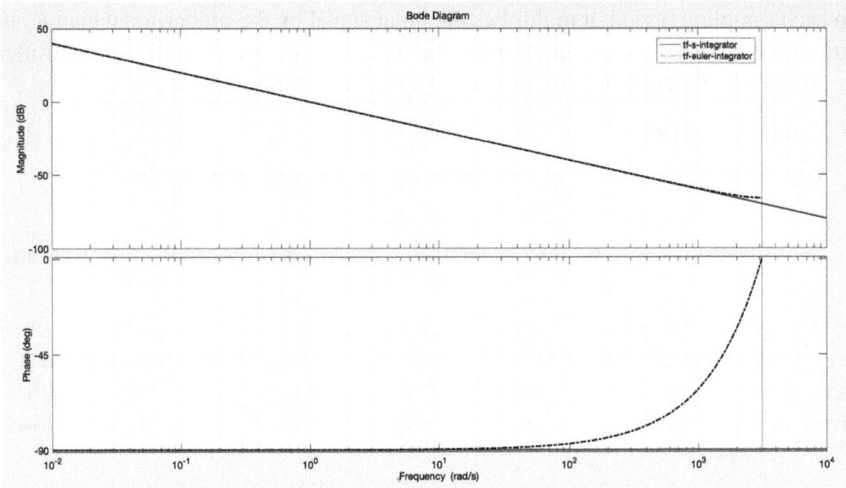

Fig. A.12 Bode plot of the analog integrator versus the Euler's integrator

```
T=0.001;

z=tf('z', T);

s=tf('s');

tf_euler_integrator = T*z/(z-1);

tf_s_integrator = 1/s;

h=bodeoptions; h.PhaseMatching='on';

h.Title.FontSize = 14;

h.XLabel.FontSize = 14;

h.YLabel.FontSize = 14;

h.TickLabel.FontSize = 14;

bodeplot(tf_s_integrator,'-b',tf_euler_integrator,'-.k',{0.01,10000},h);

legend('tf-s-integrator','tf-euler-integrator');

h = findobj(gcf,'type','line');

set(h,'linewidth',2); grid on;
```

As shown in Fig. A.12, the **Euler's integration**-based integrator closely matches the frequency-domain magnitude characteristics of the S-domain integrator when the frequency is well below half the sampling frequency (i.e., the **Nyquist frequency**, which is $\frac{2\times\pi\times1000}{2}$ (rad/s). This means the magnitude approximation is quite accurate in the low-frequency range.

In terms of **phase characteristics**, the S-domain integrator inherently introduces a phase lag of − **90°** due to its pole at the origin. However, the Euler integrator's phase lag **decreases** at higher frequencies, meaning it provides **more phase margin** in the high-frequency range. This added phase margin contributes positively to the **stability** of control systems, making the Euler integrator particularly beneficial for improving system robustness.

The second commonly used integrator is the **trapezoidal integrator**, which can be expressed as follows [2]:

$$c[n] = c[n-1] + \frac{T}{2} \times (r[n] + r[n-1]) \qquad \text{(A.4.10)}$$

where $c[n]$ is the output of the integrator at the n-th sampling period, $c[n-1]$ is the output of the integrator at the (n-1)-th sampling period, $r[n]$ is the input of the integrator at the n-th sampling period, $r[n-1]$ is the input of the integrator at the (n-1)-th sampling period, and T is the sampling period.

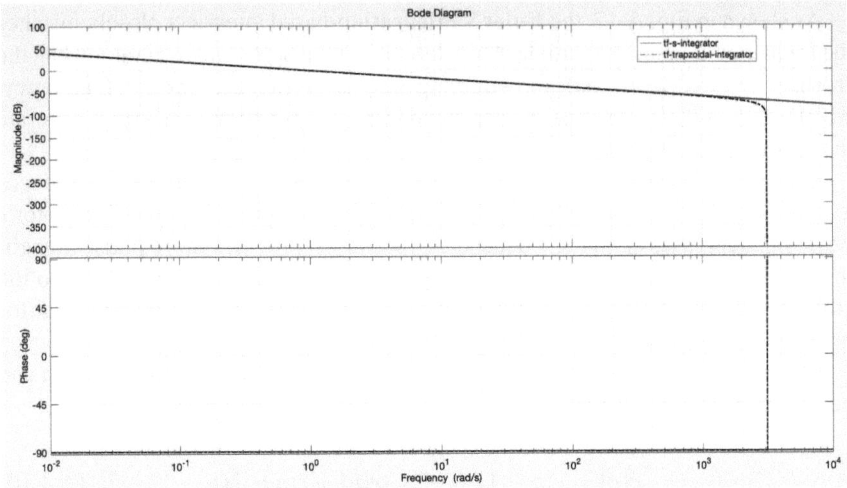

Fig. A.13 Bode plot of the analog integrator versus the Trapezoidal integrator

Applying the Z-transform to Eq. (A.4.10), we obtain the Z-transform of the trapezoidal integrator as follows:

$$\frac{C(z)}{R(z)} = \frac{T}{2}\left(\frac{z+1}{z-1}\right) \tag{A.4.11}$$

From Eqs. (A.4.11) and (A.3.5), it can be seen that the trapezoidal integrator is essentially the bilinear transformation of an integrator.

You can open the example program mA_4_2 to plot the Bode diagram of Eq. (A.4.11) and compare it with the integrator in the S-domain, as shown in Fig. A.13.

MATLAB Example Script: mA_4_2.m

```
T=0.001;

z=tf('z', T);

s=tf('s');

tf_trapezoidal_integrator = (T/2)*(z+1)/(z-1);

tf_s_integrator = 1/s;

h=bodeoptions; h.PhaseMatching='on';

h.Title.FontSize = 14;

h.XLabel.FontSize = 14;

h.YLabel.FontSize = 14;

h.TickLabel.FontSize = 14;

bodeplot(tf_s_integrator,'-b',tf_trapzoidal_integrator,'-.k',{0.01,10000},h);

legend('tf-s-integrator','tf-trapezoidal-integrator');

h = findobj(gcf,'type','line');

set(h,'linewidth',2); grid on;
```

As shown in Fig. A.13, the trapezoidal integrator exhibits a frequency-domain magnitude response that nearly overlaps with that of the S-domain integrator when the frequency is well below half the sampling frequency (i.e., the Nyquist frequency), which is $\frac{2\times\pi\times1000}{2}$ (rad/s). Therefore, its magnitude approximation is comparable to that of the Euler integrator.

In terms of phase characteristics, the trapezoidal integrator maintains a consistent -90° phase lag across the entire frequency range, matching that of the S-domain integrator. It does not provide additional phase margin at high frequencies. As a result, the trapezoidal integrator does not offer extra stability benefits to the control system.

- **Derivative controller**

We can isolate the derivative controller term from Eq. (A.4.2) as follows:

$$D\ control\ output = K_d \times s \times E(s) \tag{A.4.12}$$

As shown in Eq. (A.4.12), the derivative controller works by differentiating the error signal and then multiplying it by a derivative gain K_d. To approximate the differentiation operation in Eq. (A.4.12), the author introduces two commonly used methods. The first method is **Euler's differentiation**, which can be expressed as follows:

$$c[n] = \frac{r[n] - r[n-1]}{T} \tag{A.4.13}$$

where $c[n]$ be the output of the differentiator at the nth sampling interval, $r[n]$ the input at the nth sampling interval, $r[n-1]$ the input at the (n-1)th interval, and T the sampling period.

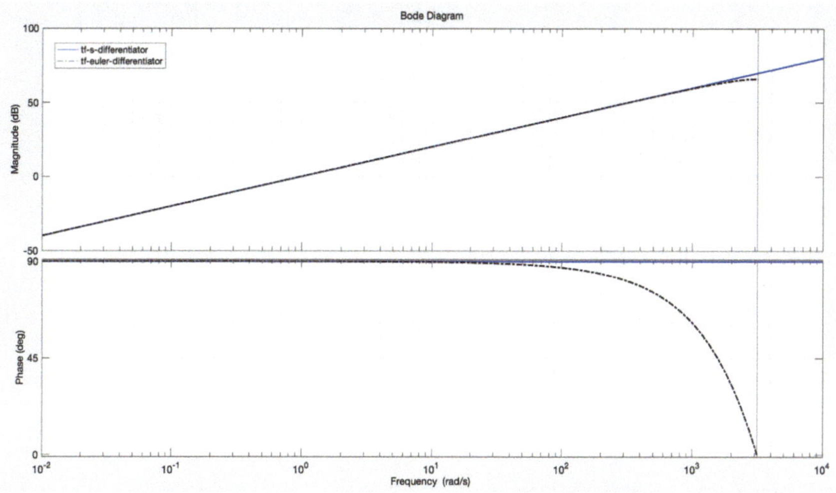

Fig. A.14 Bode plot of the analog differentiator versus the Euler differentiator

Taking the Z-transform of Eq. (A.4.13), we obtain the **Z-transform of Euler's differentiation** as:

$$\frac{C(z)}{R(z)} = \frac{z-1}{Tz} \tag{A.4.14}$$

You can open the example program **mA_4_3** to plot the Bode diagram of Eq. (A.4.14) and compare it with the differentiator in the **S-domain**, as shown in Fig. A.14.

MATLAB Example Script: mA_4_3.m

```
T=0.001;

z=tf('z', T);

s=tf('s');

tf_euler_differentiator = (z-1)/(T*z);

tf_s_differentiator = s;

h=bodeoptions; h.PhaseMatching='on';

h.Title.FontSize = 14;

h.XLabel.FontSize = 14;

h.YLabel.FontSize = 14;

h.TickLabel.FontSize = 14;

bodeplot(tf_s_differentiator,'-b',tf_euler_differentiator,'-.k',{0.01,10000},h);

legend('tf-s-differentiator','tf-euler-differentiator');

h = findobj(gcf,'type','line');

set(h,'linewidth',2); grid on;
```

As shown in Fig. A.14, the Euler differentiator exhibits a magnitude frequency response that closely matches that of the S-domain differentiator when the frequency is much lower than half of the sampling frequency (i.e., the Nyquist frequency, which is $\frac{2\times\pi\times1000}{2}$ (rad/s). This indicates that the magnitude approximation is quite accurate in the low-frequency region.

However, regarding the phase characteristics, the S-domain differentiator inherently provides a $+ 90^\circ$ phase lead due to its zero. In contrast, the Euler differentiator's phase response begins to decrease with increasing frequency in the mid-to-high-frequency range. This means that the Euler differentiator cannot maintain the full 90° phase lead in the higher-frequency regions, and its contribution to system stability diminishes as the frequency increases.

The second commonly used differentiation method is **trapezoidal differentiation**, which can be expressed as follows:

$$c[n] = -c[n-1] + \frac{2}{T}(r[n] - r[n-1]) \tag{A.4.15}$$

where

- $c[n]$ be the output of the differentiator at the n-th sampling period,
- $c[n-1]$ be the output at the (n-1)-th sampling period,
- $r[n]$ be the input of the differentiator at the n-th sampling period,
- $r[n-1]$ be the input at the (n-1)-th sampling period,
- and T be the sampling period.

Taking the Z-transform of Eq. (A.4.15), the trapezoidal differentiation in the Z-domain is given by:

$$\frac{C(z)}{R(z)} = \frac{2}{T}\left(\frac{z-1}{z+1}\right) \tag{A.4.16}$$

Equation (A.4.16) is the reciprocal of the trapezoidal integrator's Z-transform (i.e., the inverse of Eq. (A.4.11)). However, pure trapezoidal differentiation has inherent issues.

We can observe from Eq. (A.4.15) that when both $r[n]$ and $r[n-1]$ are zero, the output $c[n]$ may still change sign, which could lead to instability in the control system [2]. Therefore, Eq. (A.4.15) must be modified to:

$$c[n] = -\alpha \times c[n-1] + \frac{1+\alpha}{T}(r[n] - r[n-1]), \alpha < 1 \tag{A.4.17}$$

Taking the Z-transform of Eq. (A.4.17), we obtain the modified trapezoidal differentiation Z-transform as follows:

$$\frac{C(z)}{R(z)} = \frac{1+\alpha}{T}\left(\frac{z-1}{z+\alpha}\right), \alpha < 1 \tag{A.4.18}$$

When $\alpha = 0$, the modified trapezoidal differentiator becomes equivalent to the Euler differentiator. As the value of α increases, the phase loss caused by the differentiator in the high-frequency range becomes smaller. However, α must be less than 1.

You can open the example program **mA_4_4** to plot the Bode diagram of Eq. (A.4.18) and compare it with the differentiator in the S-domain, as shown in Fig. A.15.

MATLAB Example Script: mA_4_4.m

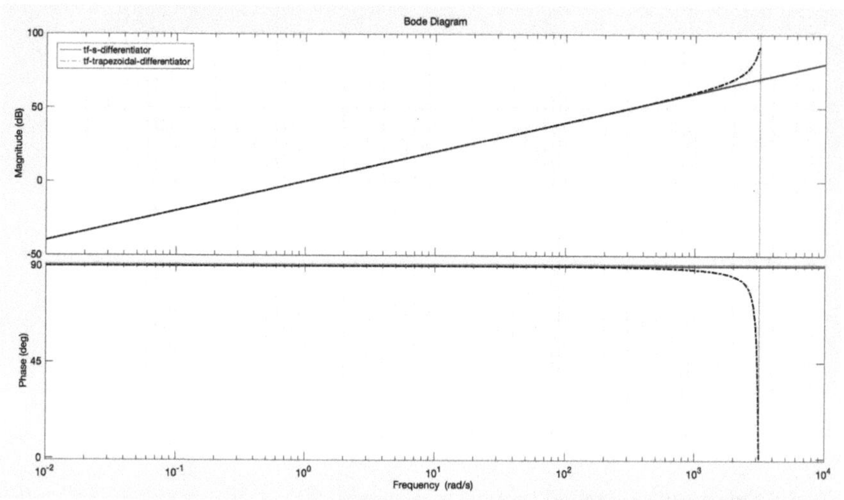

Fig. A.15 Bode plot of the analog differentiator versus the modified Trapezoidal differentiator

```
T=0.001;

Alpha=0.9;

z=tf('z', T);

s=tf('s');

tf_trapezoidal_differentiator = (1+alpha)/T*(z-1)/(z+alpha);

tf_s_differentiator = s;

h=bodeoptions; h.PhaseMatching='on';

h.Title.FontSize = 14;

h.XLabel.FontSize = 14;

h.YLabel.FontSize = 14;

h.TickLabel.FontSize = 14;

bodeplot(tf_s_differentiator,'-
b',tf_trapezoidal_differentiator,'-.k',{0.01,10000},h);

legend('tf-s-differentiator','tf-trapezoidal-differentiator');

h = findobj(gcf,'type','line');

set(h,'linewidth',2); grid on;
```

As shown in Fig. A.15, the modified trapezoidal differentiator exhibits a frequency magnitude response that closely matches the S-domain differentiator when the frequency is much lower than half the sampling frequency (i.e., the Nyquist frequency), which is $\frac{2\times\pi\times1000}{2}$ (rad/s). Therefore, the magnitude approximation is comparable to that of the Euler differentiator. Regarding the phase characteristics, by setting the α value to 0.9, the modified trapezoidal differentiator achieves a phase response in the mid-to-high frequency range that closely resembles that of the S-domain differentiator. Consequently, the modified trapezoidal differentiator can significantly improve the stability contribution to the control system.

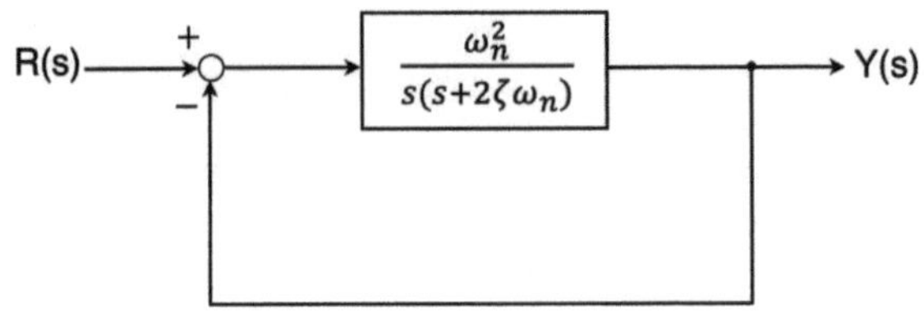

Fig. A.16 Block diagram of a standard second-order system

Summary of This Section:

- When using different approximation methods to convert an S-domain transfer function into a Z-domain transfer function, it is essential to examine the Bode plot of the Z-domain transfer function under various sampling frequencies. This allows for the observation of the phase lag introduced by the digital control system. Based on the system's stability requirements, both the sampling frequency and the approximation method should be adjusted accordingly.
- From the perspective of numerical analysis, the trapezoidal method offers better approximation accuracy than the Euler method. However, in control system design, accuracy is only one factor in evaluating controller performance—phase characteristics are another critical consideration. Since phase response directly affects system stability, the phase characteristics of the controller may be even more important in many situations.

A.5 The Nature and Characteristics of Standard Second-Order Systems

In classical control theory, aside from first-order systems, the primary focus of analysis and study is on standard second-order systems [1, 2]. A block diagram of a standard second-order control system is shown in Fig. A.16.

Here, ω_n is the natural frequency of the system, and ζ is the damping ratio. As shown in Fig. A.16, the forward gain of the standard second-order system is

$$G(s) = \frac{\omega_n^2}{s(s + 2\zeta\omega_n)}$$

and the feedback gain is

$$H(s) = 1.$$

Therefore, the closed-loop transfer function of the standard second-order system is:

$$\frac{Y(s)}{R(s)} = \frac{G(s)}{1 + G(s)H(s)} = \frac{\omega_n^2}{s^2 + 2\zeta\omega_n s + \omega_n^2} \tag{A.5.1}$$

From Eq. (A.5.1), the characteristic equation of the system is:

$$\Delta(s) = s^2 + 2\zeta\omega_n s + \omega_n^2 \tag{A.5.2}$$

The two characteristic roots of a standard second-order system can be expressed as follows:

$$s_1, s_2 = -\zeta\omega_n \pm j\omega_n\sqrt{1-\zeta^2} \tag{A.5.3}$$

From Eq. (A.5.3), we can observe the following:

- When the system damping ratio $\zeta = 0$, the two roots are $s_1, s_2 = \pm j\omega_n$, indicating that the system oscillates at the natural frequency ω_n. This case is called **undamped**, where the system remains in a state of sustained oscillation. Therefore, the natural frequency ω_n is also referred to as the **undamped frequency**.
- When $0 < \zeta < 1$, the roots are complex conjugates and lie in the left half-plane. Although the system has damping, it still exhibits overshoot, and is thus called **underdamped**.
- When $\zeta \geq 1$, the roots are real (no imaginary part) and lie on the negative real axis:
  - If $\zeta = 1$, the damping is just enough to eliminate overshoot. This case is known as **critical damping**, where the two roots are repeated: $s_1, s_2 = -\omega_n, -\omega_n$. In this scenario, the standard second-order system can be considered equivalent to two cascaded first-order low-pass filters.
  - If $\zeta > 1$, the system is **overdamped**, meaning it does not overshoot but responds more slowly.

Both critically damped and overdamped systems avoid overshoot. However, in practice, systems are often designed to be slightly **overdamped** to provide robustness against parameter variations. Designing a system to be critically damped can risk it becoming underdamped if parameters change, leading to unwanted overshoot.

Section 5.6 of this book discusses the **Butterworth low-pass filter**. From the content in Sect. 5.6, we can understand that a Butterworth low-pass filter of any order is composed of a combination of first-order and standard second-order systems.

In particular, a **second-order Butterworth low-pass filter** can be viewed as a standard second-order system with a **damping ratio** $\zeta = 0.707$. It is represented as follows:

$$T(s) = \frac{\omega_n^2}{s^2 + 2 \times 0.707 \times \omega_n s + \omega_n^2} \tag{A.5.4}$$

Additionally, if the control system can be approximated as a standard second-order system like that shown in Eq. (A.5.1), then the **system bandwidth** can be calculated using the following Eq. (A.5.5):

$$BW = \omega_n \times \sqrt{(1 - 2\zeta^2) + \sqrt{4\zeta^4 - 4\zeta^2 + 2}} \tag{A.5.5}$$

When the damping ratio $\zeta = 0.707$, the **system bandwidth** exactly equals the natural frequency ω_n, i.e., $BW = \omega_n$.

When the damping ratio $\zeta \leq 0.707$, the Bode plot of the system described by Eq. (A.5.1) will exhibit a **peak**, known as the **resonant peak** M_r.

$$M_r = \frac{1}{2\zeta\sqrt{1 - \zeta^2}} \tag{A.5.6}$$

The corresponding resonant frequency ω_r is:

$$\omega_r = \omega_n\sqrt{1 - 2\zeta^2} \tag{A.5.7}$$

From Eq. (A.5.7), it can be seen that a resonant frequency ω_r exists only when the damping ratio $\zeta \leq 0.707$. If the resonant frequency ω_r does not exist, then the resonant peak M_r naturally does not exist either.

Summary of This Section:

- For certain second-order systems, even if the denominator of the transfer function matches that of Eq. (A.5.1), the presence of a zero in the numerator means Eq. (A.5.5) cannot be used to calculate the bandwidth. Only systems that conform strictly to the standard second-order form in Eq. (A.5.1) can use Eq. (A.5.5) for bandwidth calculation.
- This section primarily reviews key concepts related to the standard second-order system from classical control theory. Since there is an abundance of related material, detailed discussion is omitted here. Interested readers are encouraged to consult classical control textbooks.
- Classical control theory techniques can systematically analyze transfer functions up to second-order. Therefore, in practical applications, real physical systems must be approximated as standard second-order systems to use classical analysis techniques effectively.
- For motor speed control systems using a PI controller, the closed-loop transfer function is shown in Eq. (2.3.26). Due to the presence of a zero in the numerator of (2.3.26), the system response may still oscillate even if designed to be overdamped. By using an IP controller instead, the zero in the numerator can be canceled, resulting in a standard second-order system as shown in Eq. (2.3.27).

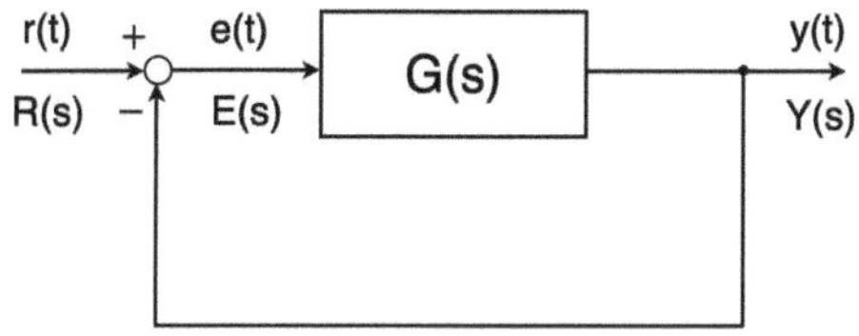

Fig. A.17 Block diagram of a typical unity-feedback negative feedback system

A.6 Steady-State Error Analysis: A Speed Control Case Study

In time-domain analysis of control systems, in addition to evaluating transient response performance, **steady-state error** is also a key indicator of system performance. Steady-state error is defined as:

The difference between the command input and the system output in steady state.

Ideally, we desire the steady-state error to be **zero**, meaning the system output closely follows the command input. However, due to system modeling inaccuracies and various nonlinear effects, **steady-state error almost always exists**. Therefore, in control system design, it is important to minimize this error or at least keep it below a certain acceptable threshold.

Moreover, **steady-state error depends on both the type of input signal and the system type**. For example, some systems may have no steady-state error for step inputs but do exhibit error when responding to ramp inputs. As a result, it is necessary to perform a **systematic and detailed analysis and modeling of steady-state error**.

In this section, we will review key concepts from classical control theory related to steady-state error and use a motor speed control system as an example to illustrate in-depth steady-state error analysis [1].

- **Definition of Steady-State Error**

A typical unity negative feedback system can be represented as shown in Fig. A.17.

Where r(t) is the time-domain function of the system command, R(s) is the Laplace transform of the system command, y(t) is the time-domain function of the system output, Y(s) is the Laplace transform of the system output, e(t) is the time-domain function of the system error, and E(s) is the Laplace transform of the system error.

The steady-state error e_{ss} of a typical unity negative feedback system is defined as:

$$e_{ss} = \lim_{t \to \infty} e(t) = \lim_{s \to 0} sE(s) \tag{A.6.1}$$

For a unity feedback system, Eq. (A.6.1) can be further expressed as:

$$e_{ss} = \lim_{s \to 0} sE(s) = \lim_{s \to 0} s\left(\frac{1}{1 + G(s)}\right)R(s) \tag{A.6.2}$$

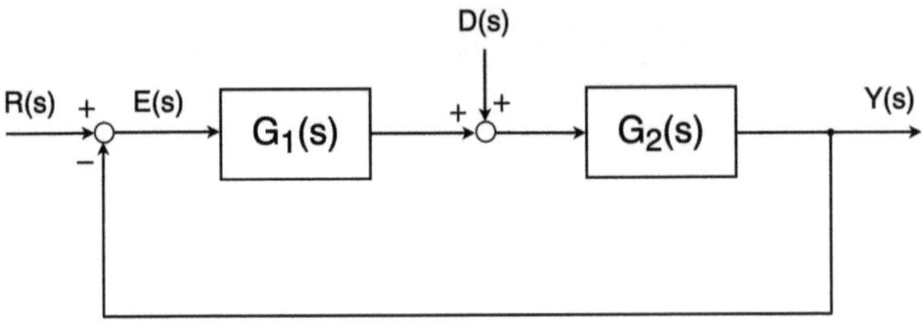

Fig. A.18 Block diagram of a typical unity-feedback negative feedback system with disturbance input

- **Steady-State Error Considering Disturbances**

Steady-state error is not solely caused by the input; disturbances can also affect it. A feedback control system that considers disturbances can be represented as shown in Fig. A.18, where $D(s)$ denotes the Laplace transform of the disturbance signal.

By applying the principle of superposition, we can determine the error signal generated by the input $R(s)$ and the disturbance $D(s)$.

$$E(s) = \left(\frac{1}{1+G_1(s)G_2(s)}\right)R(s) + \left(\frac{-G_2(s)}{1+G_1(s)G_2(s)}\right)D(s) \tag{A.6.3}$$

The steady-state error of the system, denoted as e_{ss}, is defined as:

$$e_{ss} = \lim_{s\to 0} sE(s) = \lim_{s\to 0} s\left(\frac{1}{1+G_1(s)G_2(s)}\right)R(s) + \lim_{s\to 0} s\left(\frac{-G_2(s)}{1+G_1(s)G_2(s)}\right)D(s) \tag{A.6.4}$$

From Eq. (A.6.4), it can be seen that both the input $R(s)$ and the disturbance $D(s)$ contribute to the steady-state error, affecting the overall performance of the system.

- **System Type of a Control System**

Let us first consider the steady-state error caused by the input $R(s)$ in Fig. A.17.

$$e_{ss}(t) = \lim_{s\to 0} sE(s) = \lim_{s\to 0} s\left(\frac{1}{1+G(s)}\right)R(s) \tag{A.6.5}$$

where $G(s)$ is the forward (open-loop) transfer function of the system.

Since the steady-state error of the system is related to the **type** of the forward transfer function $G(s)$, the term **type** is defined as the number of poles of G(s) located at $s = 0$. In general, the forward transfer function $G(s)$ of the system can be expressed in the following general form:

$$G(s) = \frac{K(s+z_1)(s+z_2)\ldots}{s^i(s+p_1)(s+p_2)\ldots} \tag{A.6.6}$$

Here, i represents the number of poles of $G(s)$ at $s = 0$, and this number i also defines the **type** of the system. If $i = 0$, the system is a **Type 0** system; if $i = 1$, it is a **Type 1** system, and so on.

- **Steady-State Error for a Step Input Command**

Assuming the input signal is a step command, i.e., $R(s) = R/s$, then the steady-state error caused by the input $R(s)$ is:

$$e_{ss} = \lim_{s\to 0} s\left(\frac{1}{1+G(s)}\right)\frac{R}{s} = \lim_{s\to 0}\frac{R}{1+\lim\limits_{s\to 0} G(s)} \tag{A.6.7}$$

Here, $\lim\limits_{s\to 0} G(s)$ is defined as the position error constant K_p, i.e., $K_p = \lim\limits_{s\to 0} G(s)$. Therefore, Eq. (A.6.7) can be expressed as:

$$e_{ss} = \lim_{s\to 0}\frac{R}{1+K_p} \tag{A.6.7}$$

To make the steady-state error e_{ss} equal to zero, the position error constant K_p must be infinite. This condition is satisfied only if the system type is at least 1. If the system is Type 0, then K_p cannot be infinite, and a steady-state error will exist.

- **Steady-State Error for a Ramp Input Command**

Assuming the input signal is a ramp command, i.e., $R(s) = R/s^2$, substituting it into Eq. (A.6.2) yields the steady-state error caused by the input $R(s)$ as:

$$e_{ss} = \lim_{s\to 0} s\left(\frac{1}{1+G(s)}\right)\frac{R}{s^2} = \lim_{s\to 0}\frac{R}{s+sG(s)} = \frac{R}{\lim\limits_{s\to 0} sG(s)} \tag{A.6.8}$$

Here, $\lim\limits_{s\to 0} sG(s)$ is defined as the velocity error constant K_v, i.e., $K_v = \lim\limits_{s\to 0} sG(s)$. Therefore, Eq. (A.6.8) can be expressed as:

$$e_{ss} = \lim_{s\to 0}\frac{R}{K_v} \tag{A.6.9}$$

To make the steady-state error e_{ss} equal to zero, the velocity error constant K_v must be infinite. This condition, $K_v = \infty$, is satisfied only if the system type is at least 2.

If the system type is 1, then e_{ss} becomes a finite constant, meaning that a steady-state error exists.

If the system type is 0, then e_{ss} becomes infinite, and the system's steady-state error will grow without bound.

- **Steady-State Error for a Parabolic Input Command**

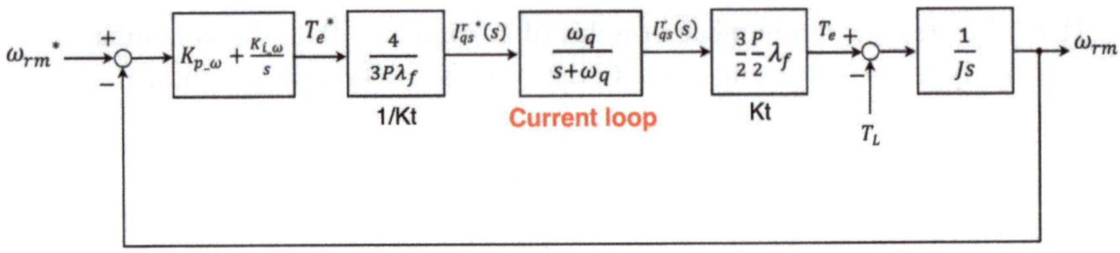

Fig. A.19 PMSM speed control block diagram with PI controller

Assuming the input signal is a parabolic command, i.e., $R(s) = R/s^3$, substituting it into Eq. (A.6.2) yields the steady-state error caused by the input $R(s)$:

$$e_{ss} = \lim_{s \to 0} s\left(\frac{1}{1+G(s)}\right)\frac{R}{s^3} = \lim_{s \to 0} \frac{R}{s^2 + s^2 G(s)} = \frac{R}{\lim_{s \to 0} s^2 G(s)} \tag{A.6.10}$$

Here, the $\lim_{s \to 0} s^2 G(s)$ is defined as the parabolic error constant K_a, that is, $K_a = \lim_{s \to 0} s^2 G(s)$. Therefore, Eq. (A.6.10) can be expressed as follows:

$$e_{ss} = \lim_{s \to 0} \frac{R}{K_a} \tag{A.6.11}$$

To make the steady-state error e_{ss} equal to zero, the parabolic error constant K_a must be infinite. For $K_a = \infty$ to hold, the system type must be at least 3. If the system type is 2, then e_{ss} is a constant, meaning a steady-state error exists. If the system type is 0 or 1, then e_{ss} is infinite, and the steady-state error will diverge.

- **Analyze the steady-state error using the motor speed control loop as an example**

In this section, the author uses the speed control loop of a permanent magnet synchronous motor (PMSM) as an example to analyze the steady-state error. Figure A.19 shows the PMSM PI controller speed loop used in Chap. 2 [same as Fig. 2.39]. Since delay effects do not affect the steady-state error, they are neglected here to simplify the mathematical derivation.

Since the bandwidth of the speed loop is much lower than that of the current loop, the current loop transfer function can be approximated as a unity gain, that is,

$$\frac{\omega_q}{s + \omega_q} \cong 1 \tag{A.6.12}$$

Next, by comparing Fig. A.19 with Fig. A.18, the following correspondence can be obtained:

$$G_1(s) = \frac{K_{p\_\omega} s + K_{i\_\omega}}{s} \tag{A.6.13}$$

$$G_2(s) = \frac{1}{Js} \tag{A.6.14}$$

$$D(s) = -T_L(s) \tag{A.6.15}$$

Using Eq. (A.6.4), the steady-state error can be obtained as:

$$\begin{aligned} e_{ss} &= \lim_{s \to 0} s\left(\frac{1}{1+G_1(s)G_2(s)}\right)\omega^*_{rm}(s) + \lim_{s \to 0} s\left(\frac{G_2(s)}{1+G_1(s)G_2(s)}\right)T_L(s) \\ &= \lim_{s \to 0} s\left(\frac{1}{1+\frac{K_{p\omega}s+K_{i\omega}}{s}\times\frac{1}{Js}}\right)\omega^*_{rm}(s) + \lim_{s \to 0} s\left(\frac{\frac{1}{Js}}{1+\frac{K_{p\omega}s+K_{i\omega}}{s}\times\frac{1}{Js}}\right)T_L(s) \\ &= \lim_{s \to 0} s\left(\frac{Js^2}{Js^2+K_{p\_\omega}s+K_{i\_\omega}}\right)\omega^*_{rm}(s) + \lim_{s \to 0} s\left(\frac{s}{Js^2+K_{p\_\omega}s+K_{i\_\omega}}\right)T_L(s) \end{aligned} \tag{A.6.16}$$

Assuming the speed command is a step signal, then $\omega^*_{rm}(s) = R/s$. *The steady-state error caused by the command* $\omega^*_{rm}(s)$ is:

$$\lim_{s \to 0} s\left(\frac{Js^2}{Js^2+K_{p\_\omega}s+K_{i\_\omega}}\right)\times\frac{R}{s} = 0 \tag{A.6.17}$$

Assuming the speed command $\omega^*_{rm}(s)$ *is a ramp signal, then* $\omega^*_{rm}(s) = R/s^2$. The steady-state error caused by the command is:

$$\lim_{s \to 0} s\left(\frac{Js^2}{Js^2+K_{p\_\omega}s+K_{i\_\omega}}\right)\times\frac{R}{s^2} = 0 \tag{A.6.18}$$

From Eqs. (A.6.17) and (A.6.18), it can be seen that whether the input is a step command or a ramp command, the steady-state error of the speed loop using a PI controller is zero. This is because the controlled plant of the motor, $\frac{1}{Js}$, is inherently a Type 1 system. With the addition of the PI controller's transfer function, the open-loop transfer function becomes a Type 2 system. As a result, the system can maintain zero steady-state error for both step and ramp commands. However, for a parabolic command, the system will produce a steady-state error.

Next, we analyze the steady-state error caused by disturbances. Assuming the load torque $T_L(s)$ is a step signal, i.e., $T_L(s) = R/s$, the steady-state error caused by the disturbance is:

$$\lim_{s \to 0} s\left(\frac{s}{Js^2+K_{p\_\omega}s+K_{i\_\omega}}\right)\times\frac{R}{s} = 0 \tag{A.6.19}$$

From Eq. (A.6.19), it can be seen that when the system encounters a step load disturbance, the steady-state error caused by the disturbance is also zero.

Based on the above analysis, the integral controller increases the type of the open-loop transfer function in the motor speed control loop. If only a proportional controller is used without an integral controller, the system will not be able to effectively eliminate the steady-state error caused by step disturbances. In contrast, the integrator increases the system type, thereby effectively eliminating the steady-state error. This is why most control loops require an integral controller.

Summary of This Section:

- In practice, due to the presence of various nonlinear effects in real-world physical systems and common modeling errors, steady-state error is generally unavoidable. Therefore, minimizing the steady-state error—or ensuring it remains below a specified tolerance—becomes an important and necessary design objective.
- A commonly adopted practical approach is to treat the friction torque caused by the motor's viscous friction coefficient B as part of the motor's load torque T_L. As a result, the motor's mechanical plant is often simplified to $\frac{1}{Js}$. This simplified mechanical model is essentially an approximation of the motor's actual mechanical equation, meaning steady-state error may still exist in reality. Nevertheless, steady-state error can be reduced to meet design specifications through the use of integral controllers or feedforward techniques.

A.7 Understanding Non-minimum Phase Systems

In classical control theory, two types of systems are often discussed: the **minimum phase system** and the **non-minimum phase system**. A minimum phase system is defined as a system that has no zeros or poles in the right-half plane (RHP) or on the imaginary axis (excluding the origin) 1. Conversely, if the transfer function has any zeros or poles in the RHP, or includes a time delay, it is referred to as a non-minimum phase system.

(Note: When a time delay element is approximated using a Padé approximation, it can introduce zeros with positive real parts. For example, the first-order Padé approximation of a delay $e^{-0.001s}$ is $\frac{-s+2000}{s+2000}$), which has a zero with a positive real part).

These definitions of minimum and non-minimum phase systems apply to both **open-loop** and **closed-loop** systems. For closed-loop systems, however, if there are poles in the RHP, the system is unstable, and such cases are generally not considered in typical analysis.

The term *non minimum phase system* originates from the frequency domain char acteristics of systems. Among systems with the same magnitude response to sinusoidal inputs, a minimum phase system exhibits the smallest phase shift, whereas a non-minimum phase system exhibits a greater phase shift. From the perspective of transfer functions, because all the poles and zeros of a minimum phase system lie in the left-half plane (LHP), its phase variation as the frequency ω increases from 0

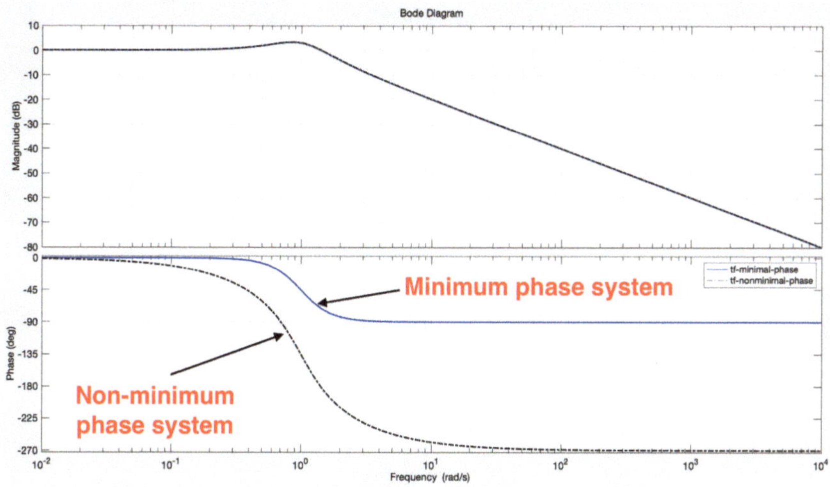

Fig. A.20 Bode plots of minimum-phase and non-minimum-phase systems

to ∞ is minimal. In contrast, a non-minimum phase system has zeros in the RHP, resulting in greater phase variation than that of a minimum phase system.

A minimum phase system is one that exhibits the smallest possible phase shift. Suppose the number of zeros in its transfer function is m and the number of poles is n; then the slope of its Bode magnitude plot follows the rule of $-20(n-m)$ dB/decade, and its phase will asymptotically approach $-90(n-m)$ degrees.

For example, consider the following two systems—one is a minimum phase system and the other is a non-minimum phase system. Both systems share the same magnitude characteristics in the frequency domain.

$$\text{minimum phase system}: \frac{s+1}{s^2+s+1} \tag{A.7.1}$$

$$\text{non-minimum phase system}: \frac{-s+1}{s^2+s+1} \tag{A.7.2}$$

Figure A.20 shows the Bode plots of these two systems. As illustrated, both systems have identical magnitude characteristics in the frequency domain, but their phase responses differ significantly. The minimum phase system exhibits the smallest phase shift, which is why it is referred to as a "minimum phase" system. In contrast, the phase shift of the non-minimum phase system is greater than that of the minimum phase system.

- **All-pass filter**

In the field of signal processing, there is a type of filter known as an **all-pass filter**. It allows signals to pass through with a constant gain across all frequencies, while its

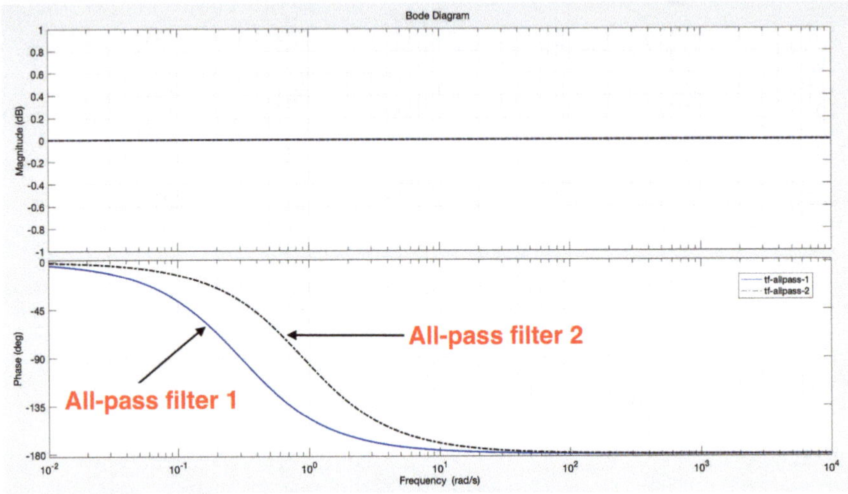

Fig. A.21 Bode plots of the two all-pass filters

phase response varies with frequency. Because an all-pass filter has a flat magnitude response in the frequency domain and its phase characteristics can be designed as needed, it is commonly used for **phase compensation** in systems. Below are two examples of all-pass filters:

$$\text{All-pass filter 1}: \frac{0.3 - s}{s + 0.3} \tag{A.7.3}$$

$$\text{All-pass filter 2}: \frac{0.9 - s}{s + 0.9} \tag{A.7.4}$$

From Eqs. (A.7.3) and (A.7.4), it can be seen that all-pass filters have zeros in the right-half plane; therefore, all-pass filters are **non-minimum phase systems**. The Bode plots of the two all-pass filters described above can be plotted using MATLAB, as shown in Fig. A.21.

As shown in Fig. A.21, All-Pass Filter 1 and All-Pass Filter 2 have identical magnitude responses in the frequency domain, but their phase responses are different.

- **The Relationship Between Magnitude and Phase in the Frequency Domain**

Suppose there is a minimum phase system with an open-loop transfer function $G_m(s)$

$$G_m(s) = \frac{s + 3}{s^2 + 2s + 1} \tag{A.7.5}$$

Next, All-Pass Filter 1 and All-Pass Filter 2 are respectively cascaded with the minimum phase system $G_m(s)$, resulting in the following two system transfer

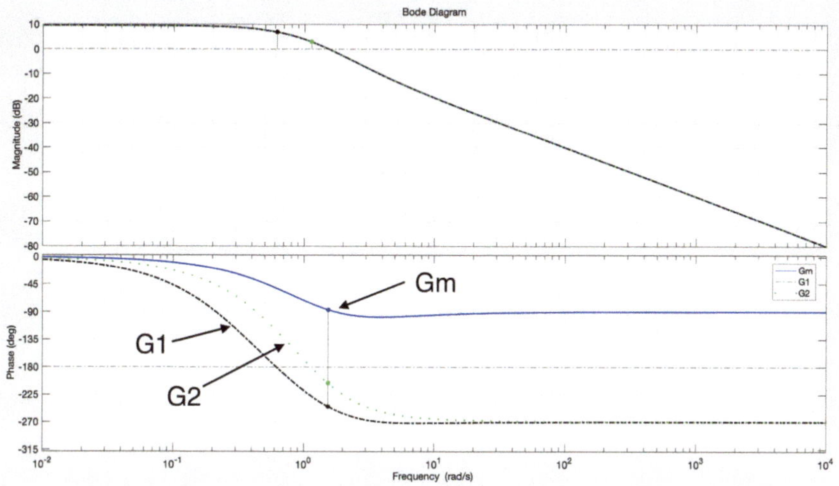

Fig. A.22 Comparison of the bode plots of three transfer functions

functions: $G_1(s)$ and $G_2(s)$.

$$G_1(s) = \left(\frac{0.3 - s}{s + 0.3}\right) \frac{s + 3}{s^2 + 2s + 1} \tag{A.7.6}$$

$$G_2(s) = \left(\frac{0.9 - s}{s + 0.9}\right) \frac{s + 3}{s^2 + 2s + 1} \tag{A.7.7}$$

Since $G_m(s)$ is a minimum phase system, it has the smallest phase shift. In contrast, both $G_1(s)$ and $G_2(s)$ have zeros in the right-half plane, making them non-minimum phase systems. As a result, the phase shifts of $G_1(s)$ and $G_2(s)$ are greater than that of $G_m(s)$. However, since both are formed by cascading $G_m(s)$ with all-pass filters, all three systems share the same magnitude characteristics in the frequency domain. The Bode plots of the three systems can be generated using MATLAB, as shown in Fig. A.21.

As shown in Fig. A.22, all three systems exhibit the same frequency-domain magnitude characteristics. However, the non-minimum phase systems $G_1(s)$ and $G_2(s)$ have greater phase shifts compared to the minimum phase system $G_m(s)$. Since all three are open-loop systems, Fig. A.22 can also be used to assess their stability. From the perspective of the Bode plot, the minimum phase system $G_m(s)$ has a phase margin of 93.2°, indicating very good stability. In contrast, the phase margins of the non-minimum phase systems $G_1(s)$ and $G_2(s)$ are − 64.8° and − 26.2°, respectively, and thus both $G_1(s)$ and $G_2(s)$ would be judged unstable. This also implies that if $G_1(s)$ and $G_2(s)$ are connected with unity feedback, the resulting systems will become unstable.

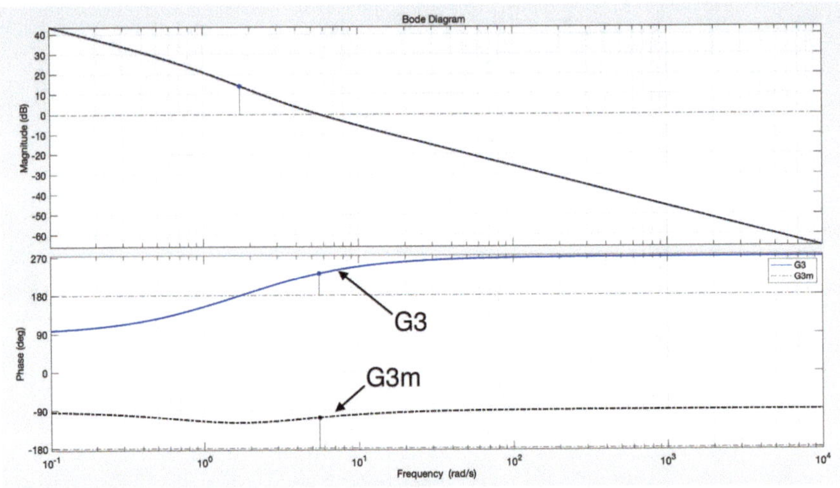

Fig. A.23 Bode plot comparison of minimum-phase and non-minimum-phase systems

However, in certain cases, relying solely on the Bode plot to assess the stability of non-minimum phase systems may lead to **misjudgment**—a topic we will explore in the next section.

For minimum phase systems, there is a **one-to-one correspondence** between magnitude and phase characteristics—once one is known, the other is uniquely determined. This unique relationship **does not exist** for non-minimum phase systems, as clearly demonstrated by Figs. A.21 and A.22.

- **Assessing the Stability of Non-Minimum Phase Systems Using Bode Plots**

Next, consider a unity feedback system with a forward transfer function $G_3(s)$.

$$G_3(s) = \frac{5(s+3)}{s(s-1)} \tag{A.7.8}$$

From Eq. (A.7.8), it can be seen that $G_3(s)$ has a pole in the right-half plane, and is therefore a non-minimum phase system. Next, consider a minimum phase system $G_{3m}(s)$ that has the same magnitude characteristics in the frequency domain.

$$G_{3m}(s) = \frac{5(s+3)}{s(s+1)} \tag{A.7.9}$$

From Eq. (A.7.9), it can be seen that all the poles and zeros of $G_{3m}(s)$ lie outside the right-half plane, making it a minimum phase system. The Bode plots of the two systems can be generated using MATLAB, as shown in Fig. A.23.

From the Bode plots in Fig. A.23, although the minimum phase system $G_{3m}(s)$ and the non-minimum phase system $G_3(s)$ share the same magnitude characteristics, their phase characteristics differ significantly. When using the Bode plot to assess system stability, the minimum phase system $G_{3m}(s)$ has a phase margin of 71.9° and a gain margin of $+\infty$, whereas the non-minimum phase system $G_3(s)$ has a phase margin of 51.6° but a gain margin of – 14 dB. According to classical control system stability criteria, a system with a negative gain margin—such as $G_3(s)$—would be judged as unstable.

However, if we derive the closed-loop transfer function $G_3(s)$ for the non-minimum phase system $G_{3c}(s)$, we obtain:

$$G_{3c}(s) = \frac{5(s+3)}{s^2 + 4s + 15} \tag{A.7.10}$$

Its two poles are both located in the left-half plane, indicating that the closed-loop system is stable. Clearly, the stability judgment based on the Bode plot contradicts the actual system behavior. Therefore, it is important to understand that in classical control theory, the use of Bode plots for evaluating system stability is suitable for minimum phase systems. This is because both phase margin and gain margin require the combined interpretation of magnitude and phase plots, and for minimum phase systems, there is a one-to-one correspondence between the magnitude and phase characteristics. Hence, using Bode plots to assess the stability of minimum phase systems is valid.

However, for non-minimum phase systems, this unique relationship between magnitude and phase does not exist. As a result, using Bode plots alone to determine the stability of non-minimum phase systems may lead to misjudgment. To accurately assess the stability of a non-minimum phase system, the Nyquist plot should be used.

Tips

You can also use MATLAB's margin command to obtain the phase margin and gain margin of the open-loop system.

- **Time-Domain Characteristics of Non-minimum Phase Systems**

Next, we consider the minimum phase and non-minimum phase systems described by Eqs. (A.7.1) and (A.7.2), respectively. Suppose both systems represent the closed-loop transfer functions of a control system. Their step responses are shown in Fig. A.24.

As shown in Fig. A.24, the minimum phase system (Eq. A.7.1), due to its minimal phase shift, exhibits time-domain behavior where the output signal quickly follows the input command. In contrast, a typical characteristic of non-minimum phase systems is the presence of a delay between the input command and the output response. As illustrated in Fig. A.24, the non-minimum phase system (Eq. A.7.2), with greater phase lag in the frequency domain, results in longer rise time and settling

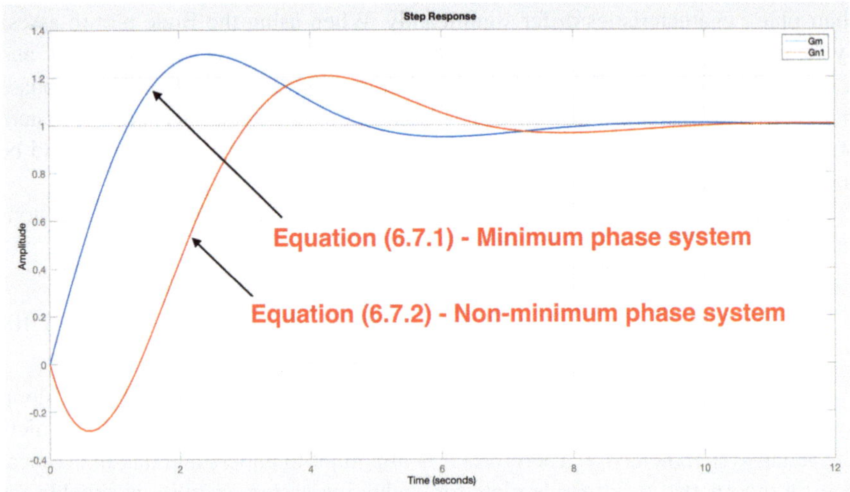

Fig. A.24 Comparison of step response waveforms between minimum-phase and non-minimum-phase systems

time in the time domain under the same step input. Excessive phase lag can even lead to an initial negative transient response.

The system can also be analyzed from its transfer function. For the minimum phase system with transfer function $\frac{s+1}{s^2+s+1}$, it can be seen as a cascade of $s+1$ and $\frac{1}{s^2+s+1}$. Assuming the input signal $R(s)$ first passes through the $s+1$ block, this corresponds in the time domain to the operation $\frac{d}{dt}r(t)+r(t)$, where $r(t)$ is the time-domain function of $R(s)$. If the input is a step command, then $\frac{d}{dt}r(t)+r(t)$ results in a positive impulse added to the original signal. Therefore, the output control signal generated by the $s+1$ operation is positive, resulting in a positive system response.

On the other hand, for the non-minimum phase system with transfer function $\frac{-s+1}{s^2+s+1}$, it can be viewed as a cascade of $-s+1$ and $\frac{1}{s^2+s+1}$. If the input signal R(s) first passes through the $-s+1$ block, this corresponds in the time domain to the operation $-\frac{d}{dt}r(t)+r(t)$. For a step command, this results in a negative impulse added to the input signal. As a result, the output of the $-s+1$ block will have a negative transient, which causes the system's transient response to exhibit an initial negative dip.

Additionally, if the open-loop transfer function of a system has a zero in the right-half plane, it is classified as a **non-minimum phase system**. According to the root locus principle, as the loop gain increases, the closed-loop poles move toward the zeros. Therefore, increasing the loop gain may cause the system to become unstable. To effectively control a non-minimum phase system, **pole-zero cancellation design methods should be avoided**, as they can create unstable internal system behavior. It is recommended to use a **predictive controller** or a **nonlinear controller** to effectively manage such systems. In general, the simplest approach is to maintain a **low loop gain** to ensure the system remains stable.

Summary of This Section:

- Both **minimum phase systems** and **non-minimum phase systems** can be represented using Bode plots. However, for minimum phase systems, there is a **one-to-one relationship** between magnitude and phase. In contrast, non-minimum phase systems do **not** have such a unique relationship—different non-minimum phase transfer functions can have the same frequency-domain magnitude response but entirely different phase responses.
- In minimum phase systems, the small phase lag allows the output signal to closely follow the input command, which is advantageous for controller design. On the other hand, non-minimum phase systems exhibit larger phase variations, which may result in longer rise time and settling time—making controller design more challenging.
- To **accurately determine the stability** of a non-minimum phase system, the **Nyquist plot** should be used. Relying solely on Bode plots may lead to incorrect conclusions.
- **Time delay** also results in a non-minimum phase system, because when the delay term e^{-sT} is linearized using a **Padé approximation**, the resulting transfer function contains a zero in the right-half plane.

A.8 Pole Placement in Modern Control: Theory and MATLAB Application

In Sect. 5.11, we previously used **pole placement** to design the characteristic roots of a **Luenberger observer**, by selecting a state feedback gain matrix $\boldsymbol{G}$ to ensure that the estimation error $\tilde{\boldsymbol{x}}$ converges to zero at the desired rate. In this section, the author will provide a comprehensive introduction to the essence of **pole placement** in modern control theory [1], and demonstrate the method through practical examples using MATLAB commands, so that readers can experience both the core concept and design philosophy of pole placement.

In classical control theory, transfer functions can only describe **single-input single-output (SISO)** linear systems, which presents limitations. If a system is **multi-input multi-output (MIMO)** or **nonlinear**, transfer functions cannot be used to describe it. Instead, modern control theory adopts the **state-space method**, where systems are represented using four matrices $\boldsymbol{A}$, $\boldsymbol{B}$, $\boldsymbol{C}$, and $\boldsymbol{D}$, as shown below:

$$\dot{\boldsymbol{x}}(t) = \boldsymbol{A}\boldsymbol{x}(t) + \boldsymbol{B}\boldsymbol{u}(t) \tag{A.8.1}$$

$$\boldsymbol{y}(t) = \boldsymbol{C}\boldsymbol{x}(t) + \boldsymbol{D}\boldsymbol{u}(t) \tag{A.8.2}$$

Equation (A.8.1) represents the **state equation** of the system, and Eq. (A.8.2) represents the **output equation**. Here, $\boldsymbol{x}(t)$ is the **state vector** of the system, with dimensions $n \times 1$; the coefficient matrix A has dimensions $n \times n$; the matrix $\boldsymbol{B}$ has

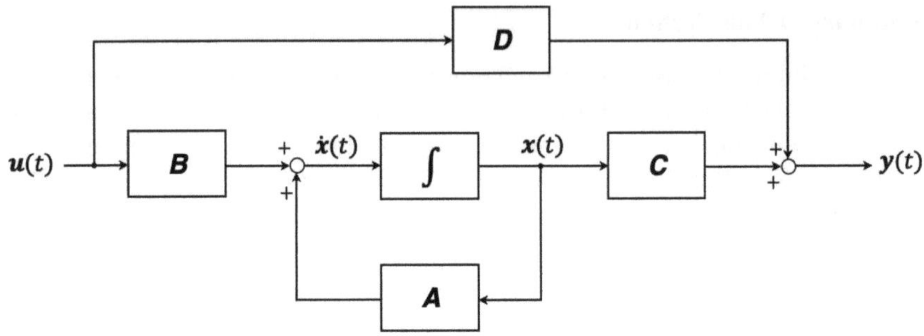

Fig. A.25 Block diagram representation of state-space equations

dimensions $n \times p$; $\boldsymbol{u}(t)$ is the **input vector**, with dimensions $p \times 1$; the matrix $\boldsymbol{C}$ has dimensions $q \times n$; $\boldsymbol{y}(t)$ is the **output vector**, with dimensions $q \times 1$; and the matrix $\boldsymbol{D}$ has dimensions $q \times p$.

(*The effect of disturbances is ignored at this point*).

The state Eq. (A.8.1) and the output Eq. (A.8.2) can be represented as a system block diagram, as shown in Fig. A.25.

In modern control theory, the definition of system stability differs from that in classical control theory, which is based on transfer functions. In modern control theory, a system's stability is determined by the **eigenvalues of matrix** A in the state equation. If all eigenvalues of matrix $\boldsymbol{A}$ have **negative real parts** (i.e., they lie in the left-half plane), the system is considered stable. If any eigenvalue of matrix $\boldsymbol{A}$ has a **positive real part**, the system is unstable. In such cases, stability can be achieved by applying **state feedback** to redesign a new matrix $\boldsymbol{A}$ by reassigning its eigenvalues. This method is known as **pole placement**. The term "pole" refers to the fact that the eigenvalues of matrix $\boldsymbol{A}$ are equivalent to the poles of the system's characteristic equation.

Figure A.26 shows the block diagram of a system with **state feedback**. As illustrated, the gain matrix $\boldsymbol{G}$ is the **state feedback matrix**. The state feedback loops the system state back to the input. In contrast to Figure A.25, where the input was $\boldsymbol{u}(t)$, the input is now modified to $\boldsymbol{r}(t) - \boldsymbol{G} \cdot \boldsymbol{x}(t)$, where $\boldsymbol{r}(t)$ is the system's reference input vector. Therefore, the state equation of the system with state feedback can be expressed as:

$$\dot{\boldsymbol{x}}(t) = \boldsymbol{A}\boldsymbol{x}(t) + \boldsymbol{B}[\boldsymbol{r}(t) - \boldsymbol{G} \cdot \boldsymbol{x}(t)] = (\boldsymbol{A} - \boldsymbol{B}\boldsymbol{G})\boldsymbol{x}(t) + \boldsymbol{B}\boldsymbol{r}(t) \tag{A.8.3}$$

From Eq. (A.8.3), it can be observed that the original matrix $\boldsymbol{A}$ is replaced by a new matrix $\boldsymbol{A} - \boldsymbol{B}\boldsymbol{G}$. By designing the gain matrix G, the system poles—i.e., the eigenvalues of the matrix $\boldsymbol{A} - \boldsymbol{B}\boldsymbol{G}$—can be freely placed to meet the desired stability requirements.

- **Implementing Pole Placement Using MATLAB**

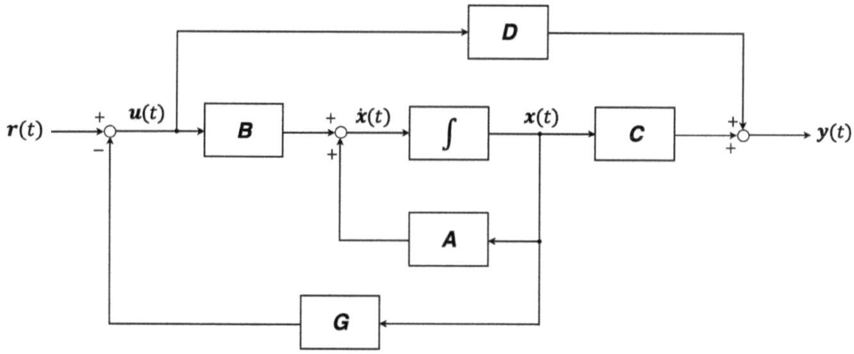

Fig. A.26 State-space block diagram with state feedback

Next, the author will guide you through the practical implementation of pole placement using MATLAB commands. Suppose the system matrices ***A***, ***B***, ***C***, and ***D*** are given as follows:

$$\boldsymbol{A} = \begin{bmatrix} -1 & -2 \\ 1 & 0 \end{bmatrix}, \boldsymbol{B} = \begin{bmatrix} 2 \\ 0 \end{bmatrix}$$

$$\boldsymbol{C} = \begin{bmatrix} 0 \; 1 \end{bmatrix}, \boldsymbol{D} = 0$$

Currently, the eigenvalues of matrix ***A*** are $-0.5 \pm j1.3229$. Suppose these eigenvalues do not meet the required stability specifications. To address this, we want to use the pole placement method to reassign the eigenvalues to two negative real roots: – 2 and – 3, so the system exhibits overdamped behavior. This can be achieved using MATLAB commands to automatically compute the corresponding state feedback matrix ***G***.

Please open the sample program **mA_8_1**. In this program, the original system matrices ***A***, ***B***, ***C***, and ***D*** are defined. The variable p1 specifies the desired pole locations, – 2 and – 3, to be used in the pole placement design. The **place** command in MATLAB is then used to calculate the corresponding state feedback gain matrix ***G*** (stored in variable G1). The variable new_A represents the new state matrix after pole placement, new_sys is the state-space system model after pole placement, and new_poles contains the new eigenvalues, which will be – 2 and – 3. Finally, the **step** command is used to plot the step response of the redesigned system.

MATLAB Example Script: MA_8_1.M

```
A = [– 1, – 2;1,0];

B = [2;0];

C = [0,1];
```

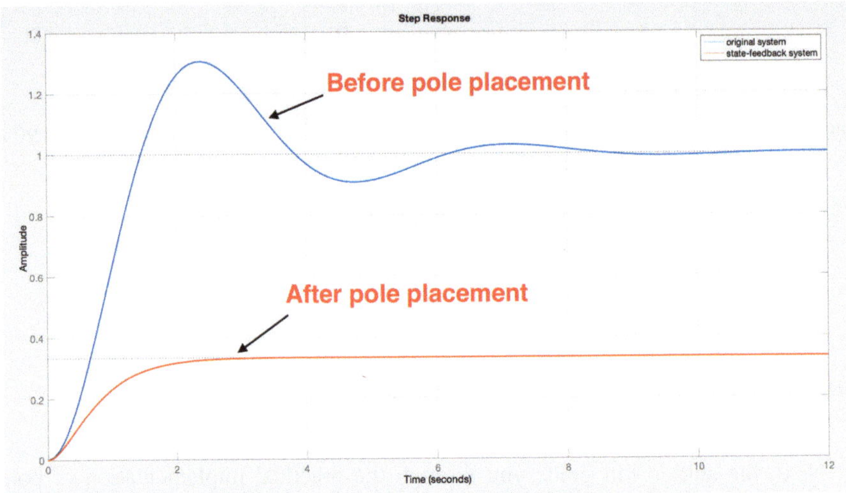

Fig. A.27 Output responses before and after pole placement

```
D = 0;
p1 = [− 2, − 3];
G1 = Place(a,B,p1);
new_A = A-B*G1;
new_sys = ss(new_A,B,C,D);
new_poles = pole(new_sys);
step(new_sys);
```

Figure A.27 shows the step response of the system after pole placement. When compared with the step response before pole placement, it can be observed that although the pole placement method allows the system to be designed as overdamped, a **steady-state error** still exists. This is because while pole placement can arbitrarily assign the locations of the poles to satisfy stability requirements, it **does not change the system type** (i.e., the Type of the system) [1].

In classical control design based on transfer functions, an **integrator** is typically introduced to increase the system type and eliminate steady-state error. Therefore, in state-space-based system design, an integrator must also be introduced to increase the system type in order to eliminate steady-state error.

We can slightly modify the block diagram in Fig. A.26. The updated system with the integrator included is shown in Fig. A.28. To eliminate the steady-state error, the system output is fed back to the input, and the error signal is passed to the integrator.

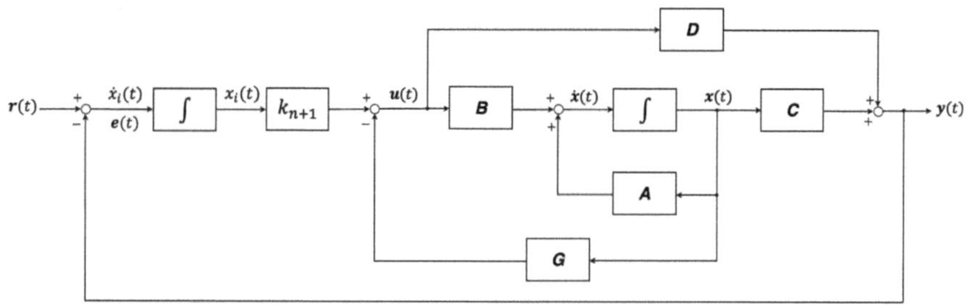

Fig. A.28 State-space block diagram with integrator [1]

The output of the integrator is defined as x_i, and the input (i.e., the error signal $\boldsymbol{e}(t)$) is denoted as $\dot{x}_i$. *The gain of the integrator is defined as* k_{n+1} [1].

Next, we present the system equations corresponding to Fig. A.28.

(*Here, we assume* D = 0).

$$\boldsymbol{u}(t) = k_{n+1} \cdot x_i - \boldsymbol{G} \cdot \boldsymbol{x}(t) = \begin{bmatrix} -\boldsymbol{G} \; k_{n+1} \end{bmatrix} \begin{bmatrix} \boldsymbol{x}(t) \\ x_i \end{bmatrix} \tag{A.8.4}$$

The resulting new state vector is $\begin{bmatrix} \boldsymbol{x}(t) \\ x_i \end{bmatrix}$, and:

$$\dot{x}_i = \boldsymbol{r}(t) - \boldsymbol{C} \cdot \boldsymbol{x}(t) \tag{A.8.5}$$

The state equation can be written as follows:

$$\begin{bmatrix} \dot{\boldsymbol{x}}(t) \\ \dot{x}_i \end{bmatrix} = \begin{bmatrix} \boldsymbol{A} & 0 \\ -\boldsymbol{C} & 0 \end{bmatrix} \begin{bmatrix} \boldsymbol{x}(t) \\ x_i \end{bmatrix} + \begin{bmatrix} \boldsymbol{B} \\ 0 \end{bmatrix} \begin{bmatrix} -\boldsymbol{G} \; k_{n+1} \end{bmatrix} \begin{bmatrix} \boldsymbol{x}(t) \\ x_i \end{bmatrix} + \begin{bmatrix} 0 \\ 1 \end{bmatrix} \boldsymbol{r}(t) \tag{A.8.6}$$

After simplification, the system equations can be expressed as:

$$\begin{bmatrix} \dot{\boldsymbol{x}}(t) \\ \dot{x}_i \end{bmatrix} = \begin{bmatrix} \boldsymbol{A} - \boldsymbol{B}\boldsymbol{G} & -\boldsymbol{B}k_{n+1} \\ -\boldsymbol{C} & 0 \end{bmatrix} \begin{bmatrix} \boldsymbol{x}(t) \\ x_i \end{bmatrix} + \begin{bmatrix} 0 \\ 1 \end{bmatrix} \boldsymbol{r}(t) \tag{A.8.7}$$

$$\boldsymbol{y}(t) = \begin{bmatrix} \boldsymbol{C} \; 0 \end{bmatrix} \begin{bmatrix} \boldsymbol{x}(t) \\ x_i \end{bmatrix} \tag{A.8.8}$$

The newly formed state matrix is:

$$\boldsymbol{A}_{new} = \begin{bmatrix} \boldsymbol{A} - \boldsymbol{B}\boldsymbol{G} & -\boldsymbol{B}k_{n+1} \\ -\boldsymbol{C} & 0 \end{bmatrix} \tag{A.8.9}$$

- **Implementing Pole Placement with Integrator Using MATLAB**

Next, we will again use MATLAB commands to implement **pole placement with an integrator**. Assume that the system matrices ***A***, ***B***, ***C***, and ***D*** are the same as in the previous example, as shown below:

$$\boldsymbol{A} = \begin{bmatrix} -1 & -2 \\ 1 & 0 \end{bmatrix}, \boldsymbol{B} = \begin{bmatrix} 2 \\ 0 \end{bmatrix}$$

$$\boldsymbol{C} = \begin{bmatrix} 0 & 1 \end{bmatrix}, \boldsymbol{D} = 0$$

Currently, the eigenvalues of matrix $\boldsymbol{A}$ are $-0.5 \pm j1.3229$. In the previous example, we used pole placement to assign the system's eigenvalues to – 2 and – 3. However, a steady-state error remained, so the system must be redesigned by introducing an **integrator**. Since adding an integrator increases the system state by one (i.e., introducing x_i, the new state matrix $\boldsymbol{A}$ becomes a 3 × 3 matrix. Therefore, we now need to assign **three poles**. Suppose we choose three negative real poles: – 6, – 9, and – 60. Using MATLAB, we can automatically compute the corresponding state feedback gain matrix $\boldsymbol{G}$ (which will now be a 1 × 3 matrix).

Please open the example program **mA_8_2**. In this script, the original matrices ***A***, ***B***, ***C***, and ***D*** are defined as before. The variable p1 is used to specify the desired pole locations: – 6, – 9, and – 60.

Since we do not yet know the values of $\boldsymbol{G}$ and k_{n+1} in Eq. (A.8.7) prior to pole placement, we temporarily substitute $\boldsymbol{G} = 0$ and $k_{n+1} = 0$ into the equation. This allows us to construct the **augmented state-space matrices** $\overline{\boldsymbol{A}}$ **and** $\overline{\boldsymbol{B}}$, which are represented in the program by the variables A_bar and B_bar.

Next, the place function is used with the augmented $\overline{\boldsymbol{A}}$, $\overline{\boldsymbol{B}}$, and desired pole locations to compute the appropriate state feedback matrix $\boldsymbol{G}$. Once $\boldsymbol{G}$ is obtained, the new system matrices ***A***, ***B***, ***C***, and ***D*** can be constructed based on Eqs. (A.8.7) and (A.8.8). Finally, the **step** command is used to plot the step response of the redesigned system, as shown in Fig. A.29.

MATLAB Example Script: mA_8_2.m

```
A = [-1,-2;1,0];
B = [2;0];
C = [0,1];
D = 0;
p1=[-6 -9 -60];
A_bar=[A zeros(2,1);-C 0];
B_bar=[B; 0];
G_new=place(A_bar,B_bar,p1);
```

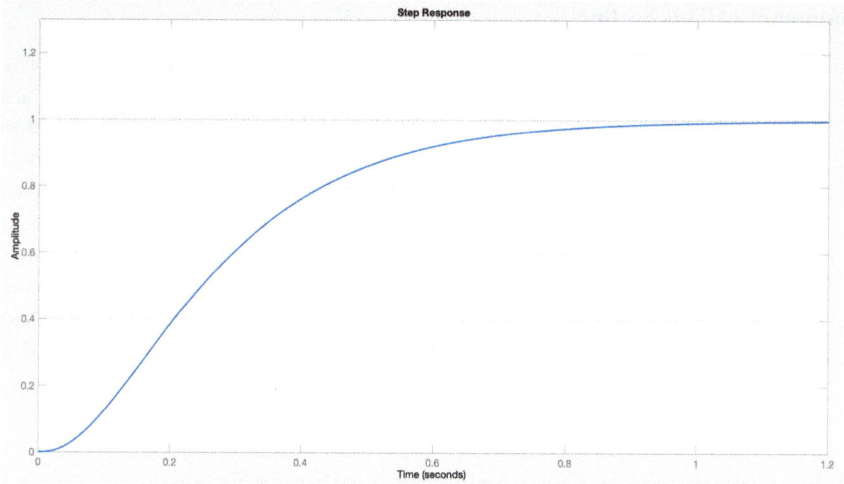

Fig. A.29 Output response waveform

```
A_new=A_bar-B_bar*G_new;
B_new=[zeros(2,1);1];
C_new=[C 0];
D_new=D;
new_sys = ss(A_new,B_new,C_new,D_new);
new_poles = pole(new_sys);
step(new_sys);
```

From the step response waveform shown in Fig. A.29, it can be observed that using the pole placement method not only allows the system to meet the desired stability requirements, but also eliminates the steady-state error after introducing the integrator. The design of control systems using **state-space pole placement** is fundamentally different from the classical control approach based on transfer functions.

In classical control theory, the design method is limited to analyzing the relationship between a single input and a single output, and relies on standard second-order system design criteria, which imposes certain limitations. In contrast, the state-space pole placement method considers **all system states simultaneously**. Through matrix operations, the required feedback gains can be computed in one step. This method can also be extended to **multi-input multi-output (MIMO)** systems, making it a more scalable and standardized design framework—particularly suitable for implementation using computers.

Summary of This Section:

- The **eigenvalues of the state matrix** A are the roots of the system's characteristic equation, i.e., the **poles** of the system. Using the **pole placement method**, these poles can be assigned arbitrarily—**provided the system is controllable**. For more information on controllability, please refer to modern control theory textbooks.
- When designing **state feedback**, the input signal is often set to zero. This approach is intended to drive the system's initial state to zero at a desired rate. This type of problem is known as a **regulator problem** in linear control theory. However, a control system that relies solely on state feedback **cannot meet tracking performance requirements** for input commands, because state feedback **does not increase the system type (Type)**. Therefore, an **integrator** must be added to improve the system's tracking performance.

A.9 Linearization Techniques for Nonlinear Systems

Classical control theory is a discipline developed based on **linear differential equations**, with the **transfer function** as its central concept, leading to the development of various linear control methods and techniques. A transfer function can be regarded as the **linearized model** of a real physical system. However, in the real world, most systems are **nonlinear**, and nonlinear systems **cannot be directly transformed** into transfer functions. Nevertheless, through **nonlinear system linearization techniques**, a transfer function model can be obtained for a nonlinear system. Therefore, this section will review the concepts and methods of nonlinear system linearization in classical control theory 1.

Nonlinear systems can be linearized using the **Taylor series expansion** technique. However, Taylor expansion requires an **operating point**, which is critically important. If this operating point is an **equilibrium point**, the system's linearized model can be obtained. The **equilibrium point** is defined as:

> An operating point at which the time derivatives of all system states are zero.

When a nonlinear system is expanded around an equilibrium point using a Taylor series, its linearized model can be obtained by retaining only the **first-order derivative terms**, while ignoring all **higher-order terms** (second-order and above).

The Taylor series formula is as follows [1]:

$$f(x) = f(x_o(t)) + \frac{f'(x_o)}{1!}(x(t) - x_o(t)) + \frac{f''(x_o)}{2!}(x(t) - x_o(t))^2 + \dots \tag{A.9.1}$$

where $f(x)$ is the nonlinear function, and $x_o(t)$ is the operating point.

The meaning of Eq. (A.9.1) is that the nonlinear function $f(x)$ is expanded around an operating point x_o using the Taylor series. If $\Delta x = x(t) - x_o(t)$ is small, then Eq. (A.9.1) will converge. In this case, the nonlinear function $f(x)$ can be approximated by the **first two terms** on the right-hand side of Eq. (A.9.1), that is:

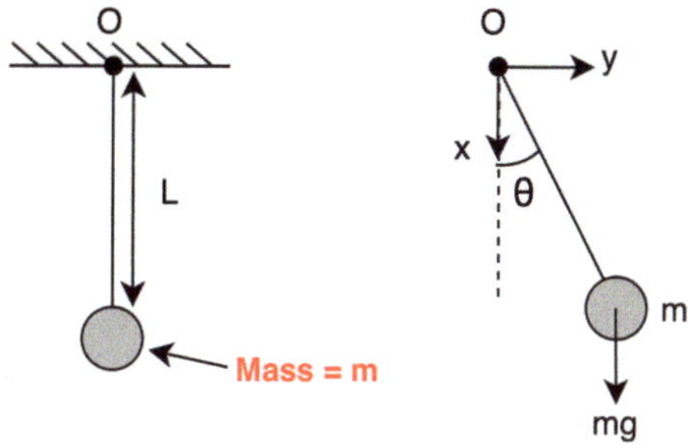

Fig. A.30 Simplified pendulum system diagram

$$f(x) \approx f(x_o(t)) + \frac{df(x_o(t))}{dt}(x(t) - x_o(t)) = c_0 + c_1 \Delta x \tag{A.9.2}$$

Next, we will go through a practical example to demonstrate the process of **linearizing a nonlinear system**.

Consider a **simple pendulum system**, as shown in Fig. A.30. The pendulum has a mass m and is attached to point O with a rigid rod of length L. The mass of the rod is negligible. Using principles from mechanics and the concept of net force, the differential equation describing the system can be derived, as shown in Eq. (A.9.3).

$$\ddot{\theta} + \frac{g}{L} sin\theta = 0 \tag{A.9.3}$$

As described in Eq. (A.9.3), when $\theta = 0$, the angular acceleration $\ddot{\theta}$ is also zero. Therefore, $\theta = 0$ is an **equilibrium point** of the system. This can also be understood from a physical perspective: when $\theta = 0$, and no external force is applied, the pendulum will remain at rest under the influence of gravity. In this state, both the velocity and acceleration of the pendulum are zero.

Next, we can define the state variables as follows:

$$x_1 = \theta, x_2 = \dot{\theta} \tag{A.9.4}$$

We can obtain

$$\dot{x}_1 = x_2, \dot{x}_2 = -\frac{g}{L} \sin \theta, \tag{A.9.5}$$

Next, expand $\dot{x}_2$ around the equilibrium point $\theta = 0$ using a Taylor series, keeping only the first-order derivative term and neglecting higher-order terms. This yields:

$$\begin{aligned} \dot{x}_2 &= f(x) = -\frac{g}{L}\sin\theta \approx f(\theta = 0) + \frac{f\prime(0)}{1!}(\theta - 0) \\ &= 0 - \frac{g}{L}\cos(0) \times \theta = -\frac{g}{L}\theta \end{aligned} \tag{A.9.6}$$

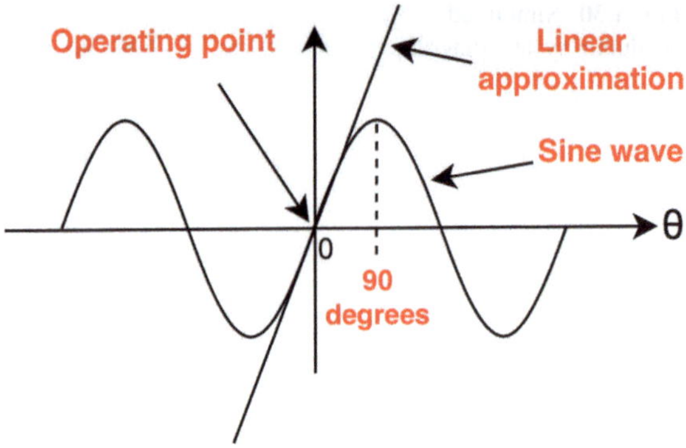

Fig. A.31 Illustration of the linearization of the sine function around $\theta = 0$

Therefore, the linearized model of the pendulum system can be expressed as:

$$\begin{bmatrix} \dot{x}_1 \\ \dot{x}_2 \end{bmatrix} = \begin{bmatrix} 0 & 1 \\ \frac{-g}{L} & 0 \end{bmatrix} \begin{bmatrix} x_1 \\ x_2 \end{bmatrix} \tag{A.9.7}$$

The state equation in Eq. (A.9.7) represents the **linearized model** of the pendulum system.

From Eq. (A.9.6), we can see that, from a functional perspective, the entire nonlinear system linearization process can be regarded as a **linear approximation of the sine function** around the operating point $\theta = 0$, as shown in Fig. A.31 [1]. When the variation $\Delta\theta = \theta - 0$ is very small, the sine function can be approximated linearly near $\theta = 0$ as:

$$\sin\theta \approx \theta \text{ when } \theta \approx 0 \tag{A.9.8}$$

You might be wondering: since the Taylor series can be expanded around any operating point, why do we usually expand around the **equilibrium point**? Is it possible to obtain a linearized model of the system if we expand around a point **other than the equilibrium**?

To explore this, let's try expanding the system around the operating point $\theta = \frac{\pi}{6}$.

$$\begin{aligned} \dot{x}_2 &= -\frac{g}{L}\sin\theta \approx f\left(\theta = \frac{\pi}{6}\right) + \frac{f'\left(\theta = \frac{\pi}{6}\right)}{1!}\left(\theta - \frac{\pi}{6}\right) \\ &= -\frac{g}{L} \times \frac{1}{2} - \frac{g}{L}\cos\left(\frac{\pi}{6}\right) \times \left(\theta - \frac{\pi}{6}\right) = -\frac{g}{L}\left\{\frac{1}{2} + \frac{\sqrt{3}}{2}\left(\theta - \frac{\pi}{6}\right)\right\} \end{aligned} \tag{A.9.9}$$

From the result in Eq. (A.9.9), we can see that the Taylor series expanded at the operating point $\theta = \frac{\pi}{6}$ contains a **constant term**. In this example, the physical meaning of that constant term is that the pendulum has an **initial velocity**. This occurs because $\theta = \frac{\pi}{6}$ is **not** an equilibrium point of the system.

As shown clearly in Fig. A.30, when the pendulum is at the position $\theta = \frac{\pi}{6}$, gravity will naturally cause it to accelerate downward, resulting in a non-zero velocity. Therefore, if the Taylor series is expanded at a **non-equilibrium point**, the term $f(x_o(t))$ will not be zero.

From a linear control system design perspective, the resulting system model—such as Eq. (A.9.9)—is **not a linear model** because of the constant term. Thus, such models must be handled using **nonlinear control techniques**, rather than linear system design methods.

Summary of This Section:

- In the real world, most systems are nonlinear, and nonlinear systems cannot be directly converted into transfer functions. However, the transfer function of a nonlinear system can be obtained through linearization techniques.
- Nonlinear systems can be linearized using the Taylor series expansion. However, Taylor series expansion requires a specific operating point, which is critically important. If the operating point is an equilibrium point, then a linearized model can be obtained. An equilibrium point is defined as:

  An operating point at which the time derivatives of all system states are zero.

References

[1] Farid Golnaraghi and Benjamin C. Kuo, Automatic Control System, McGraw-Hill Education, 2017.
[2] George Ellis, Control System Design Guide: Using Your Computer to Understand and Diagnose Feedback Controllers, Butterworth-Heinemann, 2016.
[3] Charles L. Phillips and H. Troy Nagle, Digital Control System Analysis and Design, Prentice Hall, 1994.

MIX
Papier aus verantwortungsvollen Quellen
Paper from responsible sources
FSC® C105338

If you have any concerns about our products,
you can contact us on
ProductSafety@springernature.com

In case Publisher is established outside the EU,
the EU authorized representative is:
Springer Nature Customer Service Center GmbH
Europaplatz 3, 69115 Heidelberg, Germany

Printed by Libri Plureos GmbH
in Hamburg, Germany